Accelerated Plant Breeding, Volume 5

Shabir Hussain Wani • Satbir Singh Gosal
Editors

Accelerated Plant Breeding, Volume 5

Forage Crops

Springer

Editors
Shabir Hussain Wani
National Seed Project (NSP)
Sher-e-Kashmir University of Agricultural Sciences and Technology – Kashmir (SKUAST-K)
Srinagar, Jammu and Kashmir, India

Satbir Singh Gosal
Punjab Agricultural University
Ludhiana, Punjab, India

ISBN 978-3-032-04526-3 ISBN 978-3-032-04527-0 (eBook)
https://doi.org/10.1007/978-3-032-04527-0

This Springer imprint is published by the registered company Springer Nature Switzerland AG
The registered company address is: Gewerbestrasse 11, 6330 Cham, Switzerland

Foreword

Plant breeding has witnessed a transformative shift with evolution of advanced genomic technologies, precision breeding tools, and accelerated selection strategies. Forage crops play significant role in maintaining livestock production performance to produce better quality fodder for better nutritional value and resilience to environmental stresses. Farmers need high-yielding cultivars with traits such as early maturity, higher harvest-index, better digestibility, and dual-purpose forage potential. The integration of genomic tools in forage breeding is paving the way for more efficient, sustainable, and scalable agricultural solutions. The book entitled *Accelerated Plant Breeding, Volume 5: Forage Crops* is a compilation of cutting-edge research and practical methodologies driving forage crop improvement. It provides an insightful exploration into molecular breeding, genomic-assisted selection, and modern biotechnological advancements that are shaping the production of high-performing forage varieties. In the face of climate variability and increasing global demand for quality forage, the knowledge shared will be useful for developing superior cultivars for enhanced drought tolerance, optimized nutrient composition, and compatibility with modern farming systems.

The book will serve as a comprehensive resource for researchers, academicians, and policymakers aiming to bridge fundamental genetic research with real-world agricultural applications. I congratulate the editors and authors for their vision and expertise in curating this essential reference. I hope that this volume will serve as a guiding light for future research and breeding programs worldwide.

Mangi Lal Jat
Secretary (DARE) & Director General ICAR
New Delhi, India
13 June 2025

Preface

The ever-growing global human population, coupled with the hastening effects of climate change, has raised serious concerns about food and nutritional security worldwide. Traditionally regarded as both an art and a science for modifying and enhancing plant traits, plant breeding has now become increasingly reliant on advanced scientific methodologies. As an ongoing process, it helps to develop new crop varieties tailored to meet farmers' needs boosting yields and enhancing nutritional quality. Under the current World Trade Organization (WTO) regime, there is an increasing demand for high-yielding crop varieties that cater not only to domestic consumption but also to global export markets. Conventional breeding methods alone have proven insufficient in meeting the growing need for superior cultivars. Consequently, a more efficient and strategic approach to improving existing varieties is imperative to address the challenges of rising food demand and sustainability. Forage crop improvement has customarily focused on enhancing yield potential, quality attributes, and resistance to biotic and abiotic stresses. However, modern agricultural requirements have expanded beyond these traits. Farmers and growers now seek high-yielding cultivars that incorporate durable and multiple resistance mechanisms, early maturity, higher harvest index, and dual-purpose forage potential. Additionally, superior varieties must exhibit nutrient-use and water-use efficiency, wider adaptability, and compatibility with mechanized harvesting. Enhanced shelf life, processing quality, and improved nutritional composition including optimized levels of minerals, vitamins, amino acids, proteins, antioxidants, and bioactive compounds have become essential breeding objectives. Within integrated crop–livestock systems, forage succession has emerged as a sustainable strategy for pasture utilization, particularly in regions facing seasonal rainfall deficits. It contributes to cattle feed availability while simultaneously generating biomass for no-till farming systems in subsequent cropping cycles. Therefore, identifying forage species with characteristics suitable for integrated farming systems and soil cover presents a novel opportunity for crop diversification, ensuring agricultural sustainability in changing climatic conditions.

Forage crops play a crucial role in sustaining livestock productivity, supporting food security, and enhancing agricultural sustainability worldwide. In India, where diverse agro-climatic conditions shape forage production, breeding strategies must continuously evolve to ensure resilience, nutritional quality, and improved yields. The emergence of genomics has revolutionized crop improvement, offering precision breeding tools to harness genetic diversity effectively and accelerate trait enhancement. This volume entitled *Accelerated Plant Breeding, Volume 5: Forage Crops* provides a comprehensive exploration of genomic-assisted breeding approaches applied to key forage crops, including barley, sorghum, pearl millet, finger millet, guar, Guinea grass, alfalfa, oats, and maize. Integrating molecular marker technologies, QTL mapping, genome-wide selection, and genome editing, researchers have unlocked new avenues for optimizing traits such as forage quality, stress tolerance, and productivity. With contributions from leading experts, this book presents current advancements, challenges in implementation, and future perspectives for breeding programs around the globe. Designed as a valuable resource for scientists, academicians, and policymakers, this book aims to bridge the gap between fundamental research and practical application, facilitating the development of superior forage cultivars having higher yields, improved quality and resilience to biotic as well as abiotic stresses through state-of-the-art genomic technologies.

We extend our deepest gratitude to all the contributing authors for their invaluable expertise, insightful research, and dedication in shaping this volume. Their contributions have been instrumental in advancing the understanding of genomic approaches for forage crop improvement, and their collective efforts have enriched this work.

Additionally, we sincerely thank the Springer Nature editorial team for their professional guidance, editorial support, and unwavering commitment to excellence in academic publishing. Their expertise has been vital in bringing this book to fruition, ensuring its high-quality presentation and accessibility to a global audience.

Srinagar, Jammu and Kashmir, India Shabir Hussain Wani
Ludhiana, Punjab, India Satbir Singh Gosal

Contents

About the Editors

Shabir Hussain Wani is Assistant Professor, Genetics and Plant Breeding at Sher-e-Kashmir University of Agricultural Sciences and Technology, Srinagar, J&K, India. As per Google Scholar, he has an H-index of 53 and i-10 index of 197. He has published more than 300 papers/chapters in journals and books of international and national repute. He served as guest editor and reviews editor for the journal *Frontier in Plant Science* (2015–2018). Presently, he is serving as Associate Editor for *PCTOC*, Springer, and reviews editor for *Frontiers in Plant Science*, section Abiotic Stress in Plants. He has also edited several books on current topics in crop improvement for abiotic stress tolerance published by Springer Nature and CRC Press, USA. He was elected as Fellow of Linnean Society of London 2017. Recently he was felicitated with Award for Excellence in Research 2020 from Novel Research Academy, Puducherry, India. Recently he was selected as member of Expert Working Group (EWG) Wheat Initiative for Adaptation of Wheat to Abiotic Stress (AWAS), and Wheat Breeding Methods and Strategies (WBMS), Berlin, Germany, during January 2021 and 2022, respectively. He has been judged as Fellow of International Society for development and Sustainability ISDS, Japan. He was also selected as member, The Royal Society of Biology, London, UK, and member, NASI, India. Recently in the year 2024, he has been awarded the Fellow of the Society by Society of Wheat and Barley ICAR IIWBR Karnal. He is involved as principal breeder for wheat variety Shalimar wheat 3 released in the year 2023 and associated as co-breeder for more than 10 crop varieties. He is Principal Investigator for JKDST-funded research project on wheat molecular breeding for stripe rust resistance. He is also Co-PI for several research projects funded by DBT, DST, and other funding agencies. He has been actively involved in teaching of important UG and PG courses on Genetics and Biotechnology since the year 2013. Recently he has been listed in the highly ranked scholars in the field of Genetics as per Scholar GPS ranking 2024. He has also been ranked as No. 1 researcher at SKUAST–Kashmir, as per the Scopus Elsevier 2025. Currently, he is working as Scientist in the All India Coordinated Research Project on Seed (Crops) at SKUAST K, Shalimar.

Satbir Singh Gosal possesses B.Sc. (Med.) from P U Chandigarh, India, and M.Sc. and Ph.D. (Plant breeding) from Punjab Agricultural University, Ludhiana, India. He was awarded Fellowships by the Royal Society London and the Rockefeller Foundation (USA) for his Post-Doctoral Research at the University of Nottingham, England and John Innes Centre Norwich, England. He has served Punjab Agricultural University in various capacities such as Professor Biotechnology, Director School of Agricultural Biotechnology, Additional Director Research, and Director of Research. He has also served FAO/IAEA, Vienna, Austria, and took tissue culture expert mission to Iraq during 1997. He has rigorous training on "Biosafety of GM crops" from Dan Forth Centre for Plant Science Research, St. Louis; APHIS, EPA (USDA), USTDA, Washington, DC, USA. He has been an honorary member of Board of Assessors, Australian Research Council, Canberra, President of Punjab Academy of Sciences, Elected member (Fellow) of Plant Tissue Culture Association (India), and Fellow of Indian Society of Genetics and Plant Breeding. He is a recipient of Distinction Award by Society for the promotion of plant science research, Jaipur, India (2009), Fellow of Punjab Academy of Sciences, advisory member of several universities/institutes in the area of Biotechnology, served as a member of Review Committee on Genetic Manipulation (RCGM) for 3 years at Department of Biotechnology (DBT), Government of India, New Delhi, and a member of panel of experts in area of Biotechnology for

National Fund for Strategic Research of Indian Council of Agricultural Research, New Delhi. He has participated in more than 125 national/international conferences/meetings held in India, England, Scotland, Yugoslavia, Philippines, Indonesia, Thailand, The Netherlands, Malaysia, Singapore, Austria, Iraq, P R China, Australia, Mexico, Germany, and the USA. He has guided more than 75 (M.Sc. and Ph.D.) students for theses research on various aspects of Plant Tissue culture and Plant Transformation. He has executed more than 20 externally funded research projects funded by various national and international organizations such as Punjab State Government, ICAR, DBT, DAC NATP, FAO/IAEA, and The Rockefeller Foundation, USA. He has more than 200 research papers in refereed journals of high repute, 135 research papers in conference proceedings, several TV/Radio talks, and 30 book chapters. He has co-authored five laboratory manuals, one textbook, and several edited books. Currently, he is serving as Vice Chancellor (CEO) of Punjab Agricultural University, Ludhiana, India.

2 Genetic Diversity in India

Indian subcontinent is one of the world's megacentres of crop origin and crop plant diversity, as it presents a wide spectrum of ecoclimate ranging from humid tropical to semi-arid and from temperate to alpine. The Indian gene centre possesses a rich genetic diversity in native grasses and legumes. There are reports of 245 genera and 1256 species of Gramineae, of which about 139 species belonging to 21 genera are endemic. One-third of Indian grasses are considered to have fodder value. Most of the grasses belong to the tribes Andropogoneae (30%), Paniceae (15%) and Eragrosteae (9%). Similarly, out of about 400 species of 60 genera of Leguminosae, 21 genera are reported to be useful as forage. The main centres of genetic diversity are peninsular India (for tropical types) and North Eastern Region (for subtropical types) besides some micro-centres for certain species. Major forage genera exhibiting forage biodiversity include legumes like *Desmodium*, Lablab, *Stylosanthes*, *Vigna*, *Macroptilium*, *Centrosema*, etc.; grasses like *Bothriochloa*, *Dichanthium*, *Cynodon*, *Panicum*, *Pennisetum*, *Cenchrus*, *Lasiurus*, etc.; and browse plants such as *Leucaena*, *Sesbania*, *Albizia*, *Bauhinia*, *Cassia*, *Grewia*, etc. Besides, India possesses enormous diversity of minor and under-utilized fodder species such as *Agrostis alba*, *Desmodium parvifolium*, *Leptochloa fusca*, *Potentilla fruticosa*, *Rhynchosia minima* and *Salvadora persica* (Bhag 2007).

3 Forage Crop Varieties

Many institutes are working on the genetic improvement of forage crops, viz. Indian Grassland and Fodder Research Institute (IGFRI), Jhansi; All India Coordinated Research Project on Forage Crops and Utilization with a network of 22 coordinated centres and 26 cooperating testing locations for multilocational research and testing; ICAR—Central Arid Zone Research Institute, Jodhpur; Choudhary Charan Singh Hissar Agricultural University, Sharan Singh Hissar Agriculture University, Hisar; Govind Ballabh Pant University of Agriculture and Technology, Pantnagar; Tamil Nadu Agriculture University, Coimbatore; Panjab Agriculture University, Ludhiana; Anand Agricultural University, Anand; and many more institutes and organizations. In addition to this, research work on forage aspects was intensified in All India Coordinated Projects on sorghum, maize and pearl millet. Diversity collected through exploration trips and several exotic germplasm from international institutes has been utilized to develop superior varieties through direct selection or through utilization in hybridization programme. A large number of forage varieties have been developed using exotic germplasm introduced from various countries. These include Nandi (selection from African germplasm) of *Setaria anceps*; Giant Bajra (cross between local and Australian bajra) of *Pennisetum americanum*, Punjab Guinea grass -1 (a direct introduction CPI-59985 from Australia) of *Panicum maximum*, Marwar Anjan (a clonal selection from EC014369) of *Cenchrus ciliaris*; and

Marwar Dhaman (a clonal selection from EC017655) of *Cenchrus setigerus*. Some of the introduced material has been directly released as a variety like Mescavi and Fahl (introductions from Egypt) of *Trifolium alexandrinum*. In *Vigna unguiculata*, Kohinoor, a selection from material introduced from Iran, has been released as a variety. A list of some promising forage varieties released and conserved in the National Gene Bank (NGB) at ICAR—National Bureau of Plant Genetic Resources (ICAR-NBPGR) is given below (Table 1).

4 Exotic Germplasm of Forage Genetic Resources

Thec introduction of germplasm is neccessary for catering the needs of plant breeders to boost crop improvement programes wherein ICAR-NBPGR has been actively involved since 1976. To procure these exotic germplasm, international and national obligations are followed. India being a signatory to various relevant international agreements on access to Plant Genetic Resources (PGR) and sharing of benefit arising out of their use including Convention of Biological Diversity (1992), International Treaty on Plant Genetic Resources for Food and Agriculture (2001) and the Global Plan of Action (1996), the exchange of genetic resources of the Annex 1 crop of the treaty (35 food crop genera and 29 forage crops) would be under the conditions of a Standard Material Transfer Agreement (SMTA) as per the Governing Body of Treaty. A large number of varieties of forage crops have been developed in India using exotic material introduced from various countries and distributed to different institutes for utilization in genetic improvement programmes (Table 2).

5 New Forage Crops and Wild Species

ICAR-NBPGR has been instrumental in introducing new forage crops and wild species of cultivated forage crops in India. For arid zones of the country perennial and evergeen, *Atriplex* known for salt tolerance, was introduced from Tunisia. Promising introductions include *A. canescens* (EC129768), *A. halimus* (EC1297670) and *A. nummularia* (EC129766) for drought tolerance. Fast-growing, drought tolerant tree species like subabul (*Leucaena leucocephala*) well known for its use as fodder, green manure and fuel were introduced from France, Malawi, Sierra Leone, the UK and the USA (Joshi et al. 2009). Among these notable introductions are K8 from Philippines and variety El Salvador from Australia, which are good for fodder and fuel purposes. Wild species, being rich in genetic base, generally are more variable than the corresponding crop. Plant breeders utilize wild species mainly for sources of resistance to biotic and abiotic stresses as well as for genetic enhancement. In total, 986 wild species in forage crops have been introduced from outside the country (Table 3).

Table 1 Varietal germplasm conserved at the National Gene Bank, ICAR-NBPGR

Botanical name	Donor identity	No. of varieties
Cenchrus ciliaris	IGFRI- 727, IGFRI-3108, (CAZRI-75), Marwar Anjan, (Gujarat Anand Anjan Grass-1), Bundel Anjan 4 (IGFRI 67-365)	4
Cenchrus setigerus	TNCS 265 (CO 2), Bundel Dhaman 4 (IGFRI 96-706), (GAAG-1, RCCB-2 (Bikaneri Dhaman)	3
Dichanthium annulatum	Marvel 2008-1, Marvel-09-4, Bundel Marvel Grass 2013-2 (JHD 2013-2)	**2**
Trifolium alexandrinum	Mescavi, Wardaan (S-99-1), JB-5//,BL-180,PRC-3, HFB 700 (Hisar Berseem 2), JBSC-1 (Bundel Berseem single Cut-1),BL 43 (PC 75),Jawahar Berseem 05-9 (JB 05-9), PC 91(BL 44), JHB-17-1, JHB-17-2, BL 45 (BM 12)	12
Trifolium repens	Palampur composite PWC-25 (him Palam white Clover-1)	2
Stylosanthes hamata	Phule Kranti-95	1
Cyamopsis tetragonolobus	BUNDEL GUAR-1, BUNDEL GUAR-2, BUNDEL GUAR-3	3
Vigna unguiculata	UPC-607, C-30 (GFC-4), C-26 (GFC-3), C-14 (GFC-1), (GFC-2)	5
Chrysopogon fulvus	IGC 9903 (Bundel Dharselu Grass-1)	1
Medicago sativa	Anand-2 (GFS-4), Anand Lucerne-3, GAUL-1, GAUL-2, TNLC 12 (TNAU Lucerne CO2), RRB-07-1 (Krishna), ALAMDAR-51, Lucerne CO 3 (TNLC 14), TNLC 15 (CO 4/ RCP 2-1)	9
Zea mays	African tall, BRM-1	2
Avena sativa	Sabzaar, Jawahar Oat-1, OS-7, OS-6, JHO-813, JHO- 816, JHO-2001-3, KENT, JHO 99-1(Bundel Jai-991), OS-346, JO 03-91, Kashmir Oats-1 (SKO-90), Shalimar Fodder Oats-3 (SKO-96), JO-2000-61, Bundel Jai-2013, JHO-99-1, NDO-1, JO-03-93 (Jawahar Oat 03-93), Shalimar Fodder Oats-4, OS 377, Bundel Jai-2010-1 (JHO-2010-1), OL 10 (OL 1709), NDO 10 (Narendra Jayee 10), UPO-06-1 (Pant Forage Oat-3), Bundel Jai-20,091 (JHO 2009-1), Central Oat (OS 403), Jawahar Oat 04-315 (JO 04-315), OL 1802, OL 1804, Central Oat OS 405 (OS405), OS 403, Phule Surabhi (RO-11-1), OL 1802-1, OL 1769-1, Central Fodder Oat 424 (OS 424), SKO-225 (Shalimar Fodder Oats-6), JHO 2012-2 (Bundel Jai 20,122), OL 1861, OL 1869-1 (OL 13), HFO 427, HFO 607, UPO-10-2 (Pant Forage Oat-4), OL 14 (OL 1874), JO-05-304 (MC), JO-10-506 Dual Purpose, HFO 529, HFO 611,OL 1896, OL 1876-2, HFO 707, HFO 806, OL 1874-1	53
Pennisetum glaucum	GAINT BAJRA, CO-8, CO-7, FMH-3, NDFB-3 (Narendra Chara Bajra-3)	5

(continued)

Table 1 (continued)

Botanical name	Donor identity	No. of varieties
Pennisetum pedicellatum	BUNDEL-1	1
Pennisetum typhoides	GFB-1, HHB-50, HHB-60, HHB-67, HHB-94, HC-4, HC-10, HHB-117, HC-20A, ICTP-8203, RAJKO, AHB-1666, AIB-157, J-2340, ICMV-221ICMV-88,904 (MP-221), GFB-1, MH-1307 (GHB-732) HYBRID, MH-1328 (GMB-757), MHI 192 (B-2301) HYBRID, HHB-197 (HYBRID), JKBH 676 (MH-1299), Nandi-61 MH 1548(NMH-70) Hybrid, Nandi-65 MH 1549(NMH-71) Hybrid, RHB-177(MH-1486). PAC 909 (MH 1435) hybrid,RHB-177(MH-1486) RHB-494, HHB-67	29
Sehima nervosum	IGS-9001(Bundel Sain Ghas-1)	1
Chrysopogon fulvus	IGC-9903 (Bundel Dhawalu Grass-1)	1
Vigna unguiculata	Kohinoor	5
Vigna umbellata	Bidhan Rice bean-3 (KRB-19), SURABHI	2
Clitoria ternatea	JGCT 2013-3 (Titli)	1
Sorghum bicolor	Pant Chari-6, Pusa Chari-9, PC-23, Pusa Chari hyderabad-106, CO(FS)29 IGFARI PC-23, SPV 1595(CSV 18), SPH 1567, C-10-2, S-1049, GFS-3 UTMCH1302(CSH 24 MF)H, UPMC-503, Anand Forage Sorghum-1, Jaicar Nutrigraze (CSV 43 BMR/SPV-2018), CSV 21F (SRF-286), PSC 4 (Punjab Sudax Chari 4), CSV 40 F (SPV 1797 (UTFSH 2), SPV 2317 (UTFS 85), SEJ 198	19
Sorghum vulgare	AKR 354	1
Sorghum halepense	Sej198	1
Panicum maximum	JHGG 04-01(BUNDEL GUINEA-2) FD-2190, JHGG 96-5	**3**
Heteropogon contortus	IGHC 03-4	1
Desmanthus virgatus	TND 1308 (CO 2), TSLH-1	1
Lolium perenne	RLSB-11-50 (Jaisalmeri Sewan), him Palam Rye grass 1PBRG 2 (Punjab ryegrass no. 2)	2
Setaria anceps	S-18, him Palam Setaria grass 2 (S 25)	1
Brachiaria ramosa	GPUAT 2	1
Lasiurus scindicus	RLSB-11-50 (Jaisalmeri Sewan)	1
Mucuna pruriens	Arka Dhanvantari, Arka Daks	2

Table 2 Important trait-specific introductions in forage crops

Crop/EC no. and source country	Traits
Brachiaria hybrid variety Mulato EC 549024-25/USA	Stoloniferous growth, excellent forage production, vigorous regrowth, excellent palatability, drought tolerant and produces forage round the year
Dactylis glomerata/EC871140/ USA	Late maturing, excellent forage quality and increased winter tolerance
Deschampsia antarctica/Antarctica hair grass/ EC631954/Chile	New crop, native to Antarctica
Eleusine coracana/EC915296-5299/Kenya	Highest iron (Fe) content
Eleusine coracana/EC915337-5339/Kenya	High zinc (Zn) content
Eleusine coracana EC915340-5344/Kenya	High calcium content
Elytrigia repens/EC586940/USA	Var. Eversett, advanced generation synthetic cultivar for high rhizome production ability to spread by rhizomes and used for land stabilization and reclamation
Medicago sativa/EC499771-72/ USA	Highly resistant to *Aphanomyces* root rot and northern root knot nematode, resistant to *Phytophthora*, pea aphid, spotted aphid and moderately resistant to *Verticillium* wilt, anthracnose race 1 and stem nematode, high forage yield under dry land condition
Medicago sativa/EC596671/USA	Var. OK 190, unique combination of broad genetic base, resistance to blue alfalfa aphid, spotted alfalfa aphid and *Phytophthora* rot
Medicago sativa/EC596673/USA	Var. OK 207, broad genetic base population provide resistance to blue alfalfa aphid biotype BAOK 90 and spotted alfalfa aphid
Medicago spp./EC271425-29/ USA	Produce high forage, cold tolerant and resistant to aphids
Pennisetum typhoides ssp. *montana*/EC473259/USA	Potential for increase growth rate, resistance to rust, smut and leaf spot
Pennisetum glaucum/iC802313/ USA	Genetic male sterile (ms3) stock NM-10H
Sorghum bicolor/EC750258-750459/Japan	Core set
Trifolium pratense/EC560447-448/USA	Short internodes, plants usually present a rosette appearance but flower sparingly under long day conditions
Trifolium pratense/EC578957/ USA	Var. freedom, free from pubescence, good for hay making as it permits faster drying and reduces dustiness
Trifolium pratense/EC560447-448/USA	Short internodes, plants usually present a rosette appearance but flower sparingly under long day
EC671603-EC671629/USA	High fodder yielding lines
Triticale/EC537921/Mexico	Hexaploid winter triticale, good forage quality
Triticale/EC537922/Mexico	Hexaploid intermediate triticale, good forage quality

(continued)

Table 2 (continued)

Crop/EC no. and source country	Traits
Triticale/EC534274/USA	High-yielding, superior forage quality
Triticale/EC467937/Canada	Variety bobcat, sprouting tolerant and resistant to stem rust and leaf rust
Sorghum	EC842897-843045;EC843394-843396
Pearlmillet	EC842752-842893;EC843377-843393
Pigeonpea	EC843233-843289
Finger millet	EC843290-843341, EC843342-843352
Proso millet	EC843170-843232
Little millet	EC842894
Barnyard millet	EC842374-842747

Table 3 Wild species introduced in forage crops

Genera	Species
Aeschynomene	*A.americana, A. brasiliana , A. falcata , A. scabra , A.sensitiva*
Agropyron	*A. cimmericum*
Atriplex	*A. canescens , A. halimus , A. hortensis*
Brachiaria	*B. ruziziensis*
Casuarina	*Casuarina cunninghamiana* ssp. *cunninghamiana , C. equisetifolia* ssp. *equisetifolia, C. Glauca, C. junghuhniana* ssp. *junghuhniana*
Clitoria	*C. ternetea*
Desmanthus	*Desmanthus acuminatus, D. bicornutus, D. covillei, D. fruticosus, D.illinoensis, D. leptophyllus, D. paspalaceus, D. pernambucanus, D. pubescens, D. subulatus, D. tatuhyensis, D. virgatus*
Desmodium	*D. intortum* and *D. uncinotum*
Elymus	*E. haffmannii*
Eleusine	*E. africana, E. indica, E. intermedia, Eleusine floccifolia, E. jaegeri, E.kigeziensis , E. multiflora*
Eucalyptus	*E. alba, E. bakeri, E. cholorophylla, E. curtisii, E. deglupta , E. exserta, E. gamophylla, E. gillenii, E. grandis, E. herbertiana, E. horistes, E. infera, E. longirostrata , E. mannensis* ssp. *mannensis, E. normantonensis, E.nudicaulis, E.oxymitra, E. pachyphylla, E. polybractea, E. socialis, E. urophylla,E. viridis*
Lathyrus	*L. amphicarpos, L. aphaca, L. anuus, L. cassius, L. chloranthus, L. cicera, L. cirrhosus, L. clymenum, L. gorgoni, L. hierosolymitanus, L. hirsutus, L. inconspicuus, L. latifolius, L. nissolia L. ochrus, L. odoratus, L. pratensisL. rotundifolius, L. setifolius, L. sphaericus, L. sylvestris, L. tingitanus, L. tuberosus, L. vinealis*
Lotus	*L. conimbricensis, L. coribricensis, L. ornithopodioid, L. pendunoulatus, L. purshianus, L. uliginosus*

(continued)

Table 3 (continued)

Genera	Species
Macroptilium	*M. atropurpureum, M. africanum ,M. axillare, M. lathyroides*
Medicago	*M. arabica, Medicago arborea , M. Ciliaris, M. coronata, M. disciformis , M. intertexta, M. laciniata, M. littoralis, M. lupulina, M. minima, M. orbicularis, M. polymorpha, M. radiata, M. rigidula, M. rugosa, M. scutellata, M. tornata, M. truncatula, M. turbinate, M.liciruata, M.littoralis*
Melilotus	*M. indicus, M. messanensis, M. officinalis, M. segetalis, M. sulcatus*
Pennisetum	*P. macrourum, P. mezianum, P. polystachion, P. procerum, P. purpureum, P. ramosum , P. sphacelatum , P. trachyphyllum, P. unisetum*
Raphanus	*R. raphanistrum Rapistrum rugose, R. densiflora, R. elegans, R. malacophylla, R. minima, R. reticulata*
Sorghum	*Sorghum caudatum , S. durra*
Stylosanthes	*Stylosanthes angustifolia, S. calcicola, S. capitata, S. guianensis, S. fruticosa, S. gracilis, S. guianensis, S. hamata, S. humilis, S. ingrate S.leiocarpa ,S. macrocarpa, S. macrocephala S. mexicana, S. scabra, S. subsericea, S. sympodialis, S. viscosa*
Trifolium	*T. alpestre, T. angustifolium, T. arvense, T. aureum, T. apertum, T. baccarinii, T. balansaa, T. berytheum, T. billardiari, T. campestre, T. canescens, T. caucasicum, T. cherleri, T. clypeatum, T. constantinopolitanum, T. dasyurum, T. dichroanthum, T. diffusum, T. echinatum, T. fomentosum, T. fragiferum, T. grandiflorum, T hirtum, T. incarnatum, T. lappaceum, T. leucanthum, T. ligusticum, T. meironenseT. michelianum, T. nigrescens, T. obscurum, T ochroleucon, T pannonicum, T. pilulare, T. plebeium. T. pratense, T. purpureum, T salmoneum, T. scabrum, T scutatum, T. squamosum, T. stellatum, T. striatum, T. striatum var. spinescens, T. subterraneum, T. tomentosum, T. vavilovii, T. vesiculosum*
Vicia	*V.hyaeniscyamus , V. hybrida , V johannis , V.johannis* var. *johannis , V. kalakhensis , V. melanops ,V narbonensis , V sativa , V sativa* subsp. *amphicarpa , V.sativa* subsp. *cordata, V. sativa* subsp. *marcocarpa , V.sativa* subsp. *nigra , V. serratifolia*
Zornia	*Z.brasiliensis, Z. diphylla, Z. glabra, Z. glochidiata, Z. latifolia*

6 Forage Germplasm Conservation

Forage genetic resources are conserved in a network mode using seeds and field gene bank conservation strategies at various institutes across the country. Field gene bank at IGFRI, Jhansi, and at centres of AICRP (forages). ICAR-IGFRI, being a National Active Germplasm Site (NAGS) for forage crops, presently holds approximately 10,116 accessions representing about 71 forage crops as seed in Medium-Term Storage (MTS) facility. These include cereal forages (3437 accessions), cultivated legumes (2682 accessions), range legumes (746 accessions), range grasses (3175 accessions) and 76 accessions of other species of forage value (Sahay et al. 2019).

National Genebank at ICAR-NBPGR conserves the seeds of orthodox species of forages. In total, 7427 accessions (http://genebank.nbpgr.ernet.in/SeedBank/Default.aspx) comprising forage cereals (1715 accessions belonging to 14 species), grasses (3244 accessions belonging to 173 species), range legumes (1927accessions belonging to 70 species), forage millets (458 accessions belonging to 11 species) and other species of forage value (83 accessions) have been conserved at NGB, ICAR-NBPGR.

Forage grasses (173)
Acrachne racemose, Alloteropsis cimicina, Andropogon pumilus, Apluda mutica, Aristida adscensioni, Arthraxon longifolius, Arthraxon prionodes, Arundinella spicata, Arundinella tuberculate, Atriplex hortensis, Bambusa vulgaris, Bothriochloa intermedia, Bothriochloa odorata, Bothriochlo pertusa, Brachiaria ramosa, Brachiaria reptans, Brachiaria setigera, Brachypodium distachyon, Brachypodium pinnatum, Brachypodium sylvaticum, Bromus catharticus, Bromus inermis, Bromus japonicas, Bromus unioloides, Canarium strictum, Capillipedium hugelii, Capillipedium parviflorum, Cenchrus biflorus, Cenchrus ciliaris, Cenchrus prieurii, Cenchrus setigerus, Chloris barbata, Chloris dolichostachya, Chloris gayana, Chloris virgata, Chrysopogon aciculatus, Chrysopogon fulvus, Chrysopogon montanus, Chrysopogon zeylanicus, Chrysopogon zizanioides, Coix lacryma-jobi, Cymbopogon flexuosus, Cymbopogon jwarancusa, Cymbopogon martini, Cymbopogon scheonanthus, Cynodon dactylon, Dactylis glomerata, Dactyloctenium aegyptium, Dactyloctenium aristatum, Dactyloctenium sindicum, Desmostachya bipinatata, Dichanthium annulatum, Dichanthium caricosum, Digitaria adscendens, Digitaria ciliaris, Digitaria granularis, Digitaria pennata, Digitaria spp., *Dinebra retroflexa, Echinochloa crus-pavonis, Echinochloa frumentacea, Echinochloa stagnina, Eleusine coracana, Eleusine flagellifera, Eleusine indica, Elionurus royleanus, Enneapogon elegans, Eragrostiella bifaria, Eragrostis ciliaris, Eragrostis diarrhena, Eragrostis tef, Eragrostis tenella, Eragrostis tremula, Eragrostis unioloides, Eremopgon foveolatus, Eremostachys superba, Erianthus elephantinus, Erianthus procerus, Eriochloa procera, Eriophorum* spp., *Eshne dispar, Euchlaena maricana, Eulalia argentea, Eulaliopsis* spp., *Festuca arundinacea, Festuca robra, Hackelochloa granularis, Hemarthira* spp., *Heteropogon contortus, Imperata cylindrica, Isachne disperma, Ischaemum pilosum, Ischaemum rugosum, Ischaemum* spp., *Iseilema laxum, Lasiurus hirsutus, Lasiurus scindicus, Leptochloa chinensis, Lolium multiflorum, Lolium perenne, Lolium temulentum, Melanocenchris jacquemontii, Ophiuros exaltatus, Panicum antidotale, Panicum maximum, Panicum repens, Panicum sumatrense* ssp., *Psilopodium, Panicum turgidum, Paspalidium flavidum, Paspalum dilatatum, Paspalum distichum, Paspalum* spp., *Pennisetum glaucum, Pennisetum glaucum* ssp. *violaceum, Pennisetum hohenackeri, Pennisetum orientale, Pennisetum pedicellatum, Pennisetum purpureum, Pennisetum squamulata, Pennisetum typhoides, Perotis indica, Phalaris arundinacea, Phalaris minor, Phleum pratense, Phragmites australis, Phragmites karka, Polypogon monspeliensis, Pseudarthria viscida, Pseudoraphis spinescens, Pseudosorghum fasciculare, Rhynchelytrum repens, Rhynchelytrum villosum, Saccharum arundinaceum, Saccharum munja, Saccharum officinarum, Saccharum robustum, Saccharum spontaneum, Schoenefeldia gracilis, Sehima nervosum, Spilanthes acmella, Sporobolus diander, teosinte, Teramnus labialis, Tetrapogon tenellus, Themeda* spp., *Tripogon roxburghianus, Typha angustifolia, Typha elephantine, Uraria lagopoides, Urochloa panicoides, Vetiveria zizanioide*

Forage legumes (67 species)
Acacia torta, ailanthus excels, Alysicarpus longifolius, Alysicarpus vaginalis, Atylosia platycarpa, Ciltoria ternatea, Clitoria ternatea, Coronilla varia, Desmanthus virgatus, Desmodium elegans, Desmodium gangeticum, Desmodium gyrens, Desmodium heterocarpon, Desmodium intortum, Desmodium triquetrum, Lablab purpureus, Lathyrus sativus, Leucaena pulverulenta, Lotus corniculatus, Lupinus spp., *Macroptilium atropurpureum, Medicago falcata, Medicago lupulina, Medicago monantha, Medicago polymorpha, Medicago sativa, Melilotus albus, Melilotus dentatus, Melilotus elegans, Melilotus indicus, Melilotus infestus, Melilotus italicus, Melilotus officinalis, Melilotus speciosus, Melilotus suaveolens, Mucuna pruriens, Mucuna pruriens* var. *utilis, Onobrychis vicifolia, Oplismenus burmannii, Oropetium thomaeum, Rhynchosia minima, Rottboellia exaltata, Sesbania aculeata, Sesbania bispinosa, Sesbania cannabina, Sesbania rostrata, Sesbania sesban, Sesbania tetraptera, Smithia setulosa, Stylosanthes hamata, Stylosanthes mexicana, Trifolium alexandrinum, Trifolium campestre, Trifolium fragiferum, Trifolium pratense, Trifolium repens, Trifolium resupinatum, Trigonella caerulea, Trigonella corniculata, Trigonella foenum-graecum, Vicia hirsuta, Vicia sativa, Vicia sativa* subsp. *amphicarpa, Vigna luteola, Vigna umbellata, Vigna unguiculata, Vigna unguiculata* subsp. *sesquipedalis*
Forage cereals (14 species)
Avena abyssinica, Avena brevis, Avena byzantine, Avena eriantha, Avena fatua, Avena longiglumis, *Avena magna, Avena sativa, Avena sterilis, Avena sterilis* subsp. *ludoviciana, Avena strigose, Oryza sativa,* *Zea mays, Zea mexicana*
Forage millets (11 species)
Setaria anceps, Setaria glauca, Setaria helvola, Setaria italic, Setaria speciose, Setaria verticillata, Setaria viridis, Sorghum bicolor, Sorghum bicolor subsp. *Drummondii, Sorghum halepense, Sorghum vulgare*

7 Registered Germplasm in Forage Crops

With the objective of giving credit to the scientific community involved in the development of high-yielding varieties and left with a large number of promising experimental material (including parents and inbred lines) and to facilitate the flow of germplasm among them, ICAR constituted a committee named 'Plant Germplasm Registration Committee' (PGRC). The committee chaired by DDG(CS) functions through ICAR-NBPGR with the Head of the Germplasm Conservation Division as Member-Secretary. The prime duty of this registration is to protect these germplasms from poaching. The standard guidelines are formulated in consultation with all stakeholders and are revised time to time as per the need. The details of the procedure which is fully online (Germplasm Registration Information System (GRIS)) are available on NBPGR website (https://www.nbpgr.ernet.in/). The information regarding new registrations are published in *Indian Journal of Plant Genetic Resources* and ICAR News regularly. Till date, 28 germplasms have been registered in the forage crops (Table 4).

Table 4 Trait-specific germplasm of forage crops registered by PGRC (http://www.nbpgr.ernet.in:8080/registration/InventoryofGermplasm.aspx) 2024

Botanical name	National identity/ INGR no.	Donor identity	Pedigree	Novel traits	Developing institute
Trifolium alexandrinum	IC0593646; INGR12009	IL-11-239	*T. alexandrinum* x *T. apertum* cross	Stabilized black (dark tan) seeded against normal yellow seeded	ICAR-IGFRI, Jhansi, U.P.
Trifolium alexandrinum	IC0593647; INGR12010	IL-11-238	Penta-1	Tetraploid pentafoliate	ICAR-IGFRI, Jhansi, U.P.
Trifolium alexandrinum	IC0593648; INGR12011	190-P3	Diploid Berseem	Stabilized self-incompatible tetraploid	ICAR-IGFRI, Jhansi, U.P.
Trifolium alexandrinum	EC329299; INGR12012	EC329299	Saidi ecotype	Self-compatible	ICAR-IGFRI, Jhansi, U.P.
Trifolium alexandrinum	IC0613360; INGR15026	IL-13-106	*T. alexandrinum* x *T. apertum* cross	Black seeded with pentafoliate leaf	ICAR-IGFRI, Jhansi, U.P.
Trifolium alexandrinum	IC524012; INGR5017	PBL-123	Spontaneous mutant	Purple leaf and flower	PAU, Ludhiana
Trifolium alexandrinum	IC567683; INGR9045	Penta-1	Spontaneous mutant of JHB 146	Pentafoliate mutant	ICAR-IGFRI, Jhansi, U.P.
Sesbania aculeata	IC546953; INGR6016	CSD123	Local collection from Khandauli, Agra	Very early maturing (120 days) and better tolerance to sodic soils	ICAR-CSSRI, Karnal, Haryanal
Cenchrus ciliaris.	IC0590889; INGR11062	IGFRI-CcSx –08/1	NA	Rare obligate sexual plant	ICAR-IGFRI, Jhansi, U.P.
Cenchrus ciliaris	EC397366; INGR19061	IG96-401	Selection from collected materials	Rich in sugar (more than 7%). Suitable for ensiling (silage preparation). Prostrate growth habit	ICAR-IGFRI, Jhansi, U.P.
Cenchrus ciliaris	IC0630758; INGR19062	IG 96-50	Selection from collected material	High water-soluble carbohydrate. Good for silage preparation. Leaf colour pale-low chlorophyll content	ICAR-IGFRI, Jhansi, U.P.

(continued)

Table 4 (continued)

Botanical name	National identity/ INGR no.	Donor identity	Pedigree	Novel traits	Developing institute
Cenchrus ciliaris	IC0630759; INGR19063	IG 96-96	Selection from collected germplasm	High water-soluble carbohydrate. Good for silage preparation	ICAR-IGFRI, Jhansi, U.P.
Pennisetum squamulatum	IC546955; INGR6017	Apomictic cytotype	Wild species	New apomictic cytotype with 2n = 56	ICAR-IGFRI, Jhansi, U.P.
Pennisetum pedicellatum	IC546954; INGR6018	Agros-4	Selection from germplasm	New cytotype with 2n = 72	ICAR-IGFRI, Jhansi, U.P.
Panicum maximum	IC567677; INGR9039	IG3/29/2-3(1)	M1 plant from 6× of IG 04-164	Triploid cytotype	ICAR-IGFRI, Jhansi, U.P.
Panicum maximum	IC567678; INGR9040	IG04-164	Local collection	High production progenitor of ploidy series	ICAR-IGFRI, Jhansi, U.P.
Panicum maximum	IC567679; INGR9041	IG3/29/2-5(1)	3/29/2(6×)X IG 04-164(4×)	Pentaploid cytotype	ICAR-IGFRI, Jhansi, U.P.
Panicum maximum	IC567680; INGR9042	IG3/29/2	BIII hybrid from IG 04-164	Hexaploid cytotype	ICAR-IGFRI, Jhansi, U.P.
Panicum maximum	IC567681; INGR9043	IG3/29/2-8(1)	3/29/2(6× X IG 04-164(4×) BIII	Octoploid cytotype	ICAR-IGFRI, Jhansi, U.P.
Panicum maximum	IC567682; INGR9044	IG3/29/2-9(1)	3/29/2 (6×) BIII hybrid	Monoploid cytotype	ICAR-IGFRI, Jhansi, U.P.
Panicum maximum	IC0616375; INGR15055	SRP#75	SPM92 × Riversdale	Rare self-compatible plant with obligate sexual mode of reproduction	ICAR-IGFRI, Jhansi, U.P.
Panicum maximum	IC0625987; INGR18024	GGPS 7×-1	Sexually derived progeny between facultative apomicts GGPS 8x-1 (2n = 8× = 64) and accession 3/29/2 (2n = 6x = 48)	A novel cytotype (heptaploid) (2n = 7× = 56). Member of ploidy series in Guinea grass	ICAR-IGFRI, Jhansi, U.P.

(continued)

Table 4 (continued)

Botanical name	National identity/ INGR no.	Donor identity	Pedigree	Novel traits	Developing institute
Panicum maximum	IC0625988; INGR18025	GGPS 11×-1	BIII hybrid between facultative apomicts GGPS 8x-1 (2n = 8× = 64) and accession 3/29/2 (2n = 6× = 48)	A novel cytotype (11-ploid) (2n = 11× = 88). Member of ploidy series in Guinea grass.	ICAR-IGFRI, Jhansi, U.P.
Pennisetum orientale	IC0625989; INGR18026	*P. ori* 6×-1	BIII hybrid of *P. ori* 4×-2	A novel hexaploid (6×) cytotype with 2n = 54 chromosomes. Recovered through BIII hybridization of a 4× (2n = 36) cytotype	ICAR-IGFRI, Jhansi, U.P.
Sorghum bicolor	**IC0612063; INGR15027**	SPV 2018	SPV 2018 = (SPV 462 X IS 21891)-3-1-1-1	High digestibility (IVOMD %), low lignin content; brown midrib (bmr)	ICAR-IIMR, Telengana
Sorghum bicolor	**IC0621690; INGR17016**	SGS 389	Selection from Mauranipur, Jhansi, Uttar Pradesh	Six stamens and two gynoecia against the normal sorghum genotype bearing three stamens and one gynoecium. It produces twin seeds but the spikelet bears six stamens and two gynoecia	ICAR-IGFRI, Jhansi, U.P.
Sorghum bicolor	**IC0632083; INGR19060**	S 614-13/ SPV2263	HJ 513 × IS 2389-1 p29-1-3-4 selfed up to F6 generation	IVDMD 56%. Anthracnose resistance	CCS HAU, Haryana
Sorghum bicolor	**IC0632070; INGR19064**	SM2254-8	Sorghum parent (126A) pollinated with maize line (CM208)	Sorghum forage line derived from intergeneric, sorghum x maize cross. Low HCN. High IVDMD	ICAR-IIMR, Telangana

8 Conclusions

The utilization of genetic diversity is a critical factor in enhancing productivity on a sustainable basis. Though improved varieties have been released in forage crops, attention is required to register the unique forage germplasm for further utilization in breeding programmes. Gap analysis for the unexplored areas should be given top priority. Today, the focus should be on as what type of forage crops are needed and how much diversity is required. Depending on the available information on passport and evaluation data, the development of new pools for important characteristics would be useful in easy and effective germplasm management and utilization. Awareness generation of the people at various levels (policymakers, scientific, administration, farmers, etc.) about the value of PGR wealth and its protection is essential. Further, there is an urgent need to increase the interaction between plant breeders and PGR workers. The interface among different stakeholders is likely to bring out new useful forage genetic resource management alternatives.

References

Bhag M (2007) Neglected and underutilized crop genetic resources for sustainable agriculture. Indian J Plant Genetic Resour 20:1–14

Joshi V, Tyagi V, Lal A (2009) Status of genetic resources of forage crops in India: a review. Indian J Plant Genetic Resour 22(3):243–252

Sahay G, Dikshit N, Singh T, Tyagi VC et al (2019) Forage genetic resources at ICAR-IGFRI Jhansi, accomplishments and future strategies. ICAR-Indian Grassland and Fodder Research Institute, Jhansi, p 34

Websites

http://www.nbaindia.org/ home page Biological Diversity Act https://www.nbpgr.ernet.in
http://www.nbpgr.ernet.in:8080/registration/InventoryofGermplasm.aspx
http://genebank.nbpgr.ernet.in/SeedBank/Default.aspx
http://www.biodiv.org home page CBD
http://www.biodiv.org/doc/legal/cbd-en.pdf full text of the Convention
http://www.fao.org/ag/cgrfa/itpgr.htm home page ITPGRFA
http://ext-ftp.fao.org/ag/cgrfa/it/ITPGRe.pdf full text of the treaty

Utilization of Genetic Resources Through Molecular and Genomic Approaches for Forage Barley Improvement

Gurjeet Singh, Pawan Saini, Basanagouda Gonal, Danishta Aziz, Gulab Khan Rohela, Belaghihalli N. Gnanesh, and Shabir Hussain Wani

1 Introduction

Among the grain cereal, barley (*Hordeum vulgare* L.) is a primitive food crop, which was domesticated first with other cereals, wheat, pea, lentil, cattle, sheep, and goats nearby, about 10,000 years ago in the Fertile Crescent (Smith 1998). It has been acknowledged as hardy crop, as it has been eaten by the Roman gladiators known as *hordearii*, which means barley man. It could have been used as food for mankind in the form of bread, soup, or porridges, but due to the evolution of rice,

G. Singh
Department of Plant Breeding and Genetics, Punjab Agricultural University (PAU), Ludhiana, Punjab, India

Rice Research Center, Texas A&M University, Agricultural Research and Extension Center, Beaumont, TX, USA

P. Saini (✉) · B. Gonal
CSB-Central Sericultural Research & Training Institute (CSR&TI), Pampore, Jammu and Kashmir, India
e-mail: pawansaini-coapbg@pau.edu

D. Aziz
Division of Entomology, Faculty of Agriculture, Sher-e-Kashmir University of Agricultural Sciences and Technology – Kashmir (SKUAST- K), Wadura, Sopore, Jammu and Kashmir, India

G. K. Rohela
CSB-Central Sericultural Research & Training Institute (CSR&TI), Mysuru, Karnataka, India

B. N. Gnanesh
ICAR JSS – Krishi Vigyan Kendra, Mysore, Karnataka, India

S. H. Wani
National Seed Project (NSP), Sher-e-Kashmir University of Agricultural Sciences and Technology – Kashmir (SKUAST-K), Srinagar, Jammu and Kashmir, India

S. H. Wani, S. S. Gosal (eds.), *Accelerated Plant Breeding, Volume 5*,
https://doi.org/10.1007/978-3-032-04527-0_2

wheat, and maize, the notability of barley subsided (Ullrich 2011). It is a prime crop cultivated across the globe particularly in the temperate regions. It is an inbreeding crop with ploidy level of 2× to 6× and a genome size of 5.2 GB (Mayer et al. 2012). It is hardy crop grown in extreme climatic conditions, such as high-altitude and high-latitude regions of the Himalayan nations, the African continent, and deserts (Baik and Ullrich 2008). A significant amount of barley comes from France, Spain, Turkey, the United Kingdom, Egypt, China, Denmark, and Kenya. The major utilization of this crop is in the brewing industry for the preparation of good-quality beer. Nowadays, more than 50% of barley is utilized for feed, followed by malt (30%); food (2–3%), and seed (5%) (Ullrich 2011).

It is an annual and tufted plant growing under temperate conditions with height of 0.6 to 1.2 m (Ecocrop 2011). With reference to wheat and rye, barley have three one-flowered spikelets at each rachial node, arranged alternately on opposite sides (Duke 1983; Ecocrop 2011). The barley is classified as spring and winter; two-rowed and six-rowed; hulled and hull-less; and food and feed/brewing (OECD 2004; CFIA 2005; Baik and Ullrich 2008). Barley is widely adapted to extreme environmental conditions such as drought and cold and alkaline stresses (Baum et al. 2007). However, owing to greater drought tolerance and the most tolerant of cereal grains to saline and alkaline soils, forage barley could be able to replace cool-season crops such as wheat in some cropping systems for fodder purposes (Bornare et al. 2012; Anderson et al. 2012). Although many small grain crops provide a source of high-quality forage, barley provide the best option for fall grazing due to its rapid growth when planted early (Warrick et al. 2002). The grains of barley are not only utilized for malt preparation, but also for feeding ruminant and non-ruminant animals (livestock, fish, and poultry). The barley grain is generally given as feed to cattle, poultry, goat, swine, and sheep. It is traditionally dry-rolled or ground for beef cattle rations (Blake et al. 2011). It has been reported that for the purpose of poultry production, barley is ground and supplemented with phytase and β-glucanase (Kellems and Church 2002). Barley varieties suited for fish feed have also been developed (Bregitzer et al. 2007).

Greater water use efficiency, as well as greater drought tolerance, may allow forage barley breeding to find a niche in semi-arid dryland environments or where irrigation becomes limited. Therefore, barley becomes the choice of crop in temperate regions for livestock feed as it can tolerate limited moisture, can be grown in marginal soil, and have short growing season (Blake et al. 2011). Several malting barley varieties, which are not fulfilling the malting standards, are fed to dairy animals in North America. The nutritional quality of barley is very much important for deciding the utilization of varieties for food, feed, and brewing industry. The edaphic factors like soil moisture can very much influence the grain nutritional content. The low soil moisture content leads to low accumulation of starch content and enhances the grain storage protein content, which lowers the malt extract. Also, the excess of soil moisture leads to pre-sprouting of grains and accumulation of toxins like aflatoxins, making the varieties unsuitable for malting and utilized for feed purposes (Blake et al. 2011). The primary requirement for enhancement of forage barley production is the development of climate-resilient dual-purpose barley varieties through integration of genetic resources with genomic breeding methodology.

2 Barley as a Forage Crop

Barley is mainly grown as a grain crop, although in some regions, it is grazed or harvested during vegetative stage for green forage. In the winter barley-growing regions of the United States, barley may be used for pasture, hay, or silage, making-forage quality an important trait in winter varieties (Poehlman 1985). Barley straw is an excellent candidate as a cellulosic ethanol feedstock. Barley is extensively utilized for forage purposes due to its adaptation of water use efficiency, tolerance for abiotic stresses, and cultivation under temperate environment (Winter 2005). Barley is the principal feed grain in northern and eastern Europe, Canada, and in northern United States. The largest use of barley is for livestock feed, to supply digestible energy. In most cases, the whole barley kernel is rolled, ground, or flaked prior to being fed, to improve digestibility (OECD 2004). About 15% of barley grain feed to cattle is excreted whole and undamaged, suggesting potential for livestock to disperse viably barley seed after consumption (Kaiser et al. 1997). High-protein barley is suitable for animal feed.

The harvest timing of barley is critical for the optimization of yield, which ultimately influences the milk production. Nowadays, attempts have been made for the development of forage-specific barley varieties. The awnless barley varieties are commonly used for forage/silage, as mature awns irritate the dairy animals, and if the cutting is done during or after flowering, awned varieties could be utilized for silage purposes (Anderson et al. 2012). For maximization of forage nutritious value for enhancement in milk production, the silage could be harvested between mid-boot and mid-dough stage (Stallings 1997; Jones et al. 2004). The barley silage rich in fiber with low protein content is good for feeding ruminant livestock (OECD 2004). The low temperature with longer day length at higher elevations result into less lignification, which allows for delayed harvesting and leads to increased biomass production and retains the forage quality (Yu et al. 2003). The forage barley seems to have low concentration of cell wall acid detergent fiber (ADF) and lignin than other small grain forages (Ditsch and Bitzer 2005).

Sometimes, forage barley can be directly grazed or cut in the early phases of the crop planting, without affecting grain yield. Barley forage can be used for bedding and making hats and also for cellulose pulp use (Duke 1983). Barley straw is one of the existing highest-quality cereal straws (Chriyaa and Amri 1997). For the best-quality feed and fast regrowth, potential crops should be cut or grazed at a height of 30–40 cm. The crop fertilizer should be applied after each cut to ensure high-quality regrowth. Before the formation of awns, cutting should be well finished; this will decrease the forage's palatability. The young awns are appropriate in hay; two to three cuts or grazing events are possible from an autumn plant (Hannaway 2004).

The grains of barley also have been used for animal feeding. It has been found that the phosphorus and other minerals in salts were effectively utilized by cattle, whereas swine, poultry, and fish cannot (Sugiura et al. 1999; Veum et al. 1999; Yi et al. 2001). The stomach microflora of ruminants produces essential amino acids, while the non-ruminant requires these additionally in their feed. As compared to other cereal crops, barley provides a good amount of starch, which is an excellent source of energy. The easily degradable forms like rolled or steamed barley are good for bovine rumen, as it is easily digestible (Blake et al. 2011).

With comparison to other cereals, forage barley has high nutritive value (Ross et al. 2004). The good forage barley should have high digestible of dry matter, low acid detergent fiber (ADF) and high crude protein (CP) content (Carr et al. 2004). Among the flowering and dough stages of the crop, the protein and neutral detergent fiber (NDF) content in forage barley varies significantly, declining from 12% to 9% dry matter (DM) for protein content and 63–56% DM for NDF (INRA 2007). The decline in protein and NDF is mostly due to a 20% rise in starch at the mid-dough level (Kirchgessner et al. 1989). The whole-crop silages have higher starch and lower NDF content, i.e., 29% and 47%, respectively, in Ireland (Walsh et al. 2008). It has also been reported that intercropping of barley with other crops enhances the fodder yield (Simmons et al. 1995; Ross et al. 2004; Sadeghpour et al. 2014).

3 Harnessing Available Barley Diversity

Barley (*Hordeum* subsp. *vulgare* or *spontaneum*) belongs to the tribe Triticeae, subfamily Pooideae, and Poaceae family. Bothmer and Jacobson (1985), based on morphological features, categorized the different barley species into four sections, viz., *Hordeum*, *Anisolepis*, *Critesion*, and *Stenostachys* (Table 1). The crop germplasm constitutes the available genetic diversity in the form of wild species and relatives, landraces, elite breeding material, genetic and cytogenetic stocks, and transgenic and modern cultivated varieties. Based on crossability studies and available genetic stocks, distinct gene pools have been established. Similarly, in barley also, primary, secondary, and tertiary gene pools are formed (Harlan and de Wet 1971).

4 Primary Gene Pool

The primary gene pool consists of related species, which can easily recombine with each other without facing any crossability barriers. The barley primary gene pool includes landraces, which are still cultivated in some geographical regions of the world such as Central and South Western Asia and Northern African regions, i.e., Ethiopia. Meanwhile, in some regions, the landraces disappeared rapidly, i.e.,

Table 1 Classification of different species as per Bothmer and Jacobson (1985) classification

Hordeum section	*Anisolepis* section	*Critesion* section	*Stenostachys* section
Hordeum vulgare L.	*Hordeum pusillum* Nuttall	*Hordeum pubiflorum* Hooker	*Hordeum marinum* Hudson
H. bulbosum L.	*H. intercedens* Nevski	*H. halophilum* Grisebach	*H. secalinum* Schreber
H. murinum L.	*H. euclaston* Steudel	*H. comosum* Presl	*H. capense* Thunberg
	H. flexusosum Steudel	*H. jubatum* L.	*H. bogdanii* Wilensky
	H. muticum Presl	*H. arizonicum* Covas	*H. roshevitzii* Bowden
	H. chilense Roemer & Schultes	*H. procerum* Nevski	*H. brevisubulatum* (Trinius) Link
	H. cordobense Bothmer, Jacobson, & Nicora	*H. lechleri* (Steudel) Schenck	*H. brachyantherum* Nevski
	H. stenostachys Godron		*H. depressum* (Scribner & Smith) Rydberg
			H. guatemalense Bothmer, Jacobson, & Jorgensen
			H. erectifolium Bothmer, Jacobson, & Jorgensen
			H. tetraploidium Covas
			H. fuegianum Bothmer, Jacobson, & Jorgensen
			H. parodii Covas
			H. patagonicum (Haumann) Covas

Northern and Central Europe. The wild species *Hordeum vulgare* subsp. *Spontaneum*, which is the progenitor of cultivated barley, is also grouped in the primary gene pool. The genetic analysis studies conducted using DNA markers confirmed that *H. vulgare* subsp. *spontaneum* has several key traits with respect to yield, abiotic stress tolerance, and quality parameters.

5 Secondary Gene Pool

This class of gene pool involves species which faced some difficulty during hybridization with primary gene pool members. The close relative of cultivated barley, the perennial diploid and tetraploid *H. bulbosum*, comes under secondary gene pool. It has been widely used for the development of haploid lines in wheat and barley breeding programs.

6 Tertiary Gene Pool

Tertiary gene pool involves the species which faced greater level of difficulty during hybridization with the cultivated species and all other species which have not been classified under primary and secondary gene pools. As the gene transfers from the tertiary gene pool species to the primary and secondary gene pools, special techniques such as embryo rescue, somatic hybridization, and gene transformation can be deemed useful.

The crop genetic resources played an important role in crop improvement with respect to yield, quality, and tolerance to adaptive traits. Across the globe, several crop gene bank collections are available with large number of specific crop accessions. Wheat has the largest collections of accessions followed by barley. According to the Food and Agriculture Organization (1996), more than 500,000 barley accessions are conserved in the world gene bank collections. The conservation of barley germplasm was initiated in 1981 by International Board for Plant Genetic Resources (IBPGR) with the establishment of an ad hoc Working Group on Barley Genetic Resources with the objective to monitor the existing barley germplasm collections, identification of trait-specific collections, and distribution of barley germplasm (Bockelman and Valkoun 2011). Later on, in collaboration with Europe, European Barley Core Collection (EBCC) and European Barley Database (EBDB) were established for the better coordination of barley genetic resources. This regional collaboration further extended to the global level with the formation of the International Barley Genetic Resources Network (IBGRN) and International Barley Core Collection Committee (Knüpffer and van Hintum 1995). In 2004, Global Crop Diversity Trust (GCDT) was established for the safeguarding of crop germplasm. The Svalbard Seed Vault was established in 2008, and almost all the crop germplasm were duplicated. There are several world collections that have barley genetic resources. Worldwide, barley accessions have been explored, collected, and conserved in various repositories, i.e., Plant Gene Resources of Canada (PGRC), Saskatoon (39,852); USDA-ARS National Small Grains Collection, Aberdeen, Idaho (29,870); Recursos Geneticos e Biotecnologia, Brasilia (EMBRAPA/CENARGEN), Brazil (29,227); International Center for Agricultural Research in Dry Areas (ICARDA), Aleppo (26,117); John Innes Centre, Norwich, United Kingdom (23,603); Leibniz Institute of Plant Genetics and Crop Plant Research (IPK), Gatersleben, Germany (22,106); Institute of Crop Germplasm Resources, CAAS, Beijing, China (18,818); National Institute of Agricultural Biotechnology, Suwon, Republic of Korea (19,764); N. I. Vavilov Institute of Plant Industry (VIR), St. Petersburg; Russia (17,850); Institute of Biodiversity Conservation (IBC), Addis, Ababa, Ethiopia (15,360); Research Institute for Bioresources, Okayama University, Japan (14,106); Nordic Genetic Resource Center, Alnarp, Sweden (13,435); Australian Winter Cereals Collection, Calala, New South Wales (12,600); CIMMYT, El Batan, Mexico (11,202); National Institute of Agrobiological Science, Tsukuba, Japan (8806); and ICAR—National Bureau of Plant Genetic Resources (NBPGR), New Delhi, India (8384), which are the major barley collections.

The germplasm resources/tools could be used to expand the germplasm base of high-quality forage with enhanced tolerance to biotic and abiotic stresses. Improved cultivars, landraces, and wild species of barley make up a sizable germplasm stock. Exotic germplasm has proved useful in a number of crops, including barley, with the help of traditional breeding methods as well as molecular technologies to enhance genetic diversity (Miflin 2000; Singh 2003).

Exotic germplasm refers to any germplasm that has not been hybridized or selected for adaptation to a certain area and is not widely applicable (Hallauer and Miranda 1981; Singh 2003). It is stated that exotic germplasm has been used to broaden the Nordic barley germplasm base (Veteläinen et al. 1997). Agronomically relevant genotypes can be developed through recombination using exotic germplasm, effectively incorporating novel traits into the local genetic base, with exotic germplasm contributing significantly to improved adaptation (Veteläinen 1996). Wide hybridization followed by subsequent backcrossing is an important method of introgression of useful genes from donor parent to recipient parent. Powdery mildew resistance (Pickering et al. 1995), boron toxicity tolerance (Jefferies et al. 1999), cereal cyst nematode resistance (Kretschmer et al. 1997), and net blotch resistance (Molnar et al. 2000) have all been transferred into barley from exotic germplasm resource.

7 Forage Barley Breeding

Forage grasses and legumes are grown on over large area across the world. But as the pressure increases for more grain, oilseed, and fiber crops, forages will likely be relegated to land not suited to grain or seed production. Ruminant animals can utilize forage grown on less productive areas, and because they graze, no exogenous energy source is needed for harvesting. High-quality forage cultivars are needed to improve animal yield per hectare. Much progress has been made in breeding and management to improve forage quality. Coastal bermudagrass is the most suitable example of classical plant breeding for improved forage quality. In barley, several varieties (Larker, Beacon, Dickson, Nordic, Morex, Klages, Steptoe, etc.) have been evolved through conventional breeding approaches, which lead to the enhancement in the productivity. The continuous improvement programs resulted in the evolution of high-yielding feed-type barley varieties such as Melvin, Klondike, Bedford, and Johnston. The breeding objectives of barley breeding are of short-, intermediate-, and long-term nature. The short-term objectives mainly focused on the development of high-yielding variety in quick time, while the intermediate- and long-term objectives deal with the generation of genetic resources for future breeding programs (Anderson 1985).

During the last decade, barley productivity is enhanced with the pace of 1–2% yearly due to breeding of genetically improved varieties with biotic and abiotic stress tolerance, nutritional quality improvement, and suitability to cultivation technologies. In most of the crops, visual selection is highly effective for high digestibility rate; however, measuring methods for nutritional components, such as lignin, starch, fiber, protein, acids, etc., are expansive as well as complicated (Casler 2001). Most of the forage quality traits are governed by many genes with minor effect alleles. The forage barley breeding programs did not receive much attention until the 1900s, and very few cultivars were available (Surber et al. 2011). Attempts have been made for the establishment of genetic diversity for forage-related traits along with yield (Srimali 2008).

Barley have excellent genetic resources and have a benefit of bearing mutations leading to the evolution of improved varieties with the trait of interest. Awn is an important morphological trait involved in drought stress tolerance, but it creates jaw infection when it is fed to animals. On chromosome number 4, hooded (*Kap*) gene leads to the conversion of lemma awn into a twisted floret structure with leaf, which makes varieties suitable for feeding (Muller et al. 1995). Another important mutation occurred as *vrsl* (Komatsuda et al. 2007) and *Int-c* (Komatsuda and Mano 2002), which resulted into the development of six-rowed barley with broad and fertile lateral florets. The six-rowed type barley is generally preferred by farmers in North Africa. Presence or absence of auricle is governed by a single monogenic recessive gene. The auricle-less barley species has erect leaf blades. Dami is a whole-crop forage barley variety that has *lig* (*al*, *li*, and *aur-a*) auricle-less gene derived through hybridization of BGS60 and Gangbori. The BGS60 variety is auricle-less gene, which produced high-quality whole-crop silage, and Gangbori have high biomass, winter hardiness, and barley yellow mosaic virus (BaYMV) resistance (Franckowiak et al. 2005). Early maturing and high-yielding rough form of awn barley is used for silage or hay from cultivars, namely, Yongyang, Sunwoo, Sangweon, and Soman. Barley with a smooth awn, hooded, and auricle-less awns may be ideal breeding choices for whole-crop forage (Park et al. 2008). Like other forage crops, barley also suffers from the problem of accumulation of toxic compounds. It has been reported that the nitrate level above 0.226% is not safe for livestock animals and causing nitrate poisoning. In barley also 0.23% nitrate has been reported at booting phase (Cash et al. 2002).

8 Forage Barley Varieties

Many barley varieties, both two-row and six-row, suited for forage quality have been bred by barley breeders using traditional breeding methods (Table 2). In India, several varieties have been developed by many research institutes and state agricultural university, of which few have been listed here with their characteristics (Table 3). ICAR—Indian Institute of Wheat and Barley Research (IIWBR), Karnal, also has developed several barley genetic stocks that possess desirable traits such as high yield, resistant to disease, and high quality.

Table 2 Forage barley varieties developed at the global level

Two-row barley		Six-row barley	
Variety	Feature	Variety	Feature
Busby	Feed barley, rough awned	AC Ranger	Feed barley, smooth awns
CDC Austenson	Rough awn, high feed yield	Chigwell	Feed barley, smooth awns
CDC Cowboy	Forage barley	Muskwa	Smooth awned, hulled, general purpose type
CDC Coalition	High yielding	Sundre	Feed barley for grain and fodder
CDC Maverick	Awned forage barley for swath grazing, bale grazing and silage	Trochu	Feed barley for grain and fodder
Conlon	Smooth awned	Viver	Feed barley
Gadsby	Rough awned, general purpose	SR14501	Suitable for silage green feed and swath grazing
Ponoka	Feed barley for grain or silage	Horseford	
Seebe	Feed barley for silage	Robust	
Xena	Semi-smooth awns	Bestford	
Dictator 2	Suitable for grazing, silage and hay production	Westford	
Hayes			
Haybet			
Logan			

Table 3 Barley varieties developed in India

Variety	Characteristics
Ratna	It was developed by Indian Agricultural Research Institute (IARI), New Delhi, and released for rainfed areas of eastern Uttar Pradesh, Bihar, and West Bengal, highly tolerant to saline and alkaline soil conditions. It yields about 150q/ha of green fodder after 65 days of sowing, has good tillering ability, and matures within 125–130 days
Azad (K125)	It was developed by Chandra Shekhar Azad University of Agriculture & Technology (CSAUAT), Kanpur, and suitable for cultivation in areas of eastern Uttar Pradesh, Bihar, and West Bengal, dual-purpose barley cultivar suitable for irrigated conditions, fair degree of resistance to yellow rust disease; it yields about 150 q/ha of green fodder after 65 days of sowing and matures in 115–120 days
Vijaya	It is developed by CSAUAT, Kanpur, and released for rainfed areas of western Uttar Pradesh, Delhi, and Madhya Pradesh. It matures in 120–125 days. It does well in rainfed areas of Haryana and Punjab as well. It has a yield potential of 30–35 quintals of grains per hectare
Amber	It was developed by CSAUAT, Kanpur, and suitable for growing in rainfed areas of eastern Uttar Pradesh. It matures in 130–133 days. Plants are 100–125 cm in height. Ear size is 8.5–9.5 cm. It has a yield potential of 25–30 quintals of grains per hectare
RS-6	It was developed in Rajasthan and suitable for cultivation under both rainfed and irrigated conditions and recommended for central and eastern Rajasthan. It matures in 130–135 days with a yield potential of 35 to 40 quintals of grain per hectare

(continued)

Table 3 (continued)

Variety	Characteristics
Jyoti	It was developed by CSAUAT, Kanpur. It is a six-row hulled variety, suitable for growing in irrigated areas of Haryana, Punjab, Uttar Pradesh, Delhi, north-western Rajasthan, Bihar, and West Bengal. Plant height is short. It has yield potential of about 35–40 quintals of grains per hectare. It matures in 120–125 days, under irrigated conditions, yielding 20–25 quintals of grains per hectare
Clipper	It matures in about 135–140 days. It is susceptible to yellow rust. It has yield potential of 28–30 quintals of grains per hectare under good management. It is introduced from Australia
PL-56	It is a six-row variety with broad and relatively upright leaves and a colorless leaf sheath. It has compact ears with well-spread, barbed awns. It is moderately resistant to yellow rust. It is recommended for cultivation under rainfed conditions of Punjab. It has yield potential of 30 quintals of grains per hectare
Ranjit (DL-70)	It is six-row and heavy tillering variety, recommended for commercial cultivation in Punjab only and suitable for growing under irrigated conditions. It has yield potential of —35 quintals of grains per hectare. Due to its early maturity, it fits well into multiple crop rotations, particularly for late sowing after the harvest of crops such as potato, cotton, or paddy.
Karan-201 Karan-231 Karan-264	Varieties Karan 201, 231, and 264 have been developed under the All India Coordinated Barley Improvement Project. These are high-yielding varieties and good for chappati making. All of them are huskless varieties. These varieties fit in well in the multiple and relay cropping systems. These can be grown in problematic soils too, like dry lands, alkaline and saline soils, etc. They can also be grown on lands lying fallow after a paddy crop has been taken. These are suitable for cultivation in the eastern and Bundelkhand region of Madhya Pradesh, Rajasthan, and Gurgaon and Mahendragarh district of Haryana. The average yield of Karan 201, 231, and 264 are 38, 42.5, and 46 quintals per hectare, respectively
C-164	It is a tall, six-row variety with compact ears and long awns. Its grain is bold, bright, and amber. It is resistant to yellow rust. It has stiff straw and is, therefore, more suitable for cultivation under irrigated conditions. It has yield potential of 30–32 quintals of grain per hectare
Kailash	It is a six-row hulled variety suitable for growing in rainfed areas of medium to low elevation of Himachal Pradesh hills. It matures in 145–150 days. It is resistant to yellow rust. It has yield potential of about 40 quintals of grain per hectare
Dolma	The variety Dolma is suitable for growing in rainfed areas of medium to high elevation of Himachal Pradesh and Uttar Pradesh hills. It matures in about 140–150 days. It is fairly resistant to yellow rust. It has yield potential of 35–40 quintals per hectare
Himani	Himani variety was developed at Shimla. It is suitable for growing in the medium to lower hill valley of Uttar Pradesh and Himachal Pradesh. It matures about 1 week earlier than Kailash. It has yield potential of 32–36 quintals of grains per hectare
LSB-2	LSB is a six-row naked-type barley variety suitable for growing at higher altitude of Himachal Pradesh and Uttar Pradesh hills. It matures in 145–150 days. It has yield potential of 25–30 quintals of grain per hectare
RDB-1	RDB-1 has been developed by mutation. It is suitable for growing under irrigated conditions in Rajasthan. It is susceptible to yellow rust and is suitable for rust-free areas of Rajasthan. It has yield potential of 30–35 quintals of grain per hectare
BG-25	It is developed by Chaudhary Charan Singh Haryana Agricultural University (CCSHAU), Hisar. It is suitable for normal sowing conditions in Haryana. It is resistant to yellow rust. It matures in about 120–130 days. It has yield potential of about 30 quintals of grain per hectare

(continued)

Table 3 (continued)

Variety	Characteristics
BG-108	It was developed by CCS HAU, Hisar. It is suitable for commercial cultivation under late sown conditions. It takes about 120–125 days to mature. It is fairly resistant to yellow rust. It has yield potential of 20–25 quintals per hectare under late sown conditions
Kedar	It was developed by IARI, New Delhi. It is suitable for late sown conditions. It is a dwarf variety with profuse tillering and resistant against yellow rust and insects
Neelam	It is developed by IARI, New Delhi. This variety yields up to 50 quintals per hectare. Its grains are hulled and amber in color. It is recommended for cultivation in both irrigated and rainfed conditions of Punjab, Haryana, Uttar Pradesh, and Bihar. This variety has higher contents of protein and lysine

9 Application of Available Genomics Tools

The recent advancement in technology has made feasible to genotype whole collections of crop germplasm in a very short window. A large number of genomic resources have been generated, which depicts a good spectrum of genetic diversity available in that particular crop species for desired trait of interest. The novel genomic and molecular approaches mined the useful genes from different gene pools, which are behind the different abiotic stress tolerance mechanisms in plants. Large number of genetic and genomic resources are available with respect to grain yield, but from the prospect of forage yield and quality traits, the marker technology has been utilized for the mapping of genomic regions responsible for high forage yield and its nutritive value.

In barley, Moragues et al. (2010) genotyped about 169 landraces and 171 European cultivars with 1536 Single Nucleotide Polymorphism (SNP) markers. Of these, different subsets of 384 and 96 SNP markers were selected and used to determine genetic diversity. The 384 SNP subsets characterized significant level of genetic diversity. The genetical research confirmed that ABA-independent stress response mechanism is important for imparting tolerance against several abiotic stresses such as drought, salinity, and cold (Shinozaki and Yamaguchi-Shinozaki 2000).

Under abiotic stresses, plants accumulate ABA and initiated a series of reactions like triggering of *late embryogenesis abundant* (LEA) proteins, which leads to the aggregation of *Hordeum vulgare abundant protein 1* (*HVA1*) (Marttila et al. 1996). The overexpression of *HVA1* gene has shown multiple stress resistance in plants (Nguyen and Sticklen 2013). In another study, it was revealed that the overexpression of *HvSNAC1* transcription factor resulted in the improvement of drought tolerance without affecting yield in transgenic barley (Al Abdallat et al. 2014). There are other transcription factors (TFs) involved in abiotic stresses such as *TaDREB2* and *TaDREB3*, which are known to be involved in drought tolerance in wheat and barley without affecting the plant growth and development (Morran et al. 2011). The overexpression of *TaCBF14* and *TaCBF15* transcription factors from wheat in transgenic barley enhanced the cold and freezing tolerance level, as both of these traits are important for forage (Soltész et al. 2013).

Being the temperate crop, in barley, cold stress is an important aspect and has drastic effect during reproductive period. Two frost determinants loci, i.e., *Fr-H1 and Fr-H2*, have been identified on the long arm of 5H chromosome (Francia et al. 2004; Skinner et al. 2005; Galiba et al. 2009). These two identified loci *Fr-H1* and *Fr-H2* were validated with a set of F_3 populations of Nure and Tremois using molecular markers, and of these, one Quantitative Trait Locus (QTL) was fixed at homozygous and another at heterozygous state (Francia et al. 2007). The genome-wise association mapping identified two tightly SNP associations Fr-H2 and HvBmy loci on 5H and on the long arm of 4H chromosomes in a collection of 184 barley genotypes (Visioni et al. 2013).

Winter hardiness is an important characteristic of temperate cereals, and in barley, Hayes et al. (1993) have mapped QTLs pertaining to field survival, LT_{50}, growth habit, and crown fructan content in a mapping population derived from Dicktoo and Morex. In another study, Karsai et al. (1997) dissected the genes related with the flowering process, which includes the photoperiod as well as vernalization genes. The two genes *Vrn-H1* and *Vrn-H2* were mapped on chromosomes 5 and 4, respectively. The *Vrn-H1* candidate gene (*HvBM5A*) was found linked with flowering and does not show any vernalization effect.

Several important QTLs for N, NO_3-N, non-NO_3-N, and in situ dry matter digestibility have been detected in a mapping population of Lewis x Karl on chromosomes 2, 3, 4, and 6 (Surber et al. 2001). In a similar way, Siahsar et al. (2009) identified QTLs for forage quality traits, viz., 5 QTLs for DMD and NDF, 3 QTLs for WSC and ash, and 4 QTLs for crude protein content from doubled-haploid population of 72 individuals derived from the cross between Steptoe x Morex. Abdel-Haleem et al. (2004) confirmed that DMD QTLs were found on chromosomes numbers 1H, 2H, 3H, and 6H and that the most significant DMD-QTL was found beside the *Vrs1* gene, which regulates the plant and seed growth, and that this gene may have a pleiotropic influence on the forage quality of barley. Some significant QTLs were identified for ADF, which is located on chromosomes 1H, 2H, 3H, 5H, and 6H (Abdel-Haleem et al. 2004; Siahsar et al. 2009). In the "Steptoe × Morex" population, several QTLs for influencing grain ADF content were found on chromosomes 1H, 2H, and 4H (Ullrich et al. 1996; Han et al. 2003). In a population of Falcon x Tyto, discovered QTLs for in vitro fiber digestibility were on chromosomes 1H, 2H, and 7H (Nyachiro et al. 2015). Forage quality traits such as NDFD, NDF, starch, protein, and lignin/NDF demonstrated heterogeneity in the population derived from the Falcon x Virden cross (Zantinge 2013).

In tandem with current revelations in genetic transformation and genome editing (CRISPER-Cas) tools, the discovery of novel stress resistance genes using advanced genomics and bioinformatics methods and sequence knowledge may have a significant effect on forage barley production in dry areas around the world (Lawrenson et al. 2015; Gürel et al. 2016).

Advanced genomic technologies like GWAS, MAS, and CRISPR together have a great deal of promise for speeding up barley breeding initiatives. However, converting these developments into workable solutions for barley production under climate change would require more research concentrating on the long-term effects and practical use of these technologies (Meng et al. 2023).

As molecular breeding offers precise tools to improve production, quality, and resilience to environmental challenges, it has completely changed the way feed crops are improved. Breeding processes have been greatly accelerated by technological advancements like genomic selection (GS), genome-wide association studies (GWAS), quantitative trait loci (QTL) mapping, and CRISPR-Cas9. This has resulted in cultivars with improved yield, disease resistance, stress tolerance, and nutritional quality. Breeding efficiency has been significantly enhanced by the combination of bioinformatics and high-throughput phenotyping. In addition to addressing the issues brought on by climate change and the demand for sustainable practices, this review examines the future potential of cutting-edge technologies like artificial intelligence in predictive breeding and shows the achievements of these methods in a variety of fodder crops (Chen 2024).

10 Conclusion

The next-generation breeding techniques, i.e., genomics-assisted breeding (GAB), have a wide impact on crop plants. The genome sequence information of barley is available in public domain and could be utilized for the identification of genomic regions which govern the most of the economic traits. Currently, efforts have been made for the enhancement of forage yield simultaneously with grain yield. The integration of traditional breeding approaches with the genomic tools will be useful in delivering high-yielding forage varieties with high forage nutritive value and resilience to climate change.

References

Abdel-Haleem H, Giroux M, Talbert H, Bowman J, Kanazin V, Blake T (2004) Identification of QTLs controlling the feed quality of barley. In: Plant & animal genome XII conference, San Diego, p 468

Al Abdallat AM, Ayad JY, Abu Elenein JM, Al Ajlouni Z, Harwood WA (2014) Overexpression of the transcription factor *HvSNAC1* improves drought tolerance in barley (*Hordeum vulgare* L). Mol Breed 33:401–414

Anderson MK (1985) Barley breeding. Published in Barley, Agronomy Monograph No. 26, © ASA-CSSA-SSSA, 677 South Segoe Road, Madison, WI 53711, USA. pp 231–268

Anderson V, Lardy G, Bauer M, Swanson K, Zwinger S (2012) Barley grain and forage for beef cattle. North Dakota Agricultural Experiment Station

Baik BK, Ullrich SE (2008) Barley for food: Characteristics, improvement, and renewed interest. J Cereal Sci 48(2):233–242

Baum M, van Korff M, Guo P, Lakew B, Udupa SM, Sayed H, Choumane W, Grando S, Ceccarelli S (2007) Molecular approaches and breeding strategies for drought tolerance in barley. In:

Varshney R, Tuberosa R (eds) Genomic assisted crop improvement: Genomics applications in crops, vol 2, pp 51–79
Blake T, Blake VC, Bowman JGP, Abdel-Haleem H (2011) Barley feed uses and quality improvement. In: Ullrich SE (ed) Barley: production, improvement, and uses. Wiley-Blackwell, West Sussex, pp 522–531
Bockelman HE, Valkoun J (2011) Barley Germplasm Conservation and Resources. In: Ullrich SE (ed) Barley: production, improvement, and uses. Blackwell Publishing Ltd, pp 144–159
Bornare SS, Prasad LC, Prasad R, Lal JP (2012) Perspective of barley drought tolerance; methods and mechanisms comparable to other cereals. J Progress Agric 3(2):68–70
Bregitzer P, Raboy V, Obert DE, Windes JM, Whitmore JC (2007) Registration of Herald barley. Crop Sci 47:441–442
Canadian Food Inspection Agency (CFIA) (2005) Description of cereal species (Appendix V). In: Seed program specific work instructions – cereal crop inspection procedures. Canadian Food Inspection Agency
Carr PM, Horsley RD, Poland WW (2004) Barley, oat and cereal–pea mixtures as dryland forages in the northern great plains. Agron J 96:677–684
Cash D, Funston R, King M, Wichman D (2002) Nitrate toxicity of Montana forages. Montana State University Extension Service, Bozeman
Casler MD (2001) Breeding forage crop for increased nutritive value. Adv Agron 71:51–107
Chen S (2024) Advances in Molecular Breeding of Forage Crops: Technologies, Applications and Prospects. Agriculture 14(2):279. https://doi.org/10.3390/agriculture14020279
Chriyaa A, Amri A (1997) Evaluation of straw quality of barley varieties for animal feeding. Options Méditerranéennes – Serie A 34:33–36
Ditsch DC and Bitzer MJ (2005) Managing small grains for livestock forage. University of Kentucky, Cooperative Extension Service, College of Agriculture, AGR-160
Duke JA (1983) Handbook of energy crops. NewCROPS web site. Purdue University
Ecocrop (2011) Ecocrop database. FAO
FAO (1996) State of the World's Plant genetic resources for food and agriculture. FAO, Rome
Francia E, Rizza F, Cattivelli L, Stanca AM, Galiba G, Toth B, Pecchioni N (2004) Two loci on chromosome 5H determine low-temperature tolerance in a 'Nure'(winter) × 'Tremois' (spring) barley map. Theor Appl Genet 108(4):670–680
Francia E, Barabaschi D, Tondelli A, Laido G, Rizza F, Stanca AM, Busconi M, Fogher C, Stockinger EJ, Pecchioni N (2007) Fine mapping of a *HvCBF* gene cluster at the frost resistance locus *Fr-H2* in barley. Theor Appl Genet 115(8):1083–1091
Franckowiak JD, Kleinhofs A, Lundqvist U (2005) Descriptions of barley genetic stocks for 2005. Barley Genet Newslett 35:155–210
Galiba G, Vágújfalvi A, Li CX, Soltész A, Dubcovsky J (2009) Regulatory genes involved in the determination of frost tolerance in temperate cereals. Plant Sci 176:12–19
Gürel F, Öztürk ZN, Uçarlı C, Rosellini D (2016) Barley genes as tools to confer abiotic stress tolerance in crops. Front Plant Sci 7:1137
Hallauer AR, Miranda FJB (1981) Quantitative genetics in maize breeding. Iowa State University Press, Ames
Han F, Ullrich SE, Romagosa I, Clancy JA, Froseth JA, Wesenberg DM (2003) Quantitative genetic analysis of acid detergent fiber content in barley grain. J Cereal Sci 38:167–172
Hannaway DB (2004) Barley (*x Elyhordeum Mansf. ex Zizin & Petrowa*). Oregon State University, Species Selection Information System
Harlan JR, de Wet JM (1971) Toward a rational classification of cultivated plants. Taxon 20(4):509–517
Hayes PM, Blake T, Chen TH, Tragoonrung S, Chen F, Pan A, Liu B (1993) Quantitative trait loci on barley (*Hordeum vulgare* L.) chromosome 7 associated with components of winter hardiness. Genome 36(1):66–71
Jefferies SP et al (1999) Mapping of chromosome regions conferring boron toxicity tolerance in barley (*Hordeum vulgare* L.). Theor Appl Genet 98:1293–1303

Jones CM, Heinrichs AJ, Roth GW, Ishler VA (2004) From harvest to feed: understanding silage management. Pennsylvania State University Extension

Karsai I, Meszaros K, Hayes PM, Bedo Z (1997) Effects of loci on chromosomes 2 (2H) and 7 (5H) on developmental patterns in barley (*Hordeum vulgare* L.) under different photoperiod regimes. Theor Appl Genet 94:612–618

Kellems RO, Church DC (2002) Livestock feeds and feeding, 5th edn. Prentice Hall, Upper Saddle River

Kirchgessner M, Heinzl WE, Schwarz FJ (1989) Nutritive value, for dairy cows, of silages from whole barley and wheat crops cut at different stages of maturity. 1. Chemical composition, digestibilities and energy contents. WirtschaftseigeneFütter 35(2):171–186

Knüpffer H, van Hintum TJL (1995) The Barley core collection – an international effort. In: Hodgkin T, Brown AHD, van Hintum TJL, Morales EAV (eds) Core collections of plant genetic resources. Wiley, Chichester, pp 171–178

Komatsuda T, Mano Y (2002) Molecular mapping of the intermedium spike-c (int-c) and non - brittle rachis 1 (btr1) loci in barley (*Hordeum vulgare* L.). Theor Appl Genet 105:85–90

Komatsuda T, Pourkheirandish M, He C, Azhaguvel P, Kanamori H, Perovic D, Stein N, Graner A, Wicker T, Tagiri A, Lundqvist U, Fujimura T, Matsuoka M, Matsumoto T, Yano M (2007) Six-rowed barley originated from a mutation in a homeodomain-leucine zipper I class homeoboxgene. Proc Natl Acad Sci USA 104:1424–1429

Kretschmer JM et al (1997) RFLP mapping of the Ha2 cereal cyst nematode resistance gene in barley. Theor Appl Genet 94:1060–1064

Lawrenson T, Shorinola O, Stacey N, Li C, Østergaard L, Patron N, Harwood W (2015) Induction of targeted, heritable mutations in barley and *Brassica oleracea* using RNA-guided Cas9 nuclease. Genome Biol 16:258

Marttila S, Tenhola T, Mikkonen A (1996) A barley (*Hordeum vulgare* L.) LEA3 protein, HVA1, is abundant in protein storage vacuoles. Planta 199(4):602–611

Mayer KF, Waugh R, Brown JW, Schulman A, Langridge P, Platzer M, Fincher GB, Muehlbauer GJ, Sato K et al (2012) A physical, genetic and functional sequence assembly of the barley genome. Nature 491:711–716

Meng G, Rasmussen SK, Christensen CS, Fan W, Torp AM (2023) Molecular breeding of barley for quality traits and resilience to climate change. Front Genet 13:1039996

Miflin B (2000) Crop improvement in the 21st century. J Exp Bot 51:1–8

Molnar SJ, James LE, Kasha KJ (2000) Inheritance and RAPD tagging of multiple genes for resistanceto net blotch in barley. Genome 43:224–231

Moragues M, Comadran J, Waugh R, Milne I, Flavell AJ, Russell JR (2010) Effects of ascertainment bias and marker number on estimations of barley diversity from high throughput SNP genotype data. Theor Appl Genet 120:1525–1534

Morran S, Eini O, Pyvovarenko T, Parent B, Singh R, Ismagul A, Lopato S (2011) Improvement of stress tolerance of wheat and barley by modulation of expression of DREB/CBF factors. Plant Biotech J 9(2):230–249

Muller KJ, Romano N, Gerstner O, Garcia-Marotot T, Pozzi C, Salamini F, Rohde W (1995) The barley Hooded mutation caused by a duplication in a homeobox gene caused by a duplication. Nature 374:727–730

Nguyen TX, Sticklen M (2013) Barley *HVA1* gene confers drought and salt tolerance in transgenic maize (*Zea mays* L). Adv Crop Sci Tech 1:1

Nyachiro J, Juskiw P, Swift ML, Aljarrah M, Zantinge J, Anbessa Y (2015) Estimating the heritability of barley forage fiber digestibility using a 'Falcon'/'Tyto' cross. Poster presented at 16th Australian Barley technical symposium, Melbourne, Australia

OECD (2004) Consensus document on compositional considerations for new varieties of barley (*Hordeum vulgare* L.): key food and feed nutrients and anti-nutrients. Series on the Safety of Novel Foods and Feeds No. 12, Joint meeting of the chemicals committee and the working party on chemicals, pesticides and biotechnology, OECD

Park TI, Han OK, Seo JH, Choi JS, Park KH, Kim JG (2008) New barley cultivars with improved morphological characteristics for whole crop forage in Korea. J Korean Soc Grassl forage Sci 28(3):193–202

Pickering RA et al (1995) The transfer of a powdery mildew resistance gene from *Hordeum bulbosum* L. to barley (*H. vulgare* L.) chromosome 2 (2I). Theor Appl Genet 91:1288–1292

Poehlman JM (1985) Adaptation and distribution. In: Rasmusson DC (ed) Barley. Agronomy Monograph 26. American Society of Agronomy, Madison, pp 1–17

Ross SM, King JR, O'Donovan JT, Spaner D (2004) Forage potential of intercropping berseem clover with barley, oat, or triticale. Agron J 96:1013–1020

Sadeghpour A, Jahanzad E, Lithourgidis AS, Hashemi M, Esmaeili A, Hosseini MB (2014) Forage yield and quality of barley-annual medic intercrops in semi-arid environments. Int J Plant Prod 8(1):77–90

Shinozaki K, Yamaguchi-Shinozaki K (2000) Molecular responses to dehydration and low temperature: differences and cross-talk between two stress signaling pathways. Curr Opin Plant Biol 3(3):217–223

Siahsar B, Peighambari S, Taleii A, Naghavi M, Nabipour A, Sarrafi A (2009) QTL analysis of forage quality traits in barley (*Hordeum vulgare* L.). Cereal Res Commun 37(4):479–488

Simmons SR, Sheaffer CC, Rasmusson DC, Stuthman DD, Nickel SE (1995) Alfalfa establishment with barley and oat companion crops differing in stature. Agron J 87:268–272

Singh RJ (2003) Plant cytogenetics, 2nd edn. CRC Press, Boca Raton

Skinner JS, von Zitzewitz J, Szucs P, Marquez-Cedillo L, Filichkin T, Amundsen K, Stockinger EJ, Thomashow MF, Chen THH, Hayes PM (2005) Structural, functional, and phylogenetic characterization of a large CBF gene family in barley. Plant MolBiol 59:533–551

Smith BD (1998) The emergence of agriculture. Scientific American Library, HPHLP, New York, p 231

Soltész A, Smedley M, Vashegyi I, Galiba G, Harwood W, Vágújfalvi A (2013) Transgenic barley lines prove the involvement of TaCBF14 and TaCBF15 in the cold acclimation process and infrost tolerance. J Exp Bot 64(7):1849–1862

Srimali M (2008) Prospects of barley in augmenting forage resource in arid and semi-arid regions in India. In: Proceedings 10th international Barley genetics symposium, Alexandria, Egypt, 5–10 April, 2008, pp 635–639

Stallings C (1997) Rye, Barley, and Wheat can make excellent quality silage. Virginia Cooperative Extension. Available at: http://www.sites.ext.vt.edu/newsletter-archive/dairy/1997-4/grainsilage.html

Sugiura SH, Raboy V, Young KA, Dong FM, Hardy RW (1999) Availability of phosphorus and trace elements in low-phytate varieties of barley and corn for rainbow trout (*Oncorhynchus mykiss*). Aquaculture 170:285–296

Surber LMM, Bowman JGP, Blake TK, Robinson KN, Endecott RL and Robinson BL (2001) Identification of genetic markers associated with forage quality characteristics in Lewis × Karl barley lines. In: Proceedings of the western section of the American society of animal science (vol 52)

Surber L, Abdel-Haleem H, Martin J, Hensleigh P, Cash D, Bowman J, Blake T (2011) Mapping quantitative trait loci controlling variation in forage quality traits in barley. Mol Breed 28:189–200

Ullrich SE (2011) Significance, adaptation, production, and trade of Barley. In: Ullrich SE (ed) Barley: production, improvement, and uses. Blackwell Publishing Ltd, pp 3–11

Ullrich SE, Han F, Froseth JA, Jones BL, Newman CW, Wesenberg DM (1996) Mapping of loci that affect carbohydrate content in barley grain. In: Slinkard A, Scoles G, Rossnagel B (eds) Proceedings of the V international oat conference and of the VII international Barley. Genetics symposium. Poster sessions, vol 1. University of Saskatchewan Ext. Press, Saskatoon, pp 141–143

Veteläinen M (1996) Utilization of exotic germplasm in Nordic barley breeding and its consequences for adaptation. Euphytica 92:267–273

Veteläinen M, Suominen M, Nissilä E (1997) Agronomic performance of crosses between Nordic and exotic barleys. Euphytica 93:239–248

Veum TL, Ledoux DR, Raboy V (1999) Low phytic acid barley for growing pigs: pig performance and bone strength. In: Proceedings of the 50th annual meeting of the European association for animal production, Zurich, p 270

Visioni A, Tondelli A, Francia E, Pswarayi A, Malosetti M, Russell J, Thomas W, Waugh R, Pecchioni N, Romagosa I, Comadran J (2013) Genome-wide association mapping of frost tolerance in barley (*Hordeum vulgare* L.). BMC Genomics 14(1):1–13

von Bothmer R, Jacobsen N (1985) Origin, taxonomy, and related species. Barley 26:19–56

Walsh K, O'Kiely P, Taweel HZ, McGee M, Moloney AP, Boland TM (2008) Intake, digestibility, rumen fermentation and performance of beef cattle fed diets based on whole-crop wheat or barley harvested at two cutting heights relative to maize silage or ad libitum concentrates. Anim Feed Sci Technol 144(3–4):257–278

Warrick B, Sansone C and Johnson J (2002) Small grain production in West Central Texas. Texas A&M AgriLife Research and Extension Center at San Angelo. Available at: http://sanangelo.tamu.edu/extension/agronomy/small-grain-production-in-west-central-texas/

Winter B (2005) Forage crops – winter. Queensland Government, Primary Industries and Fisheries

Yi YC, Ledoux R, Veum TL, Raboy V, Zyla K, Wikiera A (2001) Bioavailability of phosphorus in low phytic acid barley. J Appl Poult Res 10:86–91

Yu P, Christensen D, McKinnon J, Markert J (2003) Effect of variety and maturity stage on chemical composition, carbohydrate and protein subfractions, in vitro rumen degradability and energy values of timothy and alfalfa. Can J Anim Sci 83:279–290

Zantinge J (2013) The development of molecular markers for improved fibre quality in barley. Beef Cattle Research Council, Calgary

Genomics-Assisted Breeding for Fodder Quality Improvement in Forage Sorghum

Partha Pratim Behera and Ramendra Nath Sarma

1 Introduction

Sorghum [*Sorghum bicolor* (L.) Moench], commonly known as camel's crop, is the world's fifth most important and versatile dry-land cereal crop, used for food, feed, fodder, and bioethanol production (Bell et al. 2024). The versatility of forage sorghum for various livestock applications, including silage, grazing, and standing winter feed, reinforcing its value as a multipurpose forage crop in diversified agricultural systems (Tuinstra et al. 2024; Ping et al. 2018). Its small diploid genome (2 n = 20) and phenotypic diversity make it an ideal C-4 grass model (Mace et al. 2013; Baloch et al. 2023). The top sorghum producers are the USA, Sudan, Nigeria, India, Niger, Ethiopia, Australia, Brazil, and China, which contribute about 77% of world sorghum production (Aruna et al. 2018; https://apps.fas.usda.gov). In affluent nations like the USA, Canada, and Australia, it provides 36% of animal feed and industrial residues. In semi-arid and tropic regions, livestock are underproductive and feeble due to lack of feed and competition for land with other crops. Fostering a diverse crop species diversity that can endure high risk and fragile environmental circumstances should not only promote agricultural diversification but also increase marginal farmers' livelihood possibilities. The concept that various plant genetic resources (PGR) and genomic information are accessible for each crop and serve as "building blocks" of plant breeding is generally acknowledged (Varshney et al. 2011; Liu et al. 2024). Marginal farmers may have restricted cultivation prospects, especially during the lean season, where owning livestock serves as an alternative source of income. These farmers feed crop residues to their cattle, which are reported to be nutritionally deficient. In this case, forage sorghum can act as potential feed

P. P. Behera · R. N. Sarma (✉)
Department of Plant Breeding and Genetics, Assam agricultural University,
Jorhat, Assam, India
e-mail: ramendra.sarma@aau.ac.in

S. H. Wani, S. S. Gosal (eds.), *Accelerated Plant Breeding, Volume 5*,
https://doi.org/10.1007/978-3-032-04527-0_3

source as compared to other crops, as it is not only proved to be high yielding under limited water supply conditions and poor soils but also nutritionally superior in terms of fodder quality like rich protein and easy digestion (Getachew et al. 2016; Mundia et al. 2019). Sorghum is significant as a fodder, for the health and nutrition of India's enormous livestock population, which accounts for 20% of the world's cattle.

According to a recent Food and Agriculture Organization of the United Nations (FAO) assessment, the COVID-19 pandemic would add an extra 83 to 132 million undernourished people by 2020, which highlights the vulnerability of the world food production and supply and a substantial delay in the path to zero hunger (The State of Food Security and Nutrition in the World, FAO 2020). With a projected global population of 9.7 billion by 2050, an additional 70% of food and feed must be produced to keep pace with the rising demand (Tripathi et al. 2019). In present-day agricultural research, plant breeder thus requires achieving a balance between not only fodder and feed production enhancement but also fodder nutritional quality improvement in forage sorghum crops (Godfray et al. 2010). To address this issue, the focus is on the development of dual-purpose cultivars having higher feed quality but have the ability to reduce the land and water competition. To increase fodder quality in forage sorghum, breeders must understand the quality criteria, how they interact, and their genetics. Most essential agronomic characteristics are governed by quantitative trait locci (QTLs). The quality of fodder is a quantitative attribute. Finding potential candidate genes and choosing efficient molecular markers in sorghum breeding require knowledge of their genetic process. Forage digestibility and crude protein content are the primary fodder quality traits focused on breeding goals (Murray et al. 2008; Vellaichamy Gandhimeyyan et al. 2023). The contents of cellulose (CL), hemicellulose (HC), and lignin are important components of neutral detergent fiber (NDF), which regulates forage digestibility (Wang et al. 2016). ADF, on the other hand, is a sorghum fiber derived from acid detergent-treated pasture. Acid detergent fiber (ADF) content affects nutritional digestibility and nitrogen balance (Obregón-Cano et al. 2019; Miranda-Romero et al. 2020). To improve fodder quality and digestibility, a better leaf-to-stem ratio, stay-green trait (more photosynthesis, sugars, and nutritional value), stalk sweetness (more cell soluble nitrogen and sugars), and higher protein content are desired (Wang et al. 2023; Behera et al. 2024). A higher leaf-to-stem ratio is known to boost IVDMD and crude protein. The stem has the largest concentrations of IVDMD, water-soluble carbohydrates, and non-structural carbohydrates, whereas the leaf has the highest concentrations of CP, NDF, ADF, and acid detergent lignin (Bruno et al. 1992). Non-digestible sorghum components, including lignin, silica, and tannins, which vary across genotypes, deteriorate the fodder quality. Reduced lignin concentrations might be a feasible method to improve digestibility if done without impairing the plant's agronomic performance. Recent research has focused on enhancing fodder quality using bmr sorghum, which has a lower indigestible lignin concentration and higher forage digestibility than other sorghum types (Saballos et al. 2008). Additionally, HCN releases cyanogenic glycoside dhurrin, which decreases nutritional value, and its excess concentration (higher in early growth stages and decreasing during crop development) in forage sorghum may cause major health issues and even animal fatalities (Li et al. 2023a, b). Sorghum breeders are aiming to develop varieties or hybrids with good stem quality, genetic stability, digestibility, crude protein content, sugar content, dry matter yield, and low hydrocyanic acid concentration using conventional and biotechnological tools (Ping et al. 2018).

As a result of the extensive release of several high-yielding cultivars due to phenotypic selection in traditional breeding based on breeders' experiences, plant breeding approaches that rely on traditional methods are being hampered by factors including high labor costs, inefficiency, and the impact by the environment. The fast advancement of whole-genome sequencing and marker development technologies over the past two decades has permitted the use of high-density single nucleotide polymorphism (SNP) markers to investigate the entire genome at a cheap cost. The simultaneous discovery of high-throughput markers and multiplexed genotyping of germplasm has allowed us to speed up marker-aided breeding. It has been more than a decade since the completion of the sorghum reference genome sequence (Paterson et al. 2009), the construction of several sorghum association panels (SAP) (Casa et al. 2008; Morris et al. 2013; Upadhyaya et al. 2009), the establishment of mutant libraries, a sorghum gene expression atlas, and the development of a sorghum base online (Addo-Quaye et al. 2018; Jiao et al. 2016; Tao et al. 2021); the field of sorghum study has entered a new era. Next-generation sequencing (NGS) has made it easier to use DNA-based and high-throughput SNP markers to improve crop genetics and breed better crops. This has led to more genetically diverse stocks or ideal candidate genes being found and used in crops through genome-wide allele mining, which is done with the help of association mapping, primarily genome-wide association studies (GWAS), genomic predicted selection (GS), and genome engineering with CRISPR-Cas9. This, along with high-throughput phenotyping tools for precision selection and enhanced speed breeding methodologies to reduce breeding cycles, has revolutionized the scope and toolkits of crop breeding. To feed a growing global population, crop cultivars must be genetically adapted to current and future climate change scenarios. As a result, Varshney et al. (2021) proposed the genomics-assisted breeding (GAB) technique, incorporating genomics tools into the breeding programs. So far, this approach has resulted in high-yielding, stress-tolerant, and nutritionally better cultivars (Varshney et al. 2021). The conventional and genomics-assisted breeding, enabling trait targeted superior cultivars for food, nutrition, and climate resilience (Ganie and Reddy 2021). Because of high-yielding, nutritionally dense, stress-resistant, and climate-robust cultivars, the number of undernourished people has decreased from 40% in 1960 to 11% currently (Qaim 2020). These genomics-assisted breeding tools enable much more targeted and efficient exploitation of previously restored natural and/or induced variation, minor effects of poly- and oligo-genic genes and alleles, and unprecedented phenotyping and selection of large-scale breeding populations (Chen et al. 2019; Godwin et al. 2019; Xu et al. 2020). Despite the fact that the draft genome sequence of this significant crop was released in 2009 (Paterson et al. 2009) and that genomic resources were available (Mace et al. 2013), the pace of research using AM in general and GS in particular did not progress much in this sorghum crop. Sorghum, with its relatively small, diploid, and fully sequenced genome, fits itself well to genomics-assisted breeding techniques (Bekele et al. 2013). A genomics-driven breeding approach for developing cultivars with improved nutritional quality fodder in forage sorghum is a challenging task for breeders under climate change and exponential population growth scenarios. Here, in this chapter, we summarize current advances in forage sorghum germplasm enhancement, high-throughput genotyping with next-generation sequencing, high-throughput phenotyping technologies (HTPT), natural variation investigation through marker-trait relationships, genomic prediction, and genome engineering to increase fodder quality.

2 Genetic and Genomic Resources of Sorghum and Next-Generation Sequencing in GAB for Fodder Qualities

Breeding programs rely on genetic and genomic resources to assess genetic diversity and population structure. Recently, genetic diversity has become a major concern for plant breeders due to the continuous use of modern cultivars and advanced breeding lines as they have lower genetic diversity (narrow genetic base) as compared to wild species and crop wild relatives (broad genetic base). The local land races and crop wild relatives are a great reservoir of different beneficial alleles and genes associated with forage sorghum yield and quality traits. There is a need for an efficient exploitation of the genetic potential of these genetic resources in plant breeding programs for yield and quality enhancement. There is also the possibility of discovering variations in forage sorghum derived from landraces, ancestral or wild crop relatives, and these could be used in manipulating its nutritional quality. More wild and cultivated sorghums are needed for extensive genomic variation analysis, which will enable us to reconstruct the domestication and improvement program of sorghum and breed new sorghum varieties. Forage sorghum species are more adaptable and resistant to various biotic and abiotic stresses that impact productivity and feed quality. Introducing genes from wild *Sorghum* species into domesticated forage sorghum is the first step toward gaining access to these previously unused genes for biotic and abiotic stressors that regulate fodder quality and yield. However, just a few genes from wild sorghums have been documented, and those have been neglected in fodder quality crop enhancement programs. These genes may be found in crop wild relatives (CWRs). However, CWR materials may be difficult to deal with, and finding desirable alleles requires a significant effort in pre-breeding. Even though natural variations do not always result in pleiotropic effects, they might lead to diverse phenotypes, making them an important resource for GAB research. A unique nested association mapping population based on sorghum wild relatives was employed as a novel resource for GWAS trait dissection activities (Mace et al. 2021). The sorghum CWR panel is an unparalleled resource for sorghum development, allowing researchers to explore and exploit sorghum diversity, and this might be a fruitful study in GWAS mapping in crop wild relatives in the coming years. In a natural population, it is time-consuming and expensive to assess each accession in terms of genotyping and phenotyping. The core collection approach establishes a minimal selection of germplasm to retain the maximal genetic diversity of the whole collection (Frankel 1984), which might considerably boost the efficiency of genetic resources. Plant genomics has just opened a new era in the construction and exploitation of crop core collections (Jia et al. 2017).

It is hard to use large-scale DNA fingerprinting methods like restriction fragment length polymorphisms (RFLP), random amplified polymorphic DNA (RAPD), and simple sequence repeats (SSR) on a large scale in current plant breeding since they are labor-intensive and time-consuming. The lack of localization of these markers further slows down the breeding process (Collard and Mackill 2008). With their high throughput and expanding number of genome sequences, NGS technologies provide a chance to redesign genotyping approaches for more efficient genetic mapping and genome analysis in comparison to Sanger sequencing (Pareek et al. 2011). Over the past decade, next-generation sequencing has advanced at a breakneck pace, resulting in an ever-growing number of sequenced species. Genome-wide markers (SNPs) may be developed using next-generation sequencing and whole genome re-sequencing, which is advantageous for the identification, validation, and evaluation of diagnostic markers to measure genetic diversity in plants (Lai et al. 2010). Crop genome sequences may be used to investigate genome structure and provide insight into genetic diversity for fodder quality and production by re-sequencing various forage sorghum accessions (Rakshit & Wang, 2016). Rasheed et al. (2017) assessed the status, relevance, difficulty, and outlook for "crop breeding chips" and other genotyping platforms, which have considerable promise for crop breeding. Plant breeders use genomic scale analysis tools like NGS (next-generation sequencing) to next-NGS (NGS of single DNA molecules), high-throughput genotyping using new marker systems (e.g., DArT and SNPs), and new selection strategies like MARS (marker-assisted recurrent selection) and GWS (genome-wide selection) to improve complex polygenic traits. They ensure higher linkage mapping resolution, genome-wide association studies, and training of genomic selection models, which are widely used in genomic-assisted plant breeding programs. NGS has found many molecular markers like SNPs, and these resources are publicly accessible via multiple web portals for crop improvement and breeding operations (Kilian and Graner 2012). Because NGS enabled the rapid development of genome-wide markers (SNPs and DArt), whole-genome scanning can now readily study association mapping, natural population structure, and linkage disequlibrium (LD) decay (Varshney et al. 2009; Kilian and Graner 2012). As a result of the rapid growth of NGS, a significant number of molecular markers, such as SNPs, have been discovered, and SNP-based resources are publicly accessible via multiple web portals for crop improvement and breeding programs (Kilian and Graner 2012). As a result of the rapid development of genome-wide markers (SNPs and DArt) utilizing NGS technology, whole-genome scanning, patterns of natural population structure, and the decay of linkage disequilibrium (LD) can now be investigated more efficiently (Varshney et al. 2009; Kilian and Graner 2012).

Even though sorghum and maize have the same ancestry, the sorghum genome is substantially smaller (736 Mbp) and was sequenced in 2009. Two diploid *Sorghum* species, *S. bicolor* and *S. propinquum*, constitute the primary *Sorghum* gene pool (Harlan and de Wet 1971). Its secondary gene pool includes Columbus grass (*Sorghum* x almum Parodi) and Johnson grass (*S. halepense*), both of which are considered to be among the worst weeds in the world. Since Paterson et al. (2009) published the whole genome sequence of the first crop from Africa, there has been significant progress made in understanding grass genome evolution and the *S. bicolor* taxa's hidden diversity. More than 41,000 accessions of sorghum are maintained in the 2 major sorghum germplasm banks, ICRISAT (http://www.icrisat.org) and the USDA's National Plant Germplasm System (http://www.ars-grin.gov). The Sorghum Association Panel (SAP) consisted of 228 tropical sorghums. Morris et al. (2013) integrated three panels, namely, SAP, mini-core collection, and sorghum reference set, to create a diversified panel of 971 accessions. A further investigation into natural variation in dhurrin production and catabolism genes produced a 700-line SCP panel (Hayes et al. 2015). There is only one true nested association mapping (NAM) population in sorghum that was found by using both linkage analysis and association mapping in the nested association mapping population (Bouchet et al. 2017).

Sorghum is a diploid crop with a short genome that is not duplicated. The genetic diversity of sorghum has been greatly enhanced through advanced genome assemblies of 32 inbred lines, including the foundational BTx623 (Paterson et al. 2009; McCormick et al. 2018), RTx430 (Deschamps et al. 2018), Rio (Cooper et al. 2019), and Ji2731 (Tao et al. 2021), all accessible via Phytozome at JGI. These assemblies have enabled the identification of diverse genetic variations, such as SNPs, InDels, structural variations, and copy number variants. Recent telomere-to-telomere (T2T) assemblies for BTx623, the Chinese line Ji2055 (Wei et al. 2024), and two Baijiu brewing sorghum landraces (HYZ and HDN) (Bao et al. 2024) were achieved using ultra-long Oxford Nanopore, PacBio HiFi reads, and Hi-C technologies, marking a leap forward in sorghum genomics. The reference genome sequence for BTx623 was first constructed using Sanger sequencing (Paterson et al. 2009) and has since been revised by McCormick et al. (2018). It has 58.8% retrotransposons, 8.7% DNA transposons, and 34,129 identified genes. A total of 44 sorghum lines, comprising 18 landraces, 7 wild and weedy sorghums, 17 improved inbreeds, 2 guinea margaritiferums, and 2 *S. propinquum*, were analyzed using re-sequencing data to better understand the domestication and improvement history of the plant (Mace et al. 2013). Mace et al. (2013) re-sequenced the *S. propinquum* genome and found 8 million high-quality SNPs, 1.9 million indels, and specific gene loss and gain events in *S. bicolor*. Furthermore, they found that wild or weedy sorghum genotypes had more SNPs than landraces or improved inbreeds. Wild-specific alleles (34%) exceed superior inbred alleles (8%) and landrace-specific alleles (4%). Current findings using ancient sorghum DNA reveal no clear bottlenecks during domestication, but rather an overall rising prevalence of harmful mutation load (Smith et al. 2019). Using GBS SNPs to estimate genetic diversity helped discover phenotypically distinct and genetically diversified accessions, ICSV1 and ICSV700-P10 (Somegowda et al. 2021b). Genomic variation in plants may occur in several ways, including the presence or lack of genes or genomic regions across genotypes, and genes can be classified as either core or variable within a species (Golicz et al. 2016). However, a single genome sequence is insufficient to reflect a species' full genetic architecture. By sequencing many individuals from the same species, pan-genome analysis may be able to find all of the genetic variants in the entire genome library. By analyzing the pan-genome of a wide number of genotypes, it is possible to get a better understanding of species variety, domestication, and breeding history by comparing the results. To that end, an online database on large-scale genomic variation, entitled SorGSD, has been created (Luo et al. 2016). Also, this database contains more than 62.9 million single nucleotide polymorphisms that were found in the sequencing data of 48 *Sorghum* lines (landrace, improved breeding inbreeds, and wild relatives) that were mapped to the reference genome BTx623. This is an important resource for genetic and breeding research.

3 Genomics-Assisted Breeding in Forage Sorghum for Fodder Quality

As reference genomes and genome-wide surveys using comprehensive diversity panels become more widely available, it will be easier to link allelic variation to phenotypic traits. The immense genetic resources stored in gene banks may now be evaluated for their genetic worth (GEBV) and used for crop enhancement programs more efficiently. Precise genome editing tools, together with improved trait structures, provide novel approaches to engineering complex trait variation. Accurate, precise, and large-scale assessments of plant performance are becoming easier to come by thanks to high-throughput phenotyping techniques. Accurate, precise, and large-scale measures of plant performance are becoming easier to achieve because of high-throughput phenotyping technologies. Sorghum has been lagging behind the other main cereal crops when it comes to genetic improvement. In the future, genomics should be promoted or used as the foundation of sorghum breeding programs. Genomic research should be supported or used as the foundation of sorghum breeding programs in the future. At the moment, there are high-throughput markers, cost effective genotyping methods, and precision phenotyping systems that will permit the fast adoption of GAB. Conventional breeding and assessment techniques will be enhanced, but not replaced, by new genetic and genomic technologies. Genome-wide association studies (GWAS), genomic selection (GS), and genome editing (GE) with high-throughput phenotyping (HTPT) are an integral part of genomics-assisted breeding (GAB) with the help of next-generation sequencing (NGS) and genotype by sequencing (GBS) (Varshney et al. 2021). The detailed genomics-assisted breeding pipeline in forage sorghum for fodder quality is described in Fig. 1, and the novel QTLs and candidate genes associated with the fodder qualities through genomics-assisted breeding are described in Table 1.

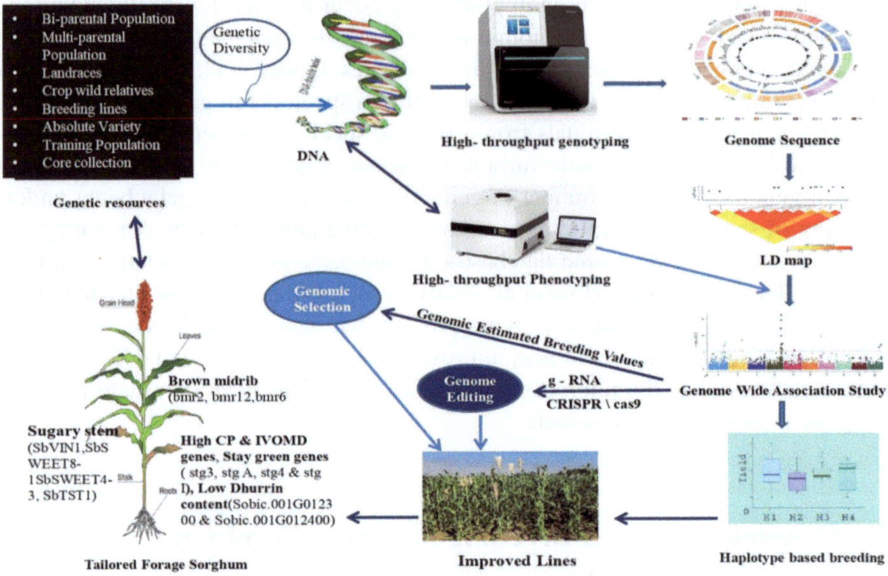

Fig. 1 Genomics-assisted breeding pipeline in forage sorghum for fodder quality

3.1 Association Mapping Through GWAS

Association mapping (AM), also termed as linkage disequilibrium (LD) mapping, is a technique for identifying marker–trait (phenotype–genotype) associations (MTA) based on historic recombinations. This method, which accounts for hundreds of polymorphisms (QTL), may be evaluated using this method, which accounts for hundreds of QTLs. It is an essential tool for finding novel genes and alleles, as well as dissecting complex traits (Baloch et al. 2023). As a result, AM is gaining popularity over traditional bi-parental QTL mapping because of its ability to take advantage of genetic diversity and search for functional variation across a wider range of germplasm, as well as its specific advantages, such as the relatively high resolution provided by its use of linkage disequilibrium (LD) structure. When using AM, it is necessary to think about factors like marker density, population, sample sizes, and population structure. The selection of germplasm, genotypic and phenotypic data quality, and the implementation of proper statistical analysis for marker–phenotype association detections and validations are critical to the success of the association study (Ibrahim et al. 2020). The candidate gene (CG) method or whole-genome AM may both be used for association mapping. It is up to the researcher to decide which method is best suited to the study objective and the available genetic information (Gupta et al. 2014). A study by Hamblin et al. (2005) found that sorghum's self-pollinating mating system and moderate amount of LD make it an excellent candidate for association mapping investigations. Sorghum genome single nucleotide diversity (Bouchet et al. 2012) and association analysis

Table 1 Major QTLs/candidate genes for important fodder quality traits in sorghum

Trait	QTL (genes)	Gene ID	Phenotype	Encoding protein	References
Sugary stem	SbVIN1	*Sobic.004G004800*	Sugar accumulation	Vacuolar invertase	McKinley et al. (2016)
	SbSWEET8-1	*Sobic.008G094000*	Sugar accumulation	Sugar transporters	Mizuno et al. (2016)
	SbSWEET4-3	*Sobic.004G136600*	Sugar accumulation	Sugar transporters	Mizuno et al. (2016)
	SbTST1	*Sb01g030430*	Sugar accumulation	Tonoplast sugar transporters	Bihmidine et al. (2016)
	SbTST2	*Sb04g008150*	Sugar accumulation	Tonoplast sugar transporters	Bihmidine et al. (2016)
Brown midrib and stem texture	bmr2	*Sb04g005210*	Brown midrib	4-Coumarate: Coenzyme A ligase	Saballos et al. (2012)
	bmr12	*Sobic.007G047300*	Brown midrib	Caffeic O-methyltransferase	Sattler et al. (2012)
	bmr6	*Sobic.004G071000*	Brown midrib	Cinnamyl alcohol dehydrogenase	Saballos et al. (2009)
Dry stem	Dry	*Sobic.006G147400*	Dry stem and midrib	NAC transcription factor	Zhang et al. (2018)

(continued)

Table 1 (continued)

Trait	QTL (genes)	Gene ID	Phenotype	Encoding protein	References
Acid detergent fiber (ADF)	8 QTLs with 91 genes	*QTL region (C2_S6746476, C5_S5473787, C6_S60364789, C7_S60623290, C8_S56789460, C9_S40246128)*	ADF rich	ADF-associated protein	Niu et al. (2021)
		SORBI_3006G272600	ADF rich	ADF biosynthesis	Niu et al. (2021)
	RSS29915	*Sobic.003G272200*	ADF rich	Transcription factor bHLH	Li et al. (2018b)
Neutral detergent fiber (NDF)	RSS7796	*Sobic.001G378300*	NDF rich	Sucrose synthase	Li et al. (2018b)
	RSS21890	*Sobic.002G390800*	NDF rich	Kinesin-like protein	Li et al. (2018b)
	RSS50197	*Sobic.005G215300*	NDF rich	Laccase-15	Li et al. (2018b)
	RSS70856	*Sobic.008G172200*	NDF rich	Transcription factor TCP	Li et al. (2018b)
	RSS76122	*Sobic.009G132000*	NDF rich	Uncharacterized protein	Li et al. (2018b)
	LG2 at 61451145	*Sobic.002G223200*	NDF rich	S26 mitochondrial inner membrane protease	Somegowda et al. (2022)
IVOMD		*Sobic.008G026400 Sobic.010G229000 Sobic.001G530300 Sobic.008G054500 Sobic.007G023400*	Low lignin and secondary carbohydrate		Somegowda et al. (2021a)
IVOMD	LG 1 at 73790038	*Sobic.001G463900*	High IVOMD	N-terminal acetyltransferase	Somegowda et al. (2022)

ME		*Sobic.003G282600*	High ME	Homogalacturonan biosynthesis	Somegowda et al. (2021a)
	64546335 on LG 1	*Sobic.001G356000*	High ME	MYB-like DNA binding protein	Somegowda et al. (2022)
	LG7 at 57595992	*Sobic.007G146200*	High ME	Starch and sucrose metabolism gene	
IVOMD, ME	LG2 at 70382979	*Sobic.002G335900*	High IVOMD, ME	Cyclin-related gene	
ADL, IVOMD, ME		*Sobic.001G354800*	High ADL, IVOMD, ME		Somegowda et al. (2021a)
Crude protein (CP)	RSS17673	*Sobic.002G217100*	High crude protein	Serine/threonine-protein kinase	Li et al. (2018b)
	RSS22092	*Sobic.002G397001*	High crude protein	Cysteine proteinase	Li et al. (2018b)
	RSS48493	*Sobic.005G171700*	High crude protein	Uncharacterized protein	Li et al. (2018b)
Hemicellulose (HC)	RSS1510	*Sobic.001G095700*	Hemicellulose rich	Transcription factor bHLH	Li et al. (2018b)
	RSS76122	*Sobic.009G132000*	Hemicellulose rich	Uncharacterized protein	Li et al. (2018b)
Cellulose (CL)	RSS21890	*Sobic.002G390800*	Cellulose rich	Kinesin-like protein	Li et al. (2018b)
Dry weight		*Sobic.007G145600 Sobic.007G146200*	Defense against pathogens	Cell wall-bound phenolic acid synthesis	Somegowda et al. (2021a)

(continued)

Table 1 (continued)

Trait	QTL (genes)	Gene ID	Phenotype	Encoding protein	References
Dhurrin	CYP79A1	*Sobic.001G012300*	Dhurrin content in plant parts	Dhurrin biosynthesis	Blomstedt et al. 2012
		3′ UTR of Sobic.001G012300	Dhurrin content in plant parts	Dhurrin biosynthesis	Wanget al. (2024)
	tcd1		Dhurrin content in plant parts	Totally cyanide deficient1 protein	Blomstedt et al. 2012
	UGT 85B1	*Sobic.001G012400*	Dhurrin content in plant parts	Cyanohydrin β-glucosyl transferase	Hayes et al. (2015)
	Putative glutathione S-transferase	*Sobic.001G012500*	Dhurrin content in plant parts	Rate-limiting for the biosynthesis of dhurrin	Hayes et al. (2015)
	Putative mat E gene	*Sobic.001G012600*	Dhurrin content in plant parts	Rate-limiting for the biosynthesis of dhurrin	Hayes et al. (2015)
	Dhurrinase-like gene	*Sobic.008G080600*	Dhurrin content in plant parts	Biosynthesis of dhurrin	Hayes et al. (2015)
	β-Glucosidases	*Sobic.008G080400 Sobic.008G079800 Sobic.008G080600*	Dhurrin content in plant parts	Rate-limiting for the biosynthesis of dhurrin	Hayes et al. (2015)
Dhurrin	NIT4B2	*SNP marker position (Chr4:57544430)*	Dhurrin content in plant parts	Dhurrin biosynthesis	Choi et al. (2020)
	CAS1	*SNP marker position Chr6:2666100 Chr6:2666092 Chr6:2666088 Chr6:2666028 Chr6:2665864 Chr6:2665840 Chr6:2665136 Chr6:2665867*	Dhurrin content in plant parts	Dhurrin biosynthesis	Choi et al. (2020)

Stay green	stg3 (Z gene)		Stay green plant	Delay in senescence	Harris et al. (2007)
	stgA (pla gene)		Stay green plant	Delay in senescence	Tao et al. (2000)
	stg4 (opr gene)		Stay green plant	Delay in senescence	Harris et al. (2007)
	stgE (bmr12)		Stay green plant	Delay in senescence	Kebede et al. (2001)
	stgI (Rs2)		Stay green plant	Delay in senescence	Kebede et al. (2001)
	RSS83457	*Sobic.010G172100*	Stay green plant	Transcription factor bHLH	Li et al. (2018b)
		Sobic.004G297400, Sobic.004G297500	Delay leaf senescence	Enoyl-CoA hydratases	Wang et al. (2023)

for variance in important agroclimatic parameters have been undertaken by Morris et al. (2013) and Nida et al. (2021). Community resources and mini-core collections in sorghum have also been produced for AM in order to boost the applicability and efficiency of the set of genotypes reflecting diversity at the genotypic and phenotypic levels. Casa et al. (2008) compiled a collection of 377 sorghum accessions from all main cultivated races, including tropical lines from varied geographic and climatic zones. Murray et al. (2009) conducted AM on 125 sorghums, mostly sweet sorghums, to see if the main QTL for total stem sugar (Brix) or any of the QTLs for height had been found in other genetic contexts.

Genome-wide association studies (GWAS) have revolutionized the mapping of quantitative characteristics because of developments in high-throughput genotyping technology and increased computational capacity, making it feasible to conduct association studies on genome-wide genetic variations (Liu and Yan 2019; Vellaichamy Gandhimeyyan et al. 2023). Genome-wide association study (GWAS) is typically used when one wants to discover all of the genomic areas that may be involved in influencing the trait of interest, taking advantage of historical recombination in diversity panels (Yano et al. 2016; Zhao et al. 2019). Sorghum is an ideal grass species for identifying natural variations underlying complex agronomic variables by GWAS, in part owing to its small genome size and modest LD (Paterson et al. 2009; McCormick et al. 2018). The advancement of next-generation sequencing (NGS) technology has permitted the establishment of a significant number of genetic markers, such as single nucleotide polymorphisms (SNP), insertion–deletion polymorphisms (InDels), and so on, even in relatively understudied crop species such as forage sorghum. The identification of new genes/alleles associated with any given characteristic might then be accomplished using genotyping-by-sequencing (GBS) techniques. Additionally, statistical connections between DNA polymorphisms and trait variations in distinct collections of germplasm that are genotyped and phenotyped for desired characteristics may be employed in genome-wide association studies (GWAS) to find the genomic areas controlling the desired qualities. By using GWAS, plant breeders hope to break through many of the limitations of traditional linkage mapping and find new ways to study complex traits. Increased mapping resolution for accurate gene/allele/QTL placement is achieved by combining NGS and GWAS, as described earlier (Varshney et al. 2014). These cutting-edge methods and technologies help to speed up the genomics-assisted breeding process and increase genetic gains (Somegowda et al. 2021a). Currently, GBS is responsible for the majority of sorghum genetic variants, which results in a substantial amount of missing data (Annicchiarico et al. 2015). As a result, a larger population and greater SNP density are required; therefore, effective statistical procedures must also be developed to deal with it. Since conventional GWAS focuses mostly on common variants with minor allele frequency (MAF) ≥ 0.05, those uncommon variants with MAF < 0.05, which may explain phenotypic variance, are generally filtered out before GWAS. To fill in the gaps in our understanding of these uncommon variations, we need to build a new statistical and association analysis strategy. They unravel complicated features by linking genotypic polymorphisms to phenotypic variation in huge panels. While unrelated individuals are desirable, this is not always the case when assembling existing samples. The unified mixed model controls both population structure and kinship in GWAS (Tibbs Cortes et al. 2021). Genome-wide association research approaches have improved computing performance by solving MLM equations more efficiently.

Ping et al. (2018) used 109 SSRs to examine the genetic diversity of 48 forage sorghum germplasm resources. Niu et al. (2021) re-sequenced 206 forage sorghum accessions and reported 14,570,430 SNPs and 1,967,033 indels, from various geographic locations in order to better investigate the genetic diversity through GWAS to identify the important genes associated with the acid detergent fiber content. They detected 8 QTLs with 91 genes on chromosomes 2, 5, 6, 7, 8 and 9 each that were linked to the amount of acid detergent fiber. Rice's OsAAE3 (*Os04g0683700*), which controls rice blast resistance, floret development, and ADF biosynthesis, and *SORBI 3006G272600* (which regulates rice floret development and ADF biosynthesis), are strongly collinear (Kikuchi et al. 2003). It is our goal in sorghum breeding for forage to choose cultivars with moderately acid detergent fiber. MTAs and materials with appropriate alleles discovered in this work should speed up the breeding of forage sorghum varieties and hybrids and raise the ADF content to an optimal level. Wang et al. (2011) identified both the Bg and Sbp genes at seven loci that were significantly associated with saccharification yield and identified possible candidate genes, including ß-tubulin, which determines the orientation of cellulose microfibrils in plant secondary cell walls, and NST1, a key transcription factor directing secondary cell wall formation in fibers. Li et al. (2018a, b) evaluated the crude protein, neutral detergent fiber, acid detergent fiber, hemicellulose, and cellulose content of 245 sorghum lines and found 42 SNPs and 14 candidate genes for forage quality parameters. The brown midrib phenotype has been exploited to find mutants that reduce lignin content and other components, potentially improving sorghum biomass for conversion. A total of four allelic classes of bmr mutant sorghums have been found, with the most extensively utilized allelic classes being bmr6 and bmr12 (Saballos et al. 2012; Sattler et al. 2012). This study also found 9 QTLs encompassing 18 genes for lignin concentration in 206 forage sorghum accessions, and 13 out of 184 sorghum lignin-related loci showed strong collinearity with other crop gene families (Niu et al. 2020). Sorghum has a huge amount of biomass and genetic variation for complex sugars, which makes it a good fodder for livestock as well as a possible source of biofuel. Few research have been done on the knowledge transfer from biofuel production to better fodder quality. With 42,926 SNPs, Burks et al. (2015) showed a significant relationship at 51.8 Mb on chromosome 6, a region that includes a dry midrib locus gene, which affects dry vs. juicy stems. Dry possesses 23 haplotypes in 42 wild sorghums and 1 *S. propinquum*, 2 haplotypes in landraces, and 4 haplotypes in cultivated sorghum (Fujimoto et al. 2018; Zhang et al. 2018). Most important digestibility factor IVOMD has low heritability under stress but high heritability in setup, which means that drought makes fodder less good (Perrier et al. 2017).

For two primary features, dry stalk weight (DSW) and IVOMD (IVOMD), GBS data was utilized to determine the best contrasting parents (Somegowda et al. 2021a, b). Sugary stem is an essential fodder quality feature in forage sorghum. *SbSWEETs*

8–1 and SbSWEETs 4–3 are critical sweet genes for sugar transport, respectively, accountable for sucrose efflux from the leaf and sucrose unloading from phloem to the stem (Mizuno et al. 2016). These tonoplast sugar transporters (SbTST1 and SbTST2) are required for significant sugar storage (Bihmidine et al. 2016). The expression of vacuolar invertase (SbVIN1) is negatively related to sucrose storage (McKinley et al. 2016). Takele et al. (2022) used 139 recombinant inbred lines of sorghum as mapping populations to genotype and phenotype. It took 1082 polymorphic SNP markers uniformly dispersed throughout sorghum's 10 linkage groups to discover and map the characteristic. It took two seasons to find four QTLs for stem sugar concentration (*qBrix1-1*, *qBrix2-1*, *qBrix4-1*, and *qBrix4-2*) (Takele et al. 2022). Stay-green expression has been shown to enhance crop residue quality (van Oosterom et al. 1996), maintain carbon fixation, and deliver starch to the sink (McBee et al. 1983). The "stay-green" phenotype is one of the most well-known and often used features in sorghum development (Rakshit et al. 2016), and stay-green sorghums like B35, SC56, and E36-1 have been widely bred for this characteristic (Reddy et al. 2018). The durra race member BTx642 (previously B35) is the most prevalent source of stay-green in sorghum. Sabadin et al. (2012) discovered four significant stay-green QTLs, namely, *Stg1*, *Stg2*, *Stg3*, and *Stg4*, as well as several smaller QTLs. They were found on *Sorghum* chromosome 3, explaining around 20% and 30% of phenotypic diversity, respectively. *Stg3* is on chromosome 2 and *Stg4* is on chromosome 5 (Harris et al. 2007). A fine-mapping population has recently found seven QTL and numerous candidate genes affecting stay-green characteristics, providing a valid experimental and theoretical basis for understanding stay-green mechanisms. They must be confirmed and incorporated into molecular tools for sorghum genetic improvement (Kiranmayee et al. 2020).

In sorghum, dhurrin and HCNp regulation are influenced by plant development and tissue type. Dhurrin [(S)-p-hydroxymandelonitrile-D-glucopyranoside] is an extensively researched cyanogenic glucoside found in sorghum. It is a precursor to prussic acid (HCN). Dhurrin in feed sorghum given to animals may induce cyanide poisoning, owing to hydrolysis of dhurrin into HCN. An analysis of 700 converted sorghum lines by Hayes et al. (2015) revealed the genetic regulation of leaf dhurrin concentration. The sorghum conversion panel showed a wide variation in leaf dhurrin concentration, with the Caudatum group having the greatest and the Guinea group having the lowest. A GWAS adopting a mixed linear model indicated significant associations around UGT 185B1 on chromosome 1 and the catabolic dhurrinase loci on chromosome 8. *SbChr 1:1160147*, *SbChr1: 1200433*, and *SbChr1:1213154* were discovered to be linked with dhurrin production in all three conditions. Cowan et al. (2022) characterized cyanogenesis in undomesticated *Sorghum* species to investigate casual variations or potential genes for dhurrin production. No HCN was detected in the leaf tissues of any of the examined *Chaetosorghum*, *Heterosorghum*, *Parasorghum*, or *Stiposorghum* species. Li et al. (2023a, b) identified a QTL most tightly linked with HCN-p, and the linked SNPs were in the 3' UTR of *Sobic.001G012300* at chromosome 1. This gene encodes CYP79A1, involved in dhurrin biosynthesis.

A haplotype is a set of alleles for distinct polymorphisms (such as SNPs, insertions/deletions, and other markers or variations) on the same chromosome that are inherited together with the least amount of recombination (Garg 2021). To overcome SNP restrictions and improve the resolution of prospective genomic areas, consider haplotypes for genome-wide analysis (Qian et al. 2017). All haplotype-defining markers may be utilized to determine the most acceptable haplotype combination for regulating a certain trait when the molecular markers defining these advantageous haplotypes are identified. It may also be possible to detect a wider range of genetic diversity using haplotypes in the future, when a huge collection of ancestral and wild germplasm of specific crop species may be re-sequenced to discover haplotypes (Bevan et al. 2017). Haplotype markers, on the other hand, have been shown to be superior to individual SNP markers when it comes to identifying complex features, as has been established by efficient gene identification and GS (Qian et al. 2017). Haplotypes, rather than single SNP markers, have been shown to be more accurate at predicting low-heritable quantitative traits (Won et al. 2020). Based on these findings, Jensen et al. (2020) found unique and superior haplotypes and also constructed a sorghum practical haplotype graph, which permits genome-wide imputation and cost-effective genomic prediction, in order to progress in sorghum fodder quality enhancement. So, sorghum research in the area of haplotype breeding is trailing behind other main cereal crops; therefore, greater attention must be paid to the crop's long-term growth and improvement as a source of high-quality fodder. Cooper et al. (2019) used comparative genomics to compare sugar metabolism genes in sweet (Rio) and grain sorghums (BTx623) and found a high proportion of non-synonymous and probable loss-of-function mutations. The comparative genomics of forage and broom types, as well as genetic pathways underpinning phenotypic divergence across sorghums, will give crucial information for future sorghum breeding. The use of genome-wide haplotypes in conjunction with GS has been shown to be viable and may substantially speed up plant breeding by providing more specificity and predictability (Spindel et al. 2018). Wang et al. (2023) studied 333 sorghum lines and identified 31 senescence linked QTL governing 148 genes. Based on haplotype analysis, 45 candidate genes extremely prolonged senescence duration. Haplotypes associated with regions *Sobic.001G230050*, *Sobic.004G295600*, *Sobic.003G052200*, and *Sobic.004G297200* were predominantly involved in delaying senescence.

3.2 *Genomic Selection*

Genomic selection (GS) is a large-scale version of MAS, where the effects of a marker are assessed concurrently over the whole genome in order to compute genomic-estimated breeding values (GEBVs) as suggested by Meuwissen et al. (2001). This is in contrast to the standard MAS, which employs a limited set of genetic markers associated with the QTL for genetic analyses. The GS procedure entails generating genomic-estimated breeding values (GEBVs) for a "breeding population" with just genotypic data using a model trained on a "training population" with genotypic and phenotypic data (Meuwissen et al. 2001). Individuals with higher GEBV scores may be employed as parents in the next breeding cycle or immediately in the variety release pipeline, as they predict their offspring's performance will be enhanced. As a result, it is required to optimize prediction accuracy, which is impacted positively by the training population's association with the selection candidates. This means that GS has at least two advantages over MAS, such as not having to find QTL that are linked to the traits that you want to improve and not having to collect phenotypes from the breeding population, which saves money.

The whole-genome genotype–phenotype link may be used to capture both large and small gene effects, lowering costs and time by enhancing genetic gain in every cycle (Guo et al. 2019). GWAS markers may be added to GS models to increase their ability to predict outcomes (Li et al. 2018a, b; Kent et al. 2023). The genome-wide map of SNP variation scores within an artificial population at the whole-genome level offers a platform for crop improvement in sorghum (Meuwissen et al. 2001; Crossa et al. 2017). GS is another potential breeding approach for quick improvement of complex traits with poor heritability, with genomic-estimated and real breeding values correlated well (Jia and Jannink 2012). Low-density markers may have a reduced accuracy in predicting GEBVs (Singh and Singh 2015). We need a high enough number of markers to ensure that the maximum number of QTL associated with a trait is in strong LD with at least one marker in order to improve prediction accuracy. Numerous statistical models and machine learning techniques for GS have been suggested, including rrBLUP, BayesA/B, random forest, and RKHS. rrBLUP is the most frequently utilized of these and is widely employed in crop breeding. Sorghum is one of the crops for which genetic selection (GS) has been used in recent years. The GS model is trained using natural populations, test-cross hybrids, and mixed populations (Dos Santos et al. 2020; Habyarimana et al. 2020). In the GS for biomass trait research, a worldwide collection of 976 sorghum accessions were genotyped using 0.72 million SNPs obtained by GBS, and 300 of the most representative accessions were phenotyped for biomass-related attributes, and prediction models were built (Yu et al. 2014). More research like this should lead to a greater understanding of the prediction models that will be utilized in the GS program in sorghum, particularly for fodder quality. More research like this, particularly for fodder quality, is likely to result in a greater knowledge of the GS approach in sorghum. There is no doubt that GS is a good technology for improving crops in the future. It could also change the way plants are bred.

The absence of data management skills and database resources is a key barrier to integrating these new developments into breeding operations on a regular basis. In terms of molecular breeding applications, NGS-based genotyping technologies like genotyping by sequencing (GbS) have the potential to be used for background selection in classic introgression breeding as well as genomic selection (GS), a new paradigm in current breeding programs. Access to high-throughput genotyping tools and the availability of high-throughput phenotyping facilities are the primary advantages and driving forces for its use. To attain its full potential, genomic selection will need to solve particular difficulties connected to GS theory, such as training population design, predicting efficacy under a modified marker and population variable scenario, and integrating GS into ongoing breeding schemes (both in scale and dimension). Despite significant progress in genomic resources, computational tools involving the selection of suitable models, a decrease in the computational complexity of analyses, the minimum size of the training population, and other factors influencing GS (Desta and Ortiz 2014; Hickey et al. 2014), the progress of GS studies in sorghum is not commendable. There is no full-length report on GS in sorghum for fodder quality traits yet. This opens up the possibility of using GS in the public and commercial sectors to boost sorghum production. The GS technique has the potential to transform plant breeding. Coupled with high-throughput molecular markers and phenomics tools, it has demonstrated significant promise for selecting superior genotypes with high accuracy and accelerating the breeding cycle in order to increase the genetic gain of interest trait (Sandhu et al. 2022).

3.3 Mutagenesis and Genome Engineering in Forage Sorghum for Fodder Quality

3.3.1 Tilling and Eco-Tilling for Fodder Quality in Sorghum

Plant breeders have been employing mutagenesis to increase crop yield and, more recently, to enhance crop plant quality. Mutagenesis approaches like tilling and eco-tilling have recently emerged as promising tools for detecting new functional genetic variants linked to our trait of interest. When it comes to genomics-assisted breeding, sequence-based tilling techniques play a critical role in the discovery of new functional alleles and genes because of the rapid advancement of NGS and GBS in crops. Tilling is a reverse-genetic approach for identifying the mutations in the mutagenized sorghum population for fodder production and quality, employing CEL I endonucleases for sequence analysis in heteroduplex areas (Bharathi Raja et al. 2016). To acquire a wide picture of gene expression changes in response to recently introduced SNPs, tilling could be used to screen phenotypes for nutritional features and define the phenotype using just a combination of whole-genome re-sequencing, linkage maps, and microarrays. In eco-tilling, the concepts of reverse genetics are applied to the natural population, rather than the mutagenized population (Bharathi Raja et al. 2016). Lignin gives mechanical strength to plant cells, but lignin lowers the digestion of feed by ruminant cattle and the efficiency of biomass for energy generation. As a result, sorghum mutants with decreased secondary cell wall lignin concentration, such as those with reddish-brown midribs and brown stems, were discovered and widely used to reduce the detrimental effects of lignin. Six additional brown midrib loci were also discovered as a result of EMS screening using novel mutants found in the previous screens (Sattler et al. 2014). However, only one of the probable CAD genes (Sobic.004G071000) was found to have an early termination codon substitution, resulting in a brown midrib null allele (bmr6) and its characteristic brown midrib coloration (Sattler et al. 2009). In the BTx623 tilling population, Scully et al. (2016) discovered six different bmr6 alleles (Xin et al. 2008). According to the standard tilling experiment, the production of acyanogenic or low-cyanogenic mutants in sorghum has been documented. Dhurrin-free mutants were produced via EMS mutagenesis, since naturally occurring non-dhurrin mutants were not found in wild or cultivated *Sorghum* species. Ten mutants with G/C–A/T substitutions in CYP79A1 (Sobic.001G012300), which is the enzyme that does the first step in making dhurrin, were found by using the tilling method in 264 M2 sorghum plants with CEL1 endonuclease.

3.3.2 Genome Editing for Fodder Quality in Sorghum

Genome-editing technologies have recently emerged as an efficient and successful method for producing novel allelic variants in cultivated cultivars' genomes. As an alternative to traditional plant breeding and transgenic approaches, targeted genome engineering has been established as a tool for improving crop plants and ensuring sustainable food production (Osakabe and Osakabe 2015; Liu et al. 2024; Ghane et al. 2024; Dubey et al. 2025). These new molecular techniques and resources allow for precise manipulation of genetic building blocks and regulatory apparatus, which underpin yield and quality enhancement as well as direct rectification of detrimental mutations by genome editing, to be used in crop plants (Hsu et al. 2014). Genome editing is currently carried out using four types of engineered nucleases, which include engineered homing endonucleases (Carroll 2014), zinc finger nucleases (Voytas 2009), transcription activator-like effector nucleases (Cermak et al. 2011), and clustered regularly interspaced short palindromic repeats (Cong et al. 2013; Chen et al. 2019); these are the technologies that will aid in the advancement of crop breeding efforts and the establishment of more direct links between mechanistic knowledge of gene function and agricultural yields (Hua et al. 2019). Clustered regularly interspaced short palindromic repeats (CRISPR/Cas9 system) have been extensively employed for precise genome editing in recent times. CRISPR was first identified as a prokaryotic immune system, but it has now evolved into a potent tool for eukaryotic genome editing. Transformation of the plant with Cas9 and production of the guide-RNA chimaera to a GN19NGG motif in the gene of interest leads to the cleavage of non-self-double-stranded DNA on both strands at a particular position near the protospacer adjacent motif (PAM) (Gasiunas et al. 2012). An enzyme from *Streptococcus* pyogenes, known as Cas9, may identify 5-NGG-3 as the PAM sequence. The Cas9–g RNA complex relies heavily on PAM for target binding and cleavage (Lei 2014). CRISPR/Cas9 offers several benefits over ZFNs and TALENs, including simple construct design and assembly and high mutation rates that equal or surpass those achieved with ZFNs and TALENs. This means that each target does not need to be made into a complicated protein, because only 20 nucleotides in g-RNA need to be changed to identify each one.

Genome editing (GE) relies mostly on the detection of target genes, transport of CRISPR machinery to the appropriate cells, selection, and regeneration of the right transformed plants. The discovery of the association between gene regulation and complex quantitative traits is equally important. Genome research has found a lot of information about crop enhancement genes, and GE methods are now coming up with new ideas that could help make climate-friendly sorghum crops that have more yield and better fodder quality (Abdallah et al. 2022; Li et al. 2023a, b). It is possible that CRISPR/Cas9 technology may be used to make these kinds of modifications, and as with many crops, feasibility studies have started with forage crops like sorghum (Jiang et al. 2013). Che et al. (2018) were the first to report effective endogenous gene editing in sorghum, and they used *Agrobacterium*-mediated transformation with a ternary vector system to achieve transformation efficiencies ranging from 21% to 25% in African grain sorghum. Li et al. (2018a) used CRISPR/Cas9 to alter an entire gene family to increase lysine content and digestibility. CRISPR/Cas9 knockout of *SbBADH2* in sorghum has yielded new varieties with a distinct aromatic scent in seeds and leaves, enhancing their appeal for animal feed (Zhang et al. 2022). Beyond editing the target gene, applying a multiplex CRISPR-Cas9 strategy could further amplify gene-editing value and utility. Brant et al. (2021) focused on the Liguleless1 (LG1) gene to see whether altered events would show a swiftly scoreable phenotype while going through a time-consuming tissue culture procedure. Several plant breeders have reported on the optimization of CRISPR/Cas9 techniques in sorghum (Liu et al. 2019; Sander 2019; Char et al. 2020). Assem et al. (2023) used CRISPR/Cas9-mediated mutagenesis to edit caffeic acid O-methyltransferase (COMT) responsible for lignin biosynthesis that limit ruminant digestion.

Furthermore, CRISPR genome editing allows for the rapid breeding of crop plants with specified gene mutations without the introduction of foreign DNA (Purugganan and Jackson 2021). This technology has been successfully used to produce specific mutations in a restricted number of sorghum genotypes. As a result, in order to increase the nutritional content of forage sorghum, more detailed investigation is needed in this area. Despite the fact that widespread use of CRISPR/Cas9 in the sorghum breeding toolbox is still a ways off, it is clear that this next-generation genome editing technique is set to make significant contributions to sorghum crop improvement in terms of fodder yield and quality.

3.4 High-Throughput Phenotyping for Fodder Quality in Sorghum

Plant genome functional analysis has progressed to a high throughput level, but plant phenotyping remains a significant barrier for modern plant breeding. Traditional phenotyping does not offer the large-scale, high-quality phenotypic data required for effective quantitative trait prediction. Traditional plant phenotyping approaches depend on manual measurements, which are arduous, imprecise, and time-consuming, making phenotypic data collection from large populations difficult. It is very important for plant breeding and the development of plant genomics to link phenotypes and genotypes to find out which genes control important traits.

Despite the availability of numerous high-throughput DNA sequencing technologies for the collection of genomic data, the ability to create high-quality phenotypic data remains far behind. High-throughput phenotyping, on the other hand, provides distinct benefits, allowing for quick, non-destructive, and high-throughput detection while also resolving the limitations of conventional approaches. In GWAS, traits derived using high-throughput phenotyping perform as well as or better than traits obtained through conventional, manual approaches. An International Plant Phenomics Network was established for high-throughput phenotyping of miniature, short-lived model plants and crops utilizing robotic, noninvasive imaging (Muraya et al. 2017). When it comes to crop breeding, high-throughput phenotyping, which allows for non-contact and dynamic measurements, has the potential to provide high-quality trait data for GWAS and GS, which will then help decipher the genetic structures of complex plant traits and become an integral part of GAB for accelerating the breeding cycle to improve genetic gains (Xiao et al. 2022). Data from high-throughput automated phenotyping has also been analyzed using GWAS. For instance, by utilizing automated feature extraction from photos in GWAS in sorghum (Zhou et al. 2019) and biomass attributes measured by aerial drones, plant breeders discovered substantial relationships between panicle architecture and biomass traits (Spindel et al. 2018). Likewise, a sophisticated instrument like "The NIRMasterTM" developed by BUCHI rapidly, accurately, and reproducibly controls the quality and nutritional facts of the feed and food samples. It improves the reproducibility of the data through polarization FT-NIR technology and the precision of the results due to topwavelength performance by an internal laser system. Here, a special software package for sorghum is needed to be installed to analyze the fodder quality simultaneously, like, ADF (acid detergent fiber), NDF (neutral detergent fiber), ADL (acid detergent lignin), ME (metabolic energy), IVODM (in vitro digestible mate), and moisture percent.

4 Conclusion and Perspectives

Although significant progress has been made in identifying the genetic loci underpinning important agronomical and adaptive traits, with the rapid deployment of GAB, high-throughput sequencing, phenotyping, pan-genomes, epigenomes, and other disciplines should be integrated to identify the genetic diversity, favorable alleles, and functional regulatory networks underlying the complex agronomical attributes, nutritional traits, and stress resilience of forage sorghum. Furthermore, effective breeding techniques are required to efficiently integrate modern breeding approaches with traditional breeding and evaluation processes in order to produce new varieties with enhanced fodder and nutritional properties. The advancements in genomics and the steadily decreasing cost of genotyping, together with new computational tools, have expanded the scope and created new paths for AM and GS to be implemented in sorghum in the future years. The new high-throughput phenotyping tools will aid in this effort. There is no doubt that AM and GS will play a significant role in the sorghum enhancement program for a wide range of uses. Optimized speed breeding procedures are essential for accelerating breeding progress, when used in conjunction with genomic breeding approaches. Fast-tracking utilization of minor effect alleles, accumulating positive alleles, and purging of harmful alleles are all goals for sustaining genomic breeding benefits.

Despite GAB's enormous potential, various bottlenecks remain in the way of its rapid implementation. None of the current cutting-edge phenome initiatives comes close to reaching the full promise of GAB, mostly due to expense and limited phenotyping capabilities. Complex inheritance phenomena, such as epigenetic events linked to fodder yield and quality features, are also of great interest and still yet to be understood. Although the third generation sequencing platforms and NGS technologies enable us to investigate crop genomes and other post-genomics in greater depth, bioinformatic tools and storage capacity for large sequence data remain a problem (Van Oosterom et al. 1996). The number of target genes, population size, and parent selection are all factors that must be taken into account based on the target use of sorghum. In sorghum, GS must additionally include genotyping and phenotyping accuracy, model selection, and improvement based on population and gene–environment interactions. Genome editing is crucial in identifying gene functions and crop genetic enhancement. However, the effectiveness of the transformation is crucial to its implementation. To encourage the use of the CRISPR/Cas9 system in sorghum breeding, further optimization of the transformation system is required. Sorghum breeding has lagged behind other important cereal crops, and forage sorghum breeding has lagged behind other sorghum types. Last but not least, it is crucial that genome-based technologies and techniques be integrated into the various end uses of sorghum breeding.

Genome-based technologies enable the option of the development of "tailored super sorghum" with ideal traits. In future decades, GAB 2.0 is predicted to play a vital role in breeding more climate-resilient forage sorghum crop cultivars with improved nutritional content in a cost-effective and timely way (Varshney et al. 2021). Although modern GAB approaches may help breeders accelerate the breeding cycle of forage sorghum and assist in the development of super varieties, their use in sorghum breeding is still in its infancy, and there is still a long way to go. Genomic breeding methods that significantly improve crop genomes via the accumulation of favorable alleles and the elimination of detrimental alleles will be critical for designing future forage sorghum crops with higher fodder qualities. These genomics-assisted breeding strategies should be encouraged or used as the cornerstone of future forage sorghum breeding programs.

References

Abdallah NA, Hamwieh A, Radwan K, Fouad N, Prakash C (2022) Genome editing techniques in plants: a comprehensive review and future prospects toward zero hunger. GM Crops Food 5:1–5

Addo-Quaye C, Tuinstra M, Carraro N, Weil C, Dilkes BP (2018) Whole-genome sequence accuracy is improved by replication in a population of mutagenized sorghum. G3 Genes Genomes Genetics 8(3):1079–1094

Annicchiarico P, Nazzicari N, Li X, Wei Y, Pecetti L, Brummer EC (2015) Accuracy of genomic selection for alfalfa biomass yield in different reference populations. BMC Genomics 16(1):1–3

Aruna C, Visarada KB, Bhat BV, Tonapi VA (eds) (2018) Breeding sorghum for diverse end uses. Woodhead Publishing

Assem SK, Basry MA, Taha TA, El-Aziz MA, Alwa T, Fouad WM (2023) Development of an in vitro regeneration system from immature inflorescences and CRISPR/Cas9-mediated gene editing in sudangrass. J Genetic Eng Biotechnol 21(1):58

Baloch FS, Altaf MT, Liaqat W, Bedir M, Nadeem MA, Cömertpay G, Çoban N, Habyarimana E, Barutçular C, Cerit I, Ludidi N (2023) Recent advancements in the breeding of sorghum crop: current status and future strategies for marker-assisted breeding. Front Genet 11(14):1150616

Bao J, Zhang H, Wang F, Li L, Zhu X, Xu J, Wang Y, Liu Z, Zhai G, Xu H, Lin F (2024) Telomere-to-telomere genome assemblies of two Chinese Baijiu-brewing sorghum landraces. Plant Commun 5:100933

Behera PP, Singode A, Bhat BV, Sarma RN (2024) Exploring genetic variability in fodder quality of forage sorghum in the North–Eastern regions of India. Electron J Plant Breed 15(3):574–584

Bekele WA, Wieckhorst S, Friedt W, Snowdon RJ (2013) High-throughput genomics in sorghum: from whole-genome resequencing to a SNP screening array. Plant Biotechnol J 11(9):1112–1125

Bell J, Xu W, Marek T, Heflin R (2024) Forage sorghum performance under drought conditions: hybrid selection and management. Texas A&M AgriLife Research and Extension. Retrieved from ogallala.tamu.edu

Bevan MW, Uauy C, Wulff BB, Zhou J, Krasileva K, Clark MD (2017) Genomic innovation for crop improvement. Nature 543(7645):346–354

Bharathi Raja R, Agasimani S, Anusheela V, Thiruvengadam V, Chibbar RN, Ganesh RS (2016) TILLING and EcoTILLING for discovery of induced and natural variations in sorghum genome. In: The sorghum genome. Springer, Cham, pp 257–267

Bihmidine S, Julius BT, Dweikat I, Braun DM (2016) Tonoplast sugar transporters (SbTSTs) putatively control sucrose accumulation in sweet sorghum stems. Plant Signal Behav 11(1):e1117721

Blomstedt CK, Gleadow RM, O'Donnell N, Naur P, Jensen K, Laursen T, Olsen CE, Stuart P, Hamill JD, Møller BL, Neale AD (2012) A combined biochemical screen and TILLING approach identifies mutations in Sorghum bicolor L. Moench resulting in acyanogenic forage production. Plant Biotechnol J 10(1):54–66

Bouchet S, Pot D, Deu M, Rami JF, Billot C, Perrier X, Rivallan R, Gardes L, Xia L, Wenzl P, Kilian A (2012) Genetic structure, linkage disequilibrium and signature of selection in sorghum: lessons from physically anchored DArT markers. PLoS One 7(3):e33470

Bouchet S, Olatoye MO, Marla SR, Perumal R, Tesso T, Yu J, Tuinstra M, Morris GP (2017) Increased power to dissect adaptive traits in global sorghum diversity using a nested association mapping population. Genetics 206(2):573–585

Brant EJ, Baloglu MC, Parikh A, Altpeter F (2021) CRISPR/Cas9 mediated targeted mutagenesis of LIGULELESS-1 in sorghum provides a rapidly scorable phenotype by altering leaf inclination angle. Biotechnol J 16(11):2100237

Bruno OA, Romero LA, Gaggiotti MC, Quaino OR (1992) Cultivars of forage sorghum for silage. 1. Dry matter yield and nutritive value. Revista Argentina de Prod Anim 12(2):157–162

Burks PS, Kaiser CM, Hawkins EM, Brown PJ (2015) Genomewide association for sugar yield in sweet sorghum. Crop Sci 55(5):2138–2148

Carroll D (2014) Genome engineering with targetable nucleases. Annu Rev Biochem 83:409–439
Casa AM, Pressoir G, Brown PJ, Mitchell SE, Rooney WL, Tuinstra MR, Franks CD, Kresovich S (2008) Community resources and strategies for association mapping in sorghum. Crop Sci 48(1):30–40
Cermak T, Doyle EL, Christian M, Wang L, Zhang Y, Schmidt C, Baller JA, Somia NV, Bogdanove AJ, Voytas DF (2011) Efficient design and assembly of custom TALEN and other TAL effector-based constructs for DNA targeting. Nucleic Acids Res 39(12):e82
Char SN, Lee H, Yang B (2020) Use of CRISPR/Cas9 for targeted mutagenesis in sorghum. Curr Prot Plant Biol 5(2):e20112
Che P, Anand A, Wu E, Sander JD, Simon MK, Zhu W, Sigmund AL, Zastrow-Hayes G, Miller M, Liu D, Lawit SJ (2018) Developing a flexible, high-efficiency agrobacterium-mediated sorghum transformation system with broad application. Plant Biotechnol J 16(7):1388–1395
Chen K, Wang Y, Zhang R, Zhang H, Gao C (2019) CRISPR/Cas genome editing and precision plant breeding in agriculture. Annu Rev Plant Biol 70:667–697
Choi SC, Chung YS, Lee YG, Kang Y, Park YJ, Park SU, Kim C (2020) Prediction of Dhurrin metabolism by transcriptome and metabolome analyses in sorghum. Plants 9(10):1390
Collard BC, Mackill DJ (2008) Marker-assisted selection: an approach for precision plant breeding in the twenty-first century. Philos Trans R Soc B Biol Sci 363(1491):557–572
Cong L, Ran FA, Cox D, Lin S, Barretto R, Habib N, Hsu PD, Wu X, Jiang W, Marraffini LA (2013) Multiplex genome engineering using CRISPR/Cas systems. Science 339(6121):819–823
Cooper EA, Brenton ZW, Flinn BS, Jenkins J, Shu S, Flowers D, Luo F, Wang Y, Xia P, Barry K, Daum C (2019) A new reference genome for Sorghum bicolor reveals high levels of sequence similarity between sweet and grain genotypes: implications for the genetics of sugar metabolism. BMC Genomics 20(1):1–3
Cowan M, Møller BL, Norton S, Knudsen C, Crocoll C, Furtado A, Henry R, Blomstedt C, Gleadow RM (2022) Cyanogenesis in the sorghum genus: from genotype to phenotype. Genes 13(1):140
Crossa J, Pérez-Rodríguez P, Cuevas J, Montesinos-López O, Jarquín D, De Los CG, Burgueño J, González-Camacho JM, Pérez-Elizalde S, Beyene Y, Dreisigacker S (2017) Genomic selection in plant breeding: methods, models, and perspectives. Trends Plant Sci 22(11):961–975
Deschamps S, Zhang Y, Llaca V, Ye L, Sanyal A, King M, May G, Lin H (2018) A chromosome-scale assembly of the sorghum genome using nanopore sequencing and optical mapping. Nat Commun 9(1):4844
Desta ZA, Ortiz R (2014) Genomic selection: genome-wide prediction in plant improvement. Trends Plant Sci 19(9):592–601
Dos Santos JP, Fernandes SB, McCoy S, Lozano R, Brown PJ, Leakey AD, Buckler ES, Garcia AA, Gore MA (2020) Novel bayesian networks for genomic prediction of developmental traits in biomass sorghum. G3 Genes Genomes Genetics 10(2):769–781
Dubey NK, Srivastava R, Singh M, Bhutani S, Ranjan A (2025) Genetic engineering for crop improvement against stresses and future agriculture. In: Omics and genome editing: revolution in crop improvement for sustainable agriculture. Springer Nature Switzerland, Cham, pp 55–67
Frankel OH (1984) Genetic perspectives of germplasm conservation. In: Genetic manipulation: impact on man and society, vol 161, p 170
Fujimoto M, Sazuka T, Oda Y, Kawahigashi H, Wu J, Takanashi H, Ohnishi T, Yoneda JI, Ishimori M, Kajiya-Kanegae H, Hibara KI (2018) Transcriptional switch for programmed cell death in pith parenchyma of sorghum stems. Proc Natl Acad Sci 115(37):E8783–E8792
Ganie SA, Reddy AS (2021) Stress-induced changes in alternative splicing landscape in Rice: functional significance of splice isoforms in stress tolerance. Biology 10(4):309
Garg S (2021) Computational methods for chromosome-scale haplotype reconstruction. Genome Biol 22(1):1–24
Gasiunas G, Barrangou R, Horvath P, Siksnys V (2012) Cas9–crRNA ribonucleoprotein complex mediates specific DNA cleavage for adaptive immunity in bacteria. Proc Natl Acad Sci 109(39):E2579–E2586

Getachew G, Putnam DH, De Ben CM, De Peters EJ (2016) Potential of sorghum as an alternative to corn forage. Am J Plant Sci 7(7):1106–1121
Ghane A, Malhotra PK, Sanghera GS, Verma SK, Jamwal NS, Kashyap L, Wani SH (2024) CRISPR/Cas technology: fueling the future of biofuel production with sugarcane. Funct Integr Genomics 24(6):205
Godfray HC, Beddington JR, Crute IR, Haddad L, Lawrence D, Muir JF, Pretty J, Robinson S, Thomas SM, Toulmin C (2010) Food security: the challenge of feeding 9 billion people. Science 327(5967):812–818
Godwin ID, Rutkoski J, Varshney RK, Hickey LT (2019) Technological perspectives for plant breeding. Theor Appl Genet 132(3):555–557
Golicz AA, Batley J, Edwards D (2016) Towards plant pangenomics. Plant Biotechnol J 14(4):1099–1105
Guo T, Yu X, Li X, Zhang H, Zhu C, Flint-Garcia S, McMullen MD, Holland JB, Szalma SJ, Wisser RJ, Yu J (2019) Optimal designs for genomic selection in hybrid crops. Mol Plant 12(3):390–401
Gupta PK, Kulwal PL, Jaiswal V (2014) Association mapping in crop plants: opportunities and challenges. Adv Genet 85:109–147
Habyarimana E, Lopez-Cruz M, Baloch FS (2020 Jan) Genomic selection for optimum index with dry biomass yield, dry mass fraction of fresh material, and plant height in biomass sorghum. Genes 11(1):61
Hamblin MT, Salas Fernandez MG, Casa AM, Mitchell SE, Paterson AH, Kresovich S (2005) Equilibrium processes cannot explain high levels of short-and medium-range linkage disequilibrium in the domesticated grass Sorghum bicolor. Genetics 171(3):1247–1256
Harlan JR, de Wet JM (1971) Toward a rational classification of cultivated plants. Taxon 20(4):509–517
Harris K, Subudhi PK, Borrell A, Jordan D, Rosenow D, Nguyen H, Klein P, Klein R, Mullet J (2007) Sorghum stay-green QTL individually reduce post-flowering drought-induced leaf senescence. J Exp Bot 58(2):327–338
Hayes CM, Burow GB, Brown PJ, Thurber C, Xin Z, Burke JJ (2015) Natural variation in synthesis and catabolism genes influences dhurrin content in sorghum. Plant Genome 8(2):plantgenome2014-09
Hickey JM, Dreisigacker S, Crossa J, Hearne S, Babu R, Prasanna BM, Grondona M, Zambelli A, Windhausen VS, Mathews K, Gorjanc G (2014) Evaluation of genomic selection training population designs and genotyping strategies in plant breeding programs using simulation. Crop Sci 54(4):1476–1488
Hsu PD, Lander ES, Zhang F (2014) Development and applications of CRISPR-Cas9 for genome engineering. Cell 157(6):1262–1278
Hua K, Zhang J, Botella JR, Ma C, Kong F, Liu B, Zhu JK (2019) Perspectives on the application of genome-editing technologies in crop breeding. Mol Plant 12(8):1047–1059
Ibrahim AK, Zhang L, Niyitanga S, Afzal MZ, Xu Y, Zhang L, Zhang L, Qi J (2020) Principles and approaches of association mapping in plant breeding. Trop Plant Biol 13(3):212–224
Jensen SE, Charles JR, Muleta K, Bradbury PJ, Casstevens T, Deshpande SP, Gore MA, Gupta R, Ilut DC, Johnson L, Lozano R (2020) A sorghum practical haplotype graph facilitates genome-wide imputation and cost-effective genomic prediction. Plant Genome 13(1):e20009
Jia Y, Jannink JL (2012) Multiple-trait genomic selection methods increase genetic value prediction accuracy. Genetics 192(4):1513–1522
Jia J, Li H, Zhang X, Li Z, Qiu L (2017) Genomics-based plant germplasm research (GPGR). Crop J 5(2):166–174
Jiang W, Zhou H, Bi H, Fromm M, Yang B, Weeks DP (2013) Demonstration of CRISPR/Cas9/sgRNA-mediated targeted gene modification in Arabidopsis, tobacco, sorghum and rice. Nucleic Acids Res 41(20):e188

Jiao Y, Burke J, Chopra R, Burow G, Chen J, Wang B, Hayes C, Emendack Y, Ware D, Xin Z (2016) A sorghum mutant resource as an efficient platform for gene discovery in grasses. Plant Cell 28(7):1551–1562

Kebede H, Subudhi PK, Rosenow DT, Nguyen HT (2001) Quantitative trait loci influencing drought tolerance in grain sorghum (Sorghum bicolor L. Moench). Theor Appl Genet 103(2):266–276

Kent MA, Fonseca JM, Klein PE, Klein RR, Hayes CM, Rooney WL (2023) Use of genomic prediction to screen sorghum B-lines in hybrid testcrosses. The Plant Genome 16(3):e20369

Kikuchi S, Satoh K, Nagata T, Kawagashira N, Doi K, Kishimoto N, Yazaki J, Ishikawa M, Yamada H, Ooka H, Hotta I (2003) Collection, mapping, and annotation of over 28,000 cDNA clones from japonica rice. Science 301(5631):376–379

Kilian B, Graner A (2012) NGS technologies for analyzing germplasm diversity in genebanks. Brief Funct Genomics 11(1):38–50

Kiranmayee KN, Hash CT, Sivasubramani S, Ramu P, Amindala BP, Rathore A, Kishor PB, Gupta R, Deshpande SP (2020) Fine-mapping of sorghum stay-green QTL on chromosome10 revealed genes associated with delayed senescence. Genes 11(9):1026

Lai J, Li R, Xu X, Jin W, Xu M, Zhao H, Xiang Z, Song W, Ying K, Zhang M, Jiao Y (2010) Genome-wide patterns of genetic variation among elite maize inbred lines. Nat Genet 42(11):1027–1030

Lei YL (2014) cRIsPRrP: a webtool for synthetic single r guide RnA design of cRIsPRr systemin plants. Mol Plant 7:1494r1496

Li A, Jia S, Yobi A, Ge Z, Sato SJ, Zhang C, Angelovici R, Clemente TE, Holding DR (2018a) Editing of an alpha-kafirin gene family increases, digestibility and protein quality in sorghum. Plant Physiol 177(4):1425–1438

Li J, Tang W, Zhang YW, Chen KN, Wang C, Liu Y, Zhan Q, Wang C, Wang SB, Xie SQ, Wang L (2018b) Genome-wide association studies for five forage quality-related traits in sorghum (Sorghum bicolor L.). Front Plant Sci 9:1146

Li J, Wang L, Bible PW, Tu W, Zheng J, Jin P, Liu Y, Du J, Zheng J, Wang YH, Zhan Q (2023a) A chromosome-scale genome sequence of sudangrass (Sorghum sudanense) highlights the genome evolution and regulation of dhurrin biosynthesis. Theor Appl Genet 136(3):60

Li X, Liu W, Wang G, Sun SS, Yuan L, Wang J (2023b) Improving digestibility of sorghum proteins by CRISPR/Cas9-based genome editing. Food Energy Secur 13:e506

Liu HJ, Yan J (2019) Crop genome-wide association study: a harvest of biological relevance. Plant J 97(1):8–18

Liu G, Li J, Godwin ID (2019) Genome editing by CRISPR/Cas9 in sorghum through biolistic bombardment. In: Sorghum. Humana Press, New York, pp 169–183

Liu F, Baye W, Zhao K, Tang S, Xie Q, Xie P (2024) Unravelling sorghum functional genomics and molecular breeding: past achievements and future prospects. J Genet Genomics 52:719

Luo H, Zhao W, Wang Y, Xia Y, Wu X, Zhang L, Tang B, Zhu J, Fang L, Du Z, Bekele WA (2016) SorGSD: a sorghum genome SNP database. Biotechnol Biofuels 9(1):1–9

Mace ES, Tai S, Gilding EK, Li Y, Prentis PJ, Bian L, Campbell BC, Hu W, Innes DJ, Han X, Cruickshank A (2013) Whole-genome sequencing reveals untapped genetic potential in Africa's indigenous cereal crop sorghum. Nat Commun 4(1):1–9

Mace ES, Cruickshank AW, Tao Y, Hunt CH, Jordan DR (2021) A global resource for exploring and exploiting genetic variation in sorghum crop wild relatives. Crop Sci 61(1):150–162

McBee GG, Waskom RM III, Miller FR, Creelman RA (1983) Effect of senescene and nonsenescence on carbohydrates in sorghum during late kernel maturity states 1. Crop Sci 23(2):372–376

McCormick RF, Truong SK, Sreedasyam A, Jenkins J, Shu S, Sims D, Kennedy M, Amirebrahimi M, Weers BD, McKinley B, Mattison A (2018) The Sorghum bicolor reference genome: improved assembly, gene annotations, a transcriptome atlas, and signatures of genome organization. Plant J 93(2):338–354

McKinley B, Rooney W, Wilkerson C, Mullet J (2016) Dynamics of biomass partitioning, stem gene expression, cell wall biosynthesis, and sucrose accumulation during development of Sorghum bicolor. Plant J 88(4):662–680

Meuwissen TH, Hayes BJ, Goddard M (2001) Prediction of total genetic value using genome-wide dense marker maps. Genetics 57(4):1819–1829

Miranda-Romero LA, Tirado-González DN, Tirado-Estrada G, Améndola-Massiotti R, Sandoval-González L, Ramírez-Valverde R, Salem AZ (2020) Quantifying non-fibrous carbohydrates, acid detergent fiber and cellulose of forage through an in vitro gas production technique. J Sci Food Agric 100(7):3099–3110

Mizuno H, Kasuga S, Kawahigashi H (2016) The sorghum SWEET gene family: stem sucrose accumulation as revealed through transcriptome profiling. Biotechnol Biofuels 9(1):1–2

Morris GP, Ramu P, Deshpande SP, Hash CT, Shah T, Upadhyaya HD, Riera-Lizarazu O, Brown PJ, Acharya CB, Mitchell SE, Harriman J (2013) Population genomic and genome-wide association studies of agroclimatic traits in sorghum. Proc Natl Acad Sci 110(2):453–458

Mundia CW, Secchi S, Akamani K, Wang G (2019) A regional comparison of factors affecting global sorghum production: the case of North America, Asia and Africa's Sahel. Sustainability 11(7):2135

Muraya MM, Chu J, Zhao Y, Junker A, Klukas C, Reif JC, Altmann T (2017) Genetic variation of growth dynamics in maize (Zea mays L.) revealed through automated non-invasive phenotyping. Plant J 89(2):366–380

Murray SC, Rooney WL, Mitchell SE, Sharma A, Klein PE, Mullet JE, Kresovich S (2008) Genetic improvement of sorghum as a biofuel feedstock: II. QTL for stem and leaf structural carbohydrates. Crop Sci 48(6):2180–2193

Murray SC, Rooney WL, Hamblin MT, Mitchell SE, Kresovich S (2009) Sweet sorghum genetic diversity and association mapping for brix and height. Plant Genome 2(1)

Nida H, Girma G, Mekonen M, Tirfessa A, Seyoum A, Bejiga T, Birhanu C, Dessalegn K, Senbetay T, Ayana G, Tesso T (2021) Genome-wide association analysis reveals seed protein loci as determinants of variations in grain mold resistance in sorghum. Theor Appl Genet 134(4):1167–1184

Niu H, Ping J, Wang Y, Lv X, Li H, Zhang F, Chu J, Han Y (2020) Population genomic and genome-wide association analysis of lignin content in a global collection of 206 forage sorghum accessions. Mol Breed 40(8):1–3

Niu H, Han Y, Wang Y, Lv X, Chu J. Genome wide association analysis of acid detergent fiber content of 206 Forage Sorghum (Sorghum Bicolor (L.) Moench) accessions; 2021

Obregón-Cano S, Moreno-Rojas R, Jurado-Millán AM, Cartea-González ME, Haro-Bailón D (2019) Analysis of the acid detergent fibre content in turnip greens and turnip tops (Brassica rapa L. Subsp. rapa) by means of near-infrared reflectance. Foods 8(9):364

Osakabe Y, Osakabe K (2015) Genome editing with engineered nucleases in plants. Plant Cell Physiol 56(3):389–400

Pareek CS, Smoczynski R, Tretyn A (2011) Sequencing technologies and genome sequencing. J Appl Genet 52(4):413–435

Paterson AH, Bowers JE, Bruggmann R, Dubchak I, Grimwood J, Gundlach H, Haberer G, Hellsten U, Mitros T, Poliakov A, Schmutz J (2009) The Sorghum bicolor genome and the diversification of grasses. Nature 457(7229):551–556

Perrier L, Rouan L, Jaffuel S, Clément-Vidal A, Roques S, Soutiras A, Baptiste C, Bastianelli D, Fabre D, Dubois C, Pot D (2017) Plasticity of sorghum stem biomass accumulation in response to water deficit: a multiscale analysis from internode tissue to plant level. Front Plant Sci 8:1516

Ping J, Zhang F, Niu H, Yang H, Lv X, Du Z, Li H, Wang Y (2018) Genetic diversity analysis of germplasm resources of forage sorghum based on SSR marker. Mol Plant Breed 14:4663–4670

Purugganan MD, Jackson SA (2021) Advancing crop genomics from lab to field. Nat Genet 53(5):595–601

Qaim M (2020) Role of new plant breeding technologies for food security and sustainable agricultural development. Appl Econ Perspect Policy 42(2):129–150

Qian L, Hickey LT, Stahl A, Werner CR, Hayes B, Snowdon RJ, Voss-Fels KP (2017) Exploring and harnessing haplotype diversity to improve yield stability in crops. Front Plant Sci 8:1534

Rakshit S, Wang YH (eds) (2016) The sorghum genome. Springer International Publishing, Singapore

Rakshit S, Swapna M, Dalal M, Sushma G, Ganapathy KN, Dhandapani A, Karthikeyan M, Talwar HS (2016) Post-flowering drought stress response of post-rainy sorghum genotypes. Indian J Plant Physiol 21(1):8–14

Rasheed A, Hao Y, Xia X, Khan A, Xu Y, Varshney RK, He Z (2017) Crop breeding chips and genotyping platforms: progress, challenges, and perspectives. Mol Plant 10(8):1047–1064

Reddy PS, Gadakh SR, Kalpande HV (2018) Genetic diversity of sorghum (Sorghum bicolor (L) Moench) for agronomic traits and shoot Fly tolerance and suitability for winter season. Indian Journal of Plant Genetic Resources 31(3):303–309

Sabadin PK, Malosetti M, Boer MP, Tardin FD, Santos FG, Guimaraes CT, Gomide RL, Andrade CL, Albuquerque PE, Caniato FF, Mollinari M (2012) Studying the genetic basis of drought tolerance in sorghum by managed stress trials and adjustments for phenological and plant height differences. Theor Appl Genet 124(8):1389–1402

Saballos A, Vermerris W, Rivera L, Ejeta G (2008) Allelic association, chemical characterization and saccharification properties of brown midrib mutants of sorghum (Sorghum bicolor (L.) Moench). Bioenergy Res 1(3):193–204

Saballos A, Ejeta G, Sanchez E, Kang C, Vermerris W (2009) A genomewide analysis of the cinnamyl alcohol dehydrogenase family in sorghum [Sorghum bicolor (L.) Moench] identifies SbCAD2 as the brown midrib6 gene. Genetics 181(2):783–795

Saballos A, Sattler SE, Sanchez E, Foster TP, Xin Z, Kang C, Pedersen JF, Vermerris W (2012) Brown midrib2 (Bmr2) encodes the major 4-coumarate: coenzyme a ligase involved in lignin biosynthesis in sorghum (Sorghum bicolor (L.) Moench). Plant J 70(5):818–830

Sander JD (2019) Gene editing in sorghum through agrobacterium. In: Sorghum. Humana Press, New York, pp 155–168

Sandhu KS, Merrick LF, Sankaran S, Zhang Z, Carter AH (2022) Prospectus of genomic selection and Phenomics in cereal, legume and oilseed breeding programs. Front Genet 12:829131. https://doi.org/10.3389/fgene.2021.829131

Sattler SE, Saathoff AJ, Haas EJ, Palmer NA, Funnell-Harris DL, Sarath G, Pedersen JF (2009) A nonsense mutation in a cinnamyl alcohol dehydrogenase gene is responsible for the sorghum brown midrib6 phenotype. Plant Physiol 150(2):584–595

Sattler SE, Palmer NA, Saballos A, Greene AM, Xin Z, Sarath G, Vermerris W, Pedersen JF (2012) Identification and characterization of four missense mutations in brown midrib 12 (Bmr12), the caffeic O-methyltranferase (COMT) of sorghum. Bioenergy Res 5(4):855–865

Sattler SE, Saballos A, Xin Z, Funnell-Harris DL, Vermerris W, Pedersen JF (2014) Characterization of novel sorghum brown midrib mutants from an EMS-mutagenized population. G3:Genes Genomes Genetics 4(11):2115–2124

Scully ED, Gries T, Funnell-Harris DL, Xin Z, Kovacs FA, Vermerris W, Sattler SE (2016) Characterization of novel Brown midrib 6 mutations affecting lignin biosynthesis in sorghum. J Integr Plant Biol 58(2):136–149

Singh BD, Singh AK (2015) Marker-assisted plant breeding: principles and practices. Springer India, New Delhi

Smith O, Nicholson WV, Kistler L, Mace E, Clapham A, Rose P, Stevens C, Ware R, Samavedam S, Barker G, Jordan D (2019) A domestication history of dynamic adaptation and genomic deterioration in sorghum. Nat Plants 5(4):369–379

Somegowda VK, Rayaprolu L, Rathore A, Deshpande SP, Gupta R (2021a) Genome-wide association studies (GWAS) for traits related to fodder quality and biofuel in sorghum: progress and prospects. Protein Pept Lett 28(8):843–854

Somegowda VK, Vemula A, Naravula J, Prasad G, Rayaprolu L, Rathore A, Blümmel M, Deshpande SP (2021b) Evaluation of fodder yield and fodder quality in sorghum and its interaction with grain yield under different water availability regimes. Curr Plant Biol 25:100191

Somegowda VK, Prasad KV, Naravula J, Vemula A, Selvanayagam S, Rathore A, Jones CS, Gupta R, Deshpande SP (2022) Genetic dissection and quantitative trait loci mapping of agronomic and fodder quality traits in sorghum under different water regimes. Front Plant Sci 13:810632

Spindel JE, Dahlberg J, Colgan M, Hollingsworth J, Sievert J, Staggenborg SH, Hutmacher R, Jansson C, Vogel JP (2018) Association mapping by aerial drone reveals 213 genetic associations for Sorghum bicolor biomass traits under drought. BMC Genomics 19(1):1–8

Takele A, Feyissa T, Disasa T (2022) Quantitative trait loci mapping of stem sugar content and stem diameter in sorghum recombinant inbred lines using genotyping-by-sequencing. Mol Biol Rep 25:10

Tao YZ, Henzell RG, Jordan DR, Butler DG, Kelly AM, McIntyre CL (2000) Identification of genomic regions associated with stay green in sorghum by testing RILs in multiple environments. Theor Appl Genet 100(8):1225–1232

Tao Y, Luo H, Xu J, Cruickshank A, Zhao X, Teng F, Hathorn A, Wu X, Liu Y, Shatte T, Jordan D (2021) Extensive variation within the pan-genome of cultivated and wild sorghum. Nat Plants 7(6):766–773

Tibbs Cortes L, Zhang Z, Yu J (2021) Status and prospects of genome-wide association studies in plants. Plant Genome 14(1):e20077

Tripathi AD, Mishra R, Maurya KK, Singh RB, Wilson DW (2019) Estimates for world population and global food availability for global health. In: The role of functional food security in global health. Academic, pp 3–24

Tuinstra M, Shelton B, Tower J, Minton N, Jackson J (2024) Evaluating forage sorghum for silage and grazing systems. Purdue University Extension. Retrieved from purdue.edu/newsroom

Upadhyaya HD, Pundir RP, Dwivedi SL, Gowda CL, Reddy VG, Singh S (2009) Developing a mini core collection of sorghum for diversified utilization of germplasm. Crop Sci 49(5):1769–1780

Van Oosterom EJ, Jayachandran R, Bidinger FR (1996) Diallel analysis of the stay-green trait and its components in sorghum. Crop Sci 36(3):549–555

Varshney RK, Nayak SN, May GD, Jackson SA (2009) Next-generation sequencing technologies and their implications for crop genetics and breeding. Trends Biotechnol 27(9):522–530

Varshney RK, Bansal KC, Aggarwal PK, Datta SK, Craufurd PQ (2011) Agricultural biotechnology for crop improvement in a variable climate: hope or hype? Trends Plant Sci 16(7):363–371

Varshney RK, Terauchi R, McCouch SR (2014) Harvesting the promising fruits of genomics: applying genome sequencing technologies to crop breeding. PLoS Biol 12(6):e1001883

Varshney RK, Bohra A, Yu J, Graner A, Zhang Q, Sorrells ME (2021) Designing future crops: genomics-assisted breeding comes of age. Trends Plant Sci 26(6):631–649

Vellaichamy Gandhimeyyan R, Chockalingam V, Muthurajan R, Raman R (2023) Accelerating gene discovery in climate-resilient and nutrient-rich major and minor millets through genome-wide association studies: Progress and prospects. Plant Breed 142(4):389–406

Voytas DF (2009) High-frequency modification of plant genes using engineered zinc-nger nucleases. Nature 459(7245):442445

Wang YH, Poudel DD, Hasenstein KH (2011) Identification of SSR markers associated with saccharification yield using pool-based genome-wide association mapping in sorghum. Genome 54(11):883–889

Wang H, Li K, Hu X, Liu Z, Wu Y, Huang C (2016) Genome-wide association analysis of forage quality in maize mature stalk. BMC Plant Biol 16(1):1–2

Wang L, Shang L, Wu X, Hao H, Jing HC (2023) Genomic architecture of leaf senescence in sorghum (Sorghum bicolor). Theor Appl Genet 136(3):45

Wang B, Xiong W, Guo Y (2024) Dhurrin in sorghum: Biosynthesis, regulation, biological function and challenges for animal production. Plants 13(16):2291

Wei C, Gao L, Xiao R, Wang Y, Chen B, Zou W, Li J, Mace E, Jordan D, Tao Y (2024) Complete telomere-to-telomere assemblies of two sorghum genomes to guide biological discovery. iMeta 3(2):e193

Won S, Park JE, Son JH, Lee SH, Park BH, Park M et al (2020) Genomic prediction accuracy using haplotypes defined by size and hierarchical clustering based on linkage disequilibrium. Front Genet 11:134

Xiao Q, Bai X, Zhang C, He Y (2022) Advanced high-throughput plant phenotyping techniques for genome-wide association studies: a review. J Adv Res 35:215–230

Xin Z, Li Wang M, Barkley NA, Burow G, Franks C, Pederson G, Burke J (2008) Applying genotyping (TILLING) and phenotyping analyses to elucidate gene function in a chemically induced sorghum mutant population. BMC Plant Biol 8(1):1–4

Xu Y, Liu X, Fu J, Wang H, Wang J, Huang C, Prasanna BM, Olsen MS, Wang G, Zhang A (2020) Enhancing genetic gain through genomic selection: from livestock to plants. Plant Commun 1(1):100005

Yano K, Yamamoto E, Aya K, Takeuchi H, Lo PC, Hu L, Yamasaki M, Yoshida S, Kitano H, Hirano K, Matsuoka M (2016) Genome-wide association study using whole-genome sequencing rapidly identifies new genes influencing agronomic traits in rice. Nat Genet 48(8):927–934

Yu X, Li X, Wu Y, Mitchell S, Roozeboom K, Wang D, Bernardo R, Wang M, Pederson G, Tesfaye T, Yu J (2014) Genomic selection of biomass traits in a global collection of 976 sorghum accessions. In: Plant and animal genome XXII conference plant and animal genome

Zhang LM, Leng CY, Luo H, Wu XY, Liu ZQ, Zhang YM, Zhang H, Xia Y, Shang L, Liu CM, Hao DY (2018) Sweet sorghum originated through selection of dry, a plant-specific NAC transcription factor gene. Plant Cell 30(10):2286–2307

Zhang D, Tang S, Xie P, Yang D, Wu Y, Cheng S, Du K, Xin P, Chu J, Yu F, Xie Q (2022) Creation of fragrant sorghum by CRISPR/Cas9. J Integr Plant Biol 64(5):961–964

Zhao Y, Qiang C, Wang X, Chen Y, Deng J, Jiang C, Sun X, Chen H, Li J, Piao W, Zhu X (2019) New alleles for chlorophyll content and stay-green traits revealed by a genome wide association study in rice (Oryza sativa). Sci Rep 9(1):1–1

Zhou Y, Srinivasan S, Mirnezami SV, Kusmec A, Fu Q, Attigala L, Salas Fernandez MG, Ganapathysubramanian B, Schnable PS (2019) Semiautomated feature extraction from RGB images for sorghum panicle architecture GWAS. Plant Physiol 179(1):24–37

Accelerated Breeding in Bajra Using Genomic Approaches

Ruchika Kashyap, Dinesh Kumar Saini, Sangram Singh Sandhu, Heena Puri, Shahbaz Ahmed, Jagmohan Singh, Karansher Sandhu, and Rahul Kapoor

1 Introduction

Pearl millet [*Pennisetum glaucum* (L.) R. Br.] is among the prime cereal crops mainly grown in Africa and South Asia. It is a C4 diploid (2n = 2× = 14) placed in the grass family *Poaceae,* which thrives well in both arid and semi-arid conditions (Serba et al. 2019). It follows a protogynous cross-pollination mechanism where the stigma is produced before anther appears, which effectively facilitates cross-pollination. The domestication of the crop dates to about 4000–5000 years ago, evidently in West Africa around 2500 BC at the southern edge of the Sahara Desert (Manning et al. 2011) (Singh et al. 2023b). It is grown over 27 million hectares and

The original version of the chapter has been revised. A correction to this chapter can be found at https://doi.org/10.1007/978-3-032-04527-0_13

R. Kashyap
Department of Agronomy, Horticulture, and Plant Sciences, South Dakota State University, Brookings, SD, USA

D. K. Saini
Department of Genetics and Plant Breeding, Chaudhary Charan Singh University, Meerut, Uttar Pradesh, India

S. S. Sandhu · R. Kapoor
College of Agriculture, Punjab Agricultural University, Ludhiana, Punjab, India

H. Puri
Insect Testing Scientist, Impetus Agriculture Inc., Saint Louis, MO, USA

S. Ahmed
Department of Crop and Soil Sciences, Washington State University, Pullman, WA, USA

J. Singh
Department of Plant Pathology, Indian Agricultural Research Institute, New Delhi, Delhi, India

K. Sandhu (✉)
Department of Crop and Soil Sciences, Washington State University, Pullman, WA, USA
e-mail: k.sandhu@wsu.edu

S. H. Wani, S. S. Gosal (eds.), *Accelerated Plant Breeding, Volume 5*,
https://doi.org/10.1007/978-3-032-04527-0_4

is a subsistence food for about 90 million poor and marginal people, making it the sixth most important cereal crop worldwide (Gulia et al. 2007; Srivastava et al. 2020).

The hardiness of the pearl millet and its potential for surviving in harsh climatic conditions make it one of the most drought-tolerant cereal crops (Rai et al. 2012). It is believed to give higher economic returns than other cereals like barley (*Hordeum vulgare* L.), maize (*Zea mays* L.), rice (*Oryzae sativa* L.), sorghum (*Sorghum bicolor* L.), and wheat (*Triticum aestivum* L.) when grown in water-deficit areas with average annual precipitation of <250 mm (Serba et al. 2019). Apart from being climate-resilient, this protogynous cross-pollinated crop has immense photosynthetic and biomass production capacity. These properties make it a highly desirable crop under changing climatic scenarios (Rajaram et al. 2013; Yadav and Rai 2013). It also produces high-quality forage, making it of use as a forage-cover crop (Burton and Forston 1966).

Pearl millet is a cost-effective energy source and a rich source of organic and inorganic nutrients (Kumar et al. 2016). It is a cheap source of iron (Fe) and zinc (Zn) and contributes to 30–40% of inorganic nutrients (Rai et al. 2012). It provides nutritional security to poor and marginal farmers, as it is a collective storehouse of fibers, amino acids, proteins, starch, and minerals (Pujar et al. 2020). Besides, it is an essential dietary component for the poultry industry as it ensures the bioavailability of micronutrients for monogastric humans (Rai et al. 2012; Kumar et al. 2016). The vigorous growth habit with larger leaves makes it excellent for forage purposes. Other properties like the ease in silage making, storage of minerals (Ca and P) and proteins, and lower levels of anti-nutritional factors (hydrocyanic and oxalic acid) make it a productive and quality forage crop. Pearl millet gives a productive pasture for grazing, especially with dwarf varieties, and silage is easily made (Kumar et al. 2012).

The cultivated pearl millet exhibits substantial variability in different agronomic characteristics such as time of flowering, length of panicle, grain and stover yield, the grain's nutrient content, and tolerance to biotic and abiotic stresses (Diack et al. 2020). This diversity is attributed to the diverse lineage, based on which pearl millet is grouped into three gene pools (GP), i.e., primary (GP1), secondary (GP2), and tertiary (GP3) (Harlan and Wet 1971). The cultivated, weedy, and wild diploids ($2n = 2\times = 14$) belong to the primary; the tetraploid *P. purpureum* (Shum.) ($2n = 4\times = 28$) belong to the secondary and distantly related *Pennisetum* species of various ploidy levels that come under the tertiary gene pool (Varshney et al. 2017).

Human selection processes have further diversified the germplasm by developing cultivars suitable to different production systems, environments, and consumer demands (Brunken 1977). The genetic improvement of pearl millet began in 1936 at Georgia Coastal Plain Experiment Station, Tifton, GA, United States (Burton and Forston 1966), followed by the development of breeding programs at Kansas State University (K-State) in 1969 (Serba et al. 2017). The breeding program at Tifton paved the way for improved breeding worldwide. The main purpose of the breeding program in the United States was to develop improved cultivars for forage use with high biomass production and high digestibility of dry matter. It helped in disseminating knowledge and breeding material, including cytoplasmic male sterility (CMS) source (A_1) (Burton and Forston 1966), and dwarf stocks of early maturity (Burton and Forston 1966) globally. In India, during the 1940s and 1950s, mass selection from several local materials was mainly used to improve grain yields (Gupta et al. 2015). Later, the CMS line's introduction from the United States led to the development of the first pearl millet hybrid—Hybrid Bajra-1 (HB-1) in 1964 (Dujardin and Hanna 1989; Andrews and Kumar 1992). This led to a striking increase in India's pearl millet production (Singh et al. 2014). The hybrid breeding programs accelerated, and seed parents developed at Hays, Kansas, were used for many such programs at International Crops Research Institute for the Semi-Arid Tropics (ICRISAT) and other parts of India (Andrews and Kumar 1992).

The breeding methods focused on obtaining thick panicles, greater yield, early maturity, larger grain, good combining ability, and better stand establishment (Andrews and Kumar 1992). Several cultivars and landraces like Iniati/Koupela—PI 185642 (landrace from the Ghana-Togo), Sereve 3A (cultivar from Uganda), and Tift A1 (CMS line) played an essential role in achieving these objectives (Andrews and Kumar 1992). The breeding material traveled worldwide from the collaborative efforts of the breeders and institutions. The germplasm exchange between K-State, ICRISAT, and International Sorghum and Millet (INTSORMIL) is just one example. The germplasm from Africa and India was used to develop early maturing populations, synthetics, and adapted seed parents in A1 and A4 (monodii) male-sterile cytoplasm (Serba et al. 2017).

New challenges in terms of biotic and abiotic stresses initiated the advancements in breeding. Even though there is a wide-germplasm collection that could aid crop improvement, only a narrow range was thought to help in quality breeding for valuable agronomic traits (Kumari et al. 2014). Following up, the pearl millet breeding saw an exemplar shift through the development of molecular markers and genetic mapping started at ICRISAT. In the 1990s, the first linkage map of pearl millet was constructed using restriction fragment-length polymorphisms (RFLP) (Liu et al. 1994). Thus, it allowed an opportunity to explore the crop's genome (Serba et al. 2017), and the breeding programs shifted toward diversification of genetically improved seed and pollinator parents to target adaptation to specific niches for better stress tolerance (Rai et al. 2012).

Furthermore, bi-parental mapping populations have been used for the allelic mining studies that have resulted in considerable crop improvement against abiotic and biotic stresses, especially drought tolerance and downy mildew resistance, in the past few years (Bidinger et al. 2007; Varshney et al. 2017). Identification of the quantitative trait loci (QTLs) associated with different valuable traits through the use of molecular markers like amplified fragment length polymorphism (AFLP), conserved intron spanning primer (CISP), diversity array technology (DArT), expressed sequence–simple sequence repeats (EST-SSRs), RFLP, and single-strand conformational polymorphism-single nucleotide polymorphism (SSCP-SNP) are nowadays in the process (Jones et al. 1995; Yadav et al. 2001; Bidinger et al. 2007).

Earlier, on one hand, where conventional linkage mapping helped in identifying stress-related QTLs (Taunk et al. 2018), association mapping, on the other hand, helped in effectively making marker-phenotype associations in a lesser amount of time and generating 100 s of recombination, opening new doors to genetic improvements (Sehgal et al. 2012; Rajaram et al. 2013; Kumari et al. 2014). Researchers and scientists are continually working towards pearl millet crop improvement (Fig. 1). This has paved the way for new and advanced technologies like high-throughput sequencing, targeted induced local lesion in genomes (tilling), gene silencing, and transgenics, which will keep the wheels of pearl millet improvement moving (Mishra et al. 2024).

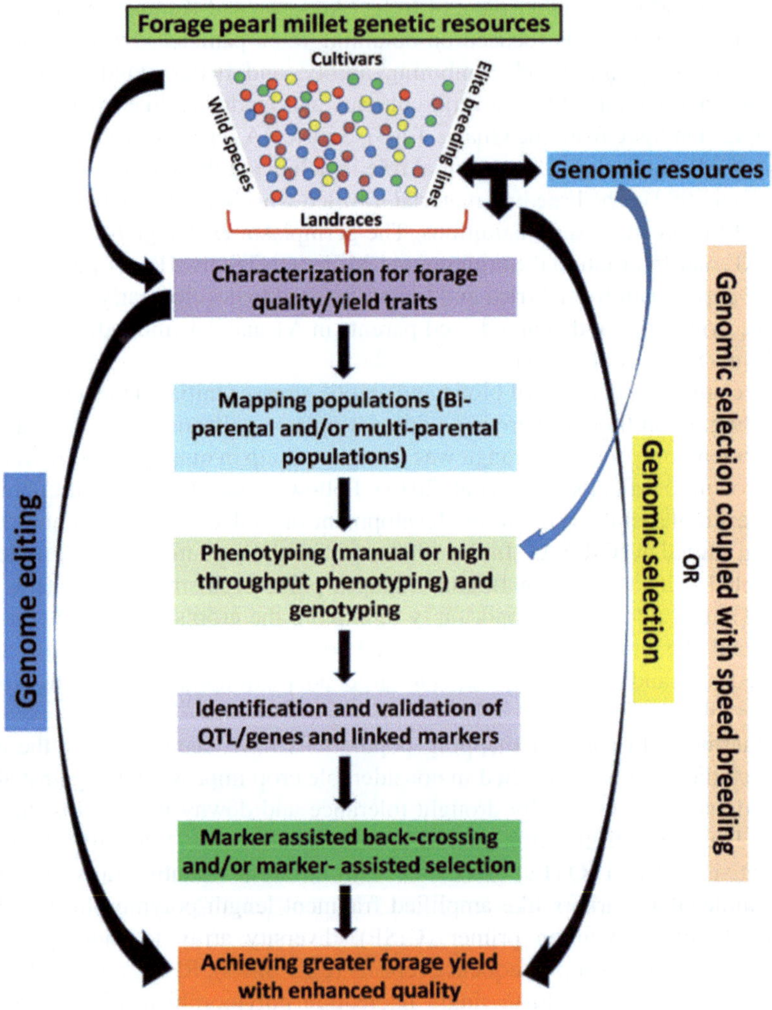

Fig. 1 The breeding strategies for genetic improvement of forage yield and quality traits that have been employed so far are represented

2 Genetic Resources in Pearl Millet

Development of forage cultivars that encompasses the ability for multi-cuts (repeated harvests), earliness to first cut (harvest), quick regeneration, and short harvesting intervals, good tillering potential, high nutritional quality factors, high green fodder, and reduced anti-nutritional factors is the primary objective for forage pearl millet breeding programs. Genetic variability is considered one of the crucial factors which determines the success of any breeding program. Pearl millet also encompasses extensive genetic variability (Serba et al. 2017; Varshney et al. 2019). Natural and artificial selection activities in pearl millet have resulted in the development and release of numerous cultivars suited for diverse environments and adapted to several production systems (Brunken 1977). To assure the availability of this extensive genetic variation for forage breeding activities, collection, characterization, conservation, assessment, and proper documentation of the wild germplasm are crucial pre-breeding activities (Rajaram et al. 2013).

Worldwide, different centers and several groups have made concerted efforts to collect and characterize pearl millet germplasm. Consequently, the ICRISAT (India) has collected and conserved as many as 22,288 accessions, while the Institute of Research and Development (France) has around 3968 accessions of pearl millet collected from 16 different countries, and the Canadian Genetic Resources Program (Canada) and US Germplasm Resource Information Network have collected and conserved around 3821 and 1283 accessions of wild species and cultivated varieties of pearl millet, respectively (Serba et al. 2017; Varshney et al. 2019). The genus *Pennisetum* is a heterogeneous collection of approximately 140 highly diverse species with contrasting basic chromosome numbers (x = 5, 7, 8, and 9) (Brunken 1977; Jauhar 1981). Ploidy level of different *Pennisetum* species varies from diploid to octaploid exhibiting both apomictic and sexual reproductive systems.

In *Pennisetum*, initially, the genetic pools were suggested based on the genetic relationship among the different species; later, these were verified based on crossability (considering the fertility level of generated hybrids) of the wild pearl millet species or accession with the cultivated ones. GP1 comprises all types of weedy, cultivated, and diploid wild relatives (2n = 2× = 14). The cultivated *P. glaucum* and its wild relatives, *P. glaucum* ssp. Monodii (having two ecotypes, namely, *P. mollissimum* and *P. violaceum*) and *P. schweinfurthii* Pilger are the major members of the GP1 (Hanna and Dujardin 1986). Meanwhile, GP2 comprises the tetraploid *P. purpureum* (Shum.) (2n = 4× = 28) and apomictic octaploid (2n = 8× = 56) *P. squamulatum* (Pattanashetti et al. 2016). GP3 consists of *Pennisetum* species of different growth habits and ploidy levels distantly associated with GP1 and GP2 (Dujardin and Hanna 1989). Members of the tertiary GPs represent strong reproductive barriers with the members of both secondary and primary GPs. The tertiary gene pool includes all species from section Brevivalvula, which comprises a total of six morphological taxa: *P. hordeoides* (Lam.) Steud., *P. atrichum* Stapf & Hubb., *P. pedicellatum* Trin., *P. setaseum* (Swartz) L. Rich., *P. polystachion* (L.) Schult., and *P. subangustum* (Schum.) Stapf & Hubb.

As aforementioned, the cultivated species and wild relatives of pearl millet have extensive genetic variation for various important traits. For the identification and mapping of the genes governing these important traits, mapping populations, generated from the cross between parents having contrasting phenotypes for the respective traits, are needed. Mostly, the parents used for crosses will be from the same species. Still, in some situations, where intraspecific variation is not adequate, related species can be utilized as one of the parents. Mapping populations aid in identifying molecular markers linked with the genes/QTLs regulating the trait (s) of interest and can be utilized for marker-assisted selection (MAS). The development of a genetic map is one of the major requisites for QTL mapping.

Several attempts have been directed toward developing linkage maps, increasing the genome coverage and density of markers, and combining the maps for better purposes. The first genetic map (based on RFLP markers) was developed utilizing an inter-varietal F2 population in the year 1994 (Liu et al. 1994), whereas the first consensus or integrated map was developed utilizing four different F2 mapping populations (derived from the crosses of parents 81B and ICMP 451; LGD and ICMP 85410; PT 732B and P1449–2; and ICMB 841 and 863B) (Li et al. 2014). Later, one more population derived from the cross of parents 841-P3 and 863B-P2 was also employed to combine a newly constructed map based on single sequence repeats (SSR) markers in earlier maps (Senthilvel et al. 2008). Recently in 2018, three F2 populations derived from the crosses ICMA 02777 × ICMR 08888, H 77/833-2 × ICMP 451, and J-2537 × ICMB-95444 were utilized for the construction of dense linkage maps, which were then employed for genetic dissection of different traits (Pucher et al. 2018; Taunk et al. 2018).

Accordingly, several RILs developed from the crosses, H 77/833-2 × PRLT 2/89-33, Tift23DB × PI536400, and H77/833-2 × PRLT 2/89-33, 81B-P6 × ICMP 451-P8, Tift 99D2B1 × Tift 454, ICMB 841-P3 × 863B-P2, ICMS 8511-S1-17-2-1-1-B-P03 × AIMP 92901-S1-183-2-2-B-0, and 89111-P6 × ICMB 90111-P6 were also employed for linkage-based mapping of pearl millet genome (Pedraza-Garcia et al. 2010; Sehgal et al. 2012; Punnuri et al. 2016; Kumar et al. 2017; Chelpuri et al. 2019). A consensus linkage map was also developed utilizing four different RILs generated from H 77/833-2 × PRLT 2/89-33, ICMB 841-P3 × 863B-P2, PT 732B-P2 × P1449-2-P1, and 81B-P6 × ICMP 451-P8 (Rajaram et al. 2013). A highly dense GBS-SNP based linkage map was developed using an F2 population derived from the cross between "116_11-(PS202–14)-121" (wild pearl millet) and "SOSAT-IBL-197" (cultivated pearl millet) (Moumouni et al. 2015). These mapping populations and linkage maps are valuable genetic resources that can be used for the genetic dissection of important forage-related traits in pearl millet.

Generally, the identification of minor effect QTLs with the traditional bi-parental populations is quite difficult as the effects of these minor QTLs are masked by major QTLs and the interactions between different QTLs. To circumvent these problems, a set of chromosome segment substitution lines (CSSLs), each of which carries one or few donor segments in the background of the recipient parent, was developed from the cross 863B x ICMB 841. These CSSLs can be efficiently used for QTL mapping, fine-mapping, etc. in pearl millet (Kumari et al. 2014; Basava et al. 2019). In the recent past, a switch from first-generation populations (bi-parental populations) to second-generation populations (multi-parental populations) has been witnessed for QTL mapping in crop plants. Multi-parental populations, namely, nested association mapping population (NAM) and multi-parent advanced generation intercross population (MAGIC) (Yu et al. 2008; Kover et al. 2009), allow the efficient exploitation of multi-parent allelic diversity and fine-mapping of QTLs for important traits by utilizing the advantages of both linkage and LD (linkage disequilibrium) based mapping (association mapping) (Bernardo and Yu 2008). The information regarding the generation and use of NAM and MAGIC in pearl millet to unveil the complex genetics underlying desirable traits is lacking.

3 Genomic Resources in Pearl Millet

The utilization of genomics tools is now considered an indispensable component of crop breeding programs. Molecular breeding strategies including MAS and marker-assisted backcross breeding may reduce the lengthy phenotypic evaluation and selection procedures and hence have a great potential on the breeding for the genetic enhancement of crops. Since the first report on the development of RFLP markers in pearl millet (Liu et al. 1994), thousands of markers of different types such as RFLP, SSR, AFLP, DArT, GBS-SNPs, etc. have been generated and utilized for the following purposes: diversity assessment, linkage mapping/QTL mapping, genome-wide association studies, and MAS in pearl millet (Pedraza-Garcia et al. 2010; Sehgal et al. 2012; Punnuri et al. 2016; Kumar et al. 2017; Chelpuri et al. 2019).

The generation of high-throughput molecular markers could only become fast and cost-effective after the development of next-generation sequencing (NGS) technologies. Moreover, the arrival of these high-throughput markers now has made genotyping of hundreds of individuals and the development of high-density genetic maps in pearl millet very easy (Diack et al. 2020). GBS is one of the most widely used NGS technology, which is being extensively used for concurrent marker discovery and genotyping in many crop plants, including pearl millet (Elshire et al. 2011; Kumar et al. 2017; Chelpuri et al. 2019).

The genome sequence, i.e., the genetic blueprint of any crop, is believed to assist the development of advanced genomics tools that may accelerate the development of improved varieties through molecular breeding. With the overall aim to sequence, the genome of pearl millet, a consortium (International Pearl Millet Genome Sequencing Consortium (IPMGSC)) was established that brought together a global team of 65 scientists from as many as 30 leading research organizations involved in pearl millet breeding research. This consortium selected a crucial ancestral genotype of pearl millet, i.e., Tift 23D2B1-P1-P5, for genome sequencing. Recently, the first draft of the whole-genome sequence ($\sim$1.79 Gb) of reference genotype containing 38,579 genes was published (Varshney et al. 2017). Besides, a total of 994 pearl millet lines were also re-sequenced and analyzed, which helped in getting insights into genetic diversity, population structure, and domestication (Varshney et al.

2017). The availability of whole-genome sequence also provides an opportunity to develop genomic resources and tools such as SNP arrays, etc. As evident in other crops, high-throughput SNP genotyping assay has the potential to revolutionize the genetic studies that involve a large number of DNA-based markers such as QTL mapping, LD-based association mapping studies, and genomic selection. We think that these valuable genomic resources should empower pearl millet researchers and breeders to improve this crucial crop for both forage and seed purposes. A boom has been observed in publications based on "pearl millet genomics" after the genome sequencing of pearl millet (highest number of papers being reported in the same year, 2017), which represent the impacts of availability of genome sequence information in the pearl millet breeding related research (Fig. 2).

4 Disease Resistance

Diseases caused by several fungi, insects, bacteria, viruses, parasitic and non-parasitic weeds, and birds happen to decrease the pearl millet productivity by extensively damaging the crop in the African and Indian subcontinent. Among the fungal diseases, downy mildew [*Sclerospora graminicola* (Sacc. Schroet.)], ergot (*Claviceps* spp.), rust (*Puccinia substriata* var. *indica*), blast (*Pyricularia grisea*), and smut (*Moesziomyces penicillariae*) cause economic damage by affecting the

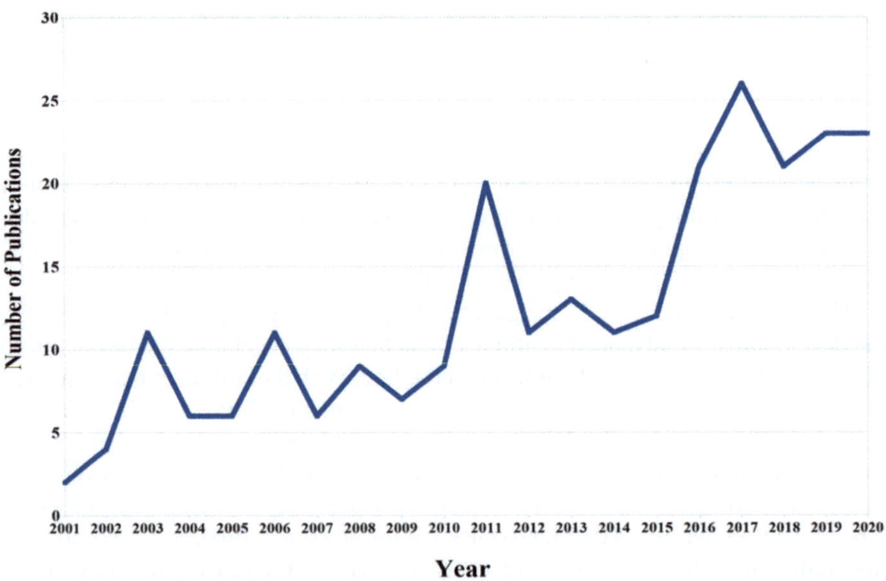

Fig. 2 The number of publications related to pearl millet genomics published from January 1, 2001 to December 31, 2020. (Source: Keywords 'pearl millet AND genomics' were used to search the number of publications in PubMed)

forage quality and yield (Kumar et al. 2012). Most of the breeding work on disease resistance is focused on downy mildew, blast, and rust.

Downy mildew or "green ear" was first reported in India and was considered a minor disease until 1970, when the disease caused heavy losses in pearl millet hybrid (HB3) (Kumari et al. 2020). The Indian hybrids based on Tift 23A1 faced the recurring problem of downy mildew, leading to production shortage in India (Kumar et al. 2012). The average approximate loss caused by the disease is 20–40% on susceptible hybrids, which increases when the disease coincides with favorable conditions. Ergot disease causes grain loss of about 58–75% (Kumar et al. 2012). Even though there is a significant reduction in ergot disease, which once caused epidemics during the 1960s and 1970s, it is still a concern, primarily because of the toxic sclerotia with the grains at the time of harvesting (Kumari et al. 2020). Another critical disease that causes heavy losses that affects both biomass and quality of forage is rust (Sharma et al. 2021). It caused about 72% yield loss in the southeastern United States, and severe losses have also been reported in Brazil and India (Yadav and Rai 2013). Thus, in the 1980s, the breeding objective shifted to diversified strategic breeding and value-addition research focused primarily on genetic diversification of CMS seed parents and restorers (Sharma et al. 2021).

Over the years, many adequate phenotypic screening methods are developed for essential diseases like downy mildew, ergot, rust, and smut at various research institutes including ICRISAT (Thakur et al. 2016). Extensive evaluation of germplasm to uncover resistance sources for developing resistant varieties against these diseases is under continuous process. So far, many improved varieties have been developed by using resistant germplasm. In 2000, Singh and Navi found 223 accessions resistant to downy mildew out of 529 wild accessions of *Pennisetum* species. *P. schweinfurthii* is thought to possess a high level of both downy-mildew and rust resistance and needs to be further explored in breeding programs.

But disease-causing pathogens are continually evolving, forming new virulent pathogens that require the development of new hybrids. Thus, it is vital to understand the changing host-pathogen interactions for developing improved and effective strategies against biotic stresses. This variability in the pathogenicity of the fungus has made the blast disease a matter of concern. Blast resistance was identified in the United States in *P. glaucum* spp. *monodii* accession (Dujardin and Hanna 1989). Apart from this, *P. violaceaum* (= *monodii* Marie) was identified as another blast resistance source against *Magnaporthe grisea* isolates in India (Chelpuri et al. 2019). Inter-simple sequence repeat (ISSR) primers have been used to characterize pathogen variability (Sudisha et al. 2009). Various researchers have identified many pathotypes of *S. graminicola*. Jogaiah et al. (2008) identified six pathotypes; Sharma et al. (2010) identified 21 pathotypes and stated pathotype P11 to be most virulent; Thakur et al., 2011 identified 20 new virulent pathotypes of *S. graminicola*. Gene pyramiding is thought to be a way to enhance disease resistance to diverse downy mildew isolates (Sharma et al. 2021).

Downy mildew resistance is generally the dominant type. However, one or more genes are involved in providing partial host plant resistance together with some modifiers (Hash and Witcombe 2001; Breese et al. 2002; Dwivedi et al. 2012).

Earlier disease-resistant breeding strategies included using the West-African germplasm as a source of downy mildew resistance (Singh et al. 1997; Thakur et al. 2006). Additionally, marker-assisted breeding and backcrossing methods played an essential role in downy mildew resistant breeding. On the other hand, open-pollinated varieties (OPVs) are preferred over hybrids in West Africa due to specific characteristics of economic seed production, relative tolerance to abiotic and biotic stresses (Kumari et al. 2014).

The first step in resistance breeding is to find sources of resistance. Researchers at ICRISAT evaluated around 3500 germplasm accessions collected from different pearl millet growing countries. There are many downy mildew resistance sources available in India, including OPVs: WC-C75, ICMS7703, ICTP8203, and ICMV155, resistant hybrids ICMH451 and Pusa23, topcross hybrid ICMH 88088, etc. (Singh 1995). The West African, Senegal, and Burkina Faso downy mildew resistant varieties include ICMV1 and ICMV2, IKMP2 and IKMP3, and IKMV8201, respectively (Yadav and Rai 2013). Over the years, several promising QTLs have been identified configuring downy mildew resistance (Jones et al. 1995; Gulia et al. 2007; Kumari et al. 2014)

Marker-assisted backcrossing breeding has been used to enhance disease resistance in many hybrids. Hybrid "HHB 67 Improved" was developed by improving the parental lines of "HHB 67." Resistant alleles on two QTLs at Linkage groups 1 and 4 were added to male parents (H 77/833-2) through selection from a recessive elite parent ICMP 451. The improved hybrid was released in drought-prone areas of India (Kumari et al. 2020).

Many cytoplasmic sterility sources are available, including A1 cytoplasm of Tift 23A, A2, A3, and A4, but A1 is the most used in India even though A4 and A5 sources are more stable (Burton and Forston 1966; Dujardin and Hanna 1989). The advancements in hybrid breeding have brought more disease epidemics, especially in the case of downy mildew, where most of the hybrids used the A1 source of male-sterile cytoplasm. However, in a study conducted by Yadav et al. (2001), it was reported that A1 cytoplasm is not associated with increased downy mildew incidence in pearl millet. No difference was found in the general combining ability of line with either A1 cytoplasm or B cytoplasm in terms of disease reaction in 40 pairs of hybrids.

More diseases appear on hybrids; however, they are still being used by farmers due to higher yield potential (15–30%) and uniformity over composites. However, utilization of A4 and A5 systems is essential to address the increasing concerns of increasing severity of downy mildew, blast, ergot, rust, and smut as it will decrease cytoplasmic uniformity. To better understand this, Yadav (1996) exploited the lesser used sources of CMS, like A2, A3, and A4, on downy mildew incidence using the disease incidence of iso-nuclear hybrids with both male sterile and fertile cytoplasm. No significant differences were found in the mean downy mildew incidence of hybrids carrying either of the cytoplasm, as the cytoplasm accounted for only 0.6% of the total variation. Since the male-sterile cytoplasm were not linked to downy mildew susceptibility, they could be exploited commercially to broaden the cytoplasmic base of the sterile male lines and, ultimately, of hybrids.

Similar studies were conducted by Yadav et al. (1992) for smut incidence. Thirty-five pairs of F1 hybrids, each carrying sterile and fertile cytoplasm, were evaluated for the effect of A1 male-sterile cytoplasm on smut severity. The results suggested that there is no link between male-sterile cytoplasm and smut severity. However, this was not the case as far as ergot disease susceptibility in sterile male hybrids was concerned. Thakur et al. (1989) compared 7 cytoplasmic male sterile lines (A-lines), their corresponding maintainer lines (B lines), 10 pollinator lines (R lines), and 48 hybrids. Moreover, 90–97% disease severity was reported in four of the seven A-lines compared to their corresponding B lines with only 54–77% severity. Significantly higher susceptibility of all A x R hybrids (except four) than their corresponding B x R hybrids is a clear indication of a positive association between a sterile cytoplasm and ergot susceptibility. But there are two significant backlogs in using other cytoplasm for hybrid development: the instability of cytoplasm across environments (A2 and A3) and limited availability of restores (A4 and A5). Even though only the ergot disease is thought to be associated with increased disease susceptibility using sterile male lines, we cannot risk using a single sterile cytoplasmic line. Apart from cytoplasmic diversification, a targeted approach for restorer development and nuclear genome diversification are equally important (Gupta et al. 2015).

Genetic and genomic resources involving use of DNA-based molecular markers, and linkage maps/mapping populations, are another ways to achieve resistance to biotic stresses (Sandhu et al. 2021c). These methods help determine the genetic differences in germplasm collections and analysis of QTLs for both abiotic and biotic stresses that aid in selecting the best parents for hybrid breeding (Sehgal et al. 2012). Moreover, SCAR markers derived from ISSR were developed for downy mildew resistance, and these were linked with the DM resistance linkage group with a linkage distance of 0.72 cM. Furthermore, QTLs for downy mildew resistance, rust and blast resistance have been identified and mapped using the pearl millet genetic linkage map through the various molecular markers (Liu et al. 1994; Jones et al. 1995; Gulia et al. 2007).

The GBS approach is lately being used to develop genetic markers and high-density linkage maps, especially for species lacking reference genomes (Punnuri et al. 2016). It is also essential to understand the physiological basis of stress, and scientists are continually working toward it. Kulkarni et al. (2016) used transcriptome analysis using next-generation sequencing to understand the mechanisms behind downy mildew resistance. They reported that the upregulation of phenylpropanoid pathway genes accelerates the pearl millet's defense systems by evoking hypersensitive and SAR pathways in resistant genotypes. Several genes have been found lately for downy mildew resistance, including Afp, NPR1, CC-NBS-LRR, and a popular downy mildew resistance (DM-QTL) (Jones et al. 1995).

Breeding for disease resistance is continually facing new challenges and thus involves ever-changing strategies. However, there is still a massive portion of resistant breeding techniques that are still undiscovered (Samantara et al. 2021). Thus, constant work and efforts are required to explore new strategies for the management of pearl millet diseases. Table 1 provided the information of different forage pearl

Table 1 Database for forage pearl millet obtained from All India Coordinated Research Project on Forage Crops and Utilization

Sr. no.	Cultivar	Nature of cultivar	Pedigree	Method of breeding/ selection	Method of propagation	If cross-pollinated, extent of natural out-crossing (specify in %)	Potential yield (green forage yield q/ ha)
1	Anand forage Bajra-3 (AFB-3)	Single cut	Developed from the local material from the North Gujarat region	Population improvement	Seed	Cross-pollinated	500–525
2.	APFB-2	Multi-cut variety	Not traceable	Recurrent selection in randomly mated population	Through seed	Highly cross-pollinated (above 85%)	350.0
3.	Avika Bajra Chari (AVKB-19)	Single-cut variety	Local material of Rajasthan	Selection from indigenous material collected from Nagore, Rajasthan	Through seed	Cross-pollinated (out crossing above 85%)	380.0
4.	BAIF Bajra-1	Single as well as multi-cut	Selection from base population of Giant Bajra	Pure line selection	By seed	100%	38 to 40 t/ha
5.	CO-8	Variety	Derivate of the cross no. 732 A × sweet Giant Bajra	Hybridization followed by pedigree method	Through seed	Cross-pollinated (out-crossing above 65%)	445.0
6.	FBC-16	Single-cut composite variety	A derivative of the five late maturing lines, namely, L5, L-7, L-13, L-14, and L-30 introduced from ICRISAT, Hyderabad	Synthesized by compositing five selected late maturing lines selected after one cycle of random mating	Through seed	Cross-pollinated (out-crossing above 80%)	575.0

(continued)

Table 1 (continued)

Sr. no.	Cultivar	Nature of cultivar	Pedigree	Method of breeding/ selection	Method of propagation	If cross-pollinated, extent of natural out-crossing (specify in %)	Potential yield (green forage yield q/ ha)
7.	KBH 108	Cross-pollinated (hybrid)	KMPS-74A × KPR-9288R	Pedigree and backcross breeding for parental NA lines and heterosis breeding for single cross hybrid	Sexually propagated	>75%	35.5 q/ha as per release proposal
8.	MH-1610	Cross-pollinated (hybrid)	M002A × M004R	Pedigree and backcross breeding for parental lines and heterosis breeding for single-cross hybrid	Sexually propagated	>75%	32.8 q/ha as per release proposal
9.	MH 1617	Cross-pollinated (hybrid)	M124F × M118R	Pedigree and backcross breeding for parental lines and heterosis breeding for single cross hybrid	Sexually propagated	>75%	33.0 q/ha as per release proposal
10.	MH 1747	Cross-pollinated (hybrid)	M163F × M164R	Pedigree and backcross breeding for parental lines and heterosis breeding for single-cross hybrid	Sexually propagated	>75%	36.5 q/ha as per release proposal
11.	Nandi-70 (MSH 224)	Cross-pollinated (hybrid)	NMS 31A × NMP 55	Pedigree and backcross breeding for parental lines and heterosis breeding for single-cross hybrid	Sexually propagated	>75%	54.4 q/ha as per release proposal

(continued)

Table 1 (continued)

Sr. no.	Cultivar	Nature of cultivar	Pedigree	Method of breeding/ selection	Method of propagation	If cross-pollinated, extent of natural out-crossing (specify in %)	Potential yield (green forage yield q/ ha)
12.	Nandi-72	Cross-pollinated (hybrid)	NMS 31A × NMP 75	Pedigree and backcross breeding for parental lines and heterosis breeding for single-cross hybrid	Sexually propagated	>75%	52.7 q/ha as per release proposal
13.	NDFB-2/ Narendra Chara Bajra-2	Single-cut system	Local selection from Kunda, Pratapgarh (U.P.)	Mass selection method	By seed	Often cross-pollinated	380–425
14.	NDFB-3/ Narendra Chara Bajra-3	Single-cut system	Selection from NDFB-202	Mass selection method	By seed	Often cross-pollinated	450–500
15.	NDFB-5/ Narendra Chara Bajra-5	Double cut/dual purpose	Selection from NDFB-1	Mass selection method	By seed	Often cross-pollinated	450–500
16.	NDFB-11/ Narendra Chara Bajra-11	Single-cut system	Selection from NDFB-202	Mass selection method	By seed	Often cross-pollinated	525–540
17.	Giant Bajra	Multi-cut variety	Derivative of the cross between Australian Bajra x local Bajra (from Dhule district)	Several selections were made from F2 to F8 generations and those were allowed to cross-pollinate freely	Through seed	Cross-pollinated (out crossing 80–90%)	645.0
18.	Gujarat forage Bajra-1 (GFB-1)	Multi-cut variety	Variety was developed from the local material collected from North Gujarat region	Plants were selected from collected material followed by population improvement program	Through seed	Cross-pollinated (out-crossing above 90%)	1595.0

(continued)

Table 1 (continued)

Sr. no.	Cultivar	Nature of cultivar	Pedigree	Method of breeding/ selection	Method of propagation	If cross-pollinated, extent of natural out-crossing (specify in %)	Potential yield (green forage yield q/ ha)
19.	Narendra Chara Bajra-3 (NDFB-3)	Single-cut variety	Local collection from Saidpur, Ghaziapur, U.P.	Mass selection	Through seed	Cross-pollinated (out-crossing above 65%)	483.0
20.	PAC-981 (Nutrifeed)	Hybrid	981F × 981 M	Heterosis breeding	Seed	–	584.4
21.	PCB-141	Single-cut variety	Composite variety	Composite selection	Through seed	Cross-pollinated (out-crossing above 80%)	545.0
22.	PCB-164	Single-cut variety composite variety	Composite variety	Composite selection	Through seed	Cross-pollinated (out-crossing above 80%)	310 0.0
23.	PHBF-1	Hybrid	Pb 408 A × PIB 213	Hybrid breeding involving CMS lines	Seed	Purely cross-pollinated	640
24.	Proagro no. 1 (FMH-3)	Forage hybrid variety	Derivative of the cross PSP-21 × PP-23	Hybridization followed by modified mass selection	Through seed	Fully cross-pollinated (out-crossing above 95%)	750.0 q/ ha (in multi-cut system)
25.	Raj Bajra 1 (RBB-1)	Composite forage Bajra variety	Composite developed by selection of superior plants from RBC-2, NDFB-3, NDFB-5 and Giant Bajra	Composite	By seed	Cross-pollinated; 80% cross-pollination	507.90

(continued)

Table 1 (continued)

Sr. no.	Cultivar	Nature of cultivar	Pedigree	Method of breeding/ selection	Method of propagation	If cross-pollinated, extent of natural out-crossing (specify in %)	Potential yield (green forage yield q/ ha)
26.	Raj Bajra Chari-2	Dual-purpose variety	Full sib selection of the population created through random mating among 20 crosses of four inbred lines (originating from West Africa), viz., MPP-7105-1, 7108-3, 7113-1-2, 7145-5 with five local inbred, viz., JB- 12-1-1, JB-19-1-2-2, JB-20-19-Br, JB-154-1-2 and JB- 203-4	Full sib selection in a population created through random mating among 20 F1 s (between 4 exotic × 5 Rajasthan inbred lines as given in 4 of the above)	Through seed	Cross-pollinated (out-crossing above 80%)	375.0
27.	APFB 09-1	Multi-cut variety	Derived from crossing between APFB-2, Giant Bajra, and Baif Bajra	Recurrent selection	Through seed	Highly cross-pollinated (above 85%)	302.0
28.	Moti Bajra	Cross-pollinated	Poly cross population of APFB-2, Giant Bajra and Baif Bajra	Recurrent selection	Seed	80–90%	811 q/ha in 3 cuts

(continued)

Table 1 (continued)

Sr. no.	Cultivar	Nature of cultivar	Pedigree	Method of breeding/ selection	Method of propagation	If cross-pollinated, extent of natural out-crossing (specify in %)	Potential yield (green forage yield q/ ha)
29.	TSFB 15-4	Single-cut variety	IP 20485, IP 20550, IP 20555, IP 20584	Recurrent selection	Through seed	Highly cross-pollinated	426
30.	TSFB 15-8	Single-cut variety	IP 22269, IP 15352, HTGP	Recurrent selection	Through seed	Highly cross-pollinated	420
31.	GAFB 4 (AFB 37)	Single-cut variety	Derived from cross between Bajra Bawal × MRB-8	Mass selection	Through seed	Highly cross-pollinated	550–580

millet genomic resources present in India and with their detailed information as obtained from the All India Coordinated Research Project on Forage Crops and Utilization.

5 Breeding for Forage Quality Traits

Fodder quality depends on its acceptability or palatability to the animal and its positive effect on the animal's health and performance. The crucial components for good forage quality are protein content, lignin content, fiber content, sugars, mineral content, and anti-nutritional components such as hydrocyanic acid (HCN). Pearl millet forage is preferred over maize and sorghum forage due to high succulence and higher crude protein content. Unlike sorghum, pearl millet does not contain HCN that has adverse effects on animal health, milk production, and fat quality of milk (Harinarayana et al. 2005). However, pearl millet also contains an anti-nutritional compound called oxalic acid. Even in silage production, pearl millet has an edge over sorghum and maize in terms of crude protein content.

Pearl millet is used as green fodder, stover (dry fodder), and silage, especially in tropical areas. The multi-cut varieties of pearl millet are explicitly used for green fodder production and interspecific hybrid of Bajra x Napier grass, being a perennial crop, provides the green fodder throughout the year. Major forage-related traits include forage yield, days to 50% flowering, plant height, number of tillers per plant, stem thickness, number of nodes, internode length, number of leaves, leaf length, leaf width, spike length, spike thickness, and 1000-seed mass. The pearl millet accessions present at ICAR-NBPGR, New Delhi, were evaluated for forage quality-related traits to create a database of pre-breeding material. The prime objectives of forage quality improvement in pearl millet are as follows: (a) to improve feed intake and digestibility, (b) to improve the amount of nutrients, and (c) to decrease anti-nutritional characteristics (Harinarayana et al. 2005). The feed quality of pearl millet is one of the most enviable traits considered for forage production. The following forage quality traits are considered while assessing the nutritive value: crude protein (CP), neutral detergent fiber (NDF), acid detergent fiber (ADF), lignin content (LC), sugars, and in vitro dry matter digestibility (IVDMD). These traits are under genetic control and linked with each other. Forage quality as ascertained by CP, NDF, ADF, LC, and IVDMD reflecting non-degradable and degradable proteins, non-structural and structural carbohydrates, celluloses and lignin, and anti-nutritional characteristics such as hydrocyanic acid (HCN), tannins, oxalic acid, and phenolics have attained much research attention throughout the world (Harinarayana et al. 2005).

Eight diverse parents and 28 F1 hybrids of pearl millet were genetically analyzed for yield and quality traits of green fodder. The results indicated that three parents Giant Bajra, RHRB-282, and RHRB-278 had a positive GCA (general combining ability) effect, and five F1 hybrids had good SCA (specific combining ability) effects for fodder yield and quality traits (Kumari et al. 2020). Pearl millet is grown for fodder purposes, whereas Napier grass is mainly cultivated as a forage crop. Napier–Bajra is a hybrid of Bajra and Napier grass with traits from both the parents. This hybrid possesess significant increase in the number of tillers, thus improving the yield and quality of forage. Napier–Bajra hybrids behave as perennials which give yield up to 2–3 years after single sowing. In a study, 24 Bajra × Napier hybrids were developed and assessed in comparison with 4 checks for forage yield, dry matter yield, number of tillers/plant, crude protein, and number of internodes/tiller. The PYT-13-391 gave the highest green fodder yield (8.26 kg/plant), which is 10.11% higher than the best check (6.88 kg/plant). The genotypes RBN-17-35 and RBN-16-55 were estimated as significantly better in crude protein over the best check by 10.11% and 9.98%, respectively (Shalini et al., 2019).

Divergent selection for IVDMD improvement can be opted for pearl millet fodder breeding. It is reported that divergent selection gives a 1.0–4.7% increase in IVDMD in various fodder crops, including legumes and cereals (Casler 2001). Conventional breeding techniques are highly convenient for improving protein quality. Moreover, improving degradable proteins in forage crops is considered more accessible and less expensive than nondegradable proteins. Breeding for fodder purposes in pearl millet is done by pure line, pedigree, and backcross breeding or through population improvement.

Dwarfing genes are important for reducing internode length and enhancing L:S (leaf/stem) ratio. The four dwarfing genes isolated from pearl millet are found recessive in the nature. The incorporation of d2 dwarfing genes leads to an 11% increase in leafiness and a 17–21% increase in IVDMD and a 30% reduction in forage yield (Burton and Forston 1966). Pearl millet is considered an important forage crop for grazing and silage in the United States, whereas it is used as a fodder crop in India.

Hence, the exploitation of dwarfing genes is not focused on pearl millet fodder breeding programs in India. The breeding programs for fodder breeding are focusing on normal height with good digestibility and wider acceptability.

Trichomes present on the leaf outer surface also determine the forage quality. Trichomeless genes are recessive in nature, which produces the phenotype that has high palatability and low digestibility. Stay-green is another forage quality trait which is an asset for both stover and forage crop. It is also governed by recessive genes that produce pleiotropic effects such as slowing down senescence and preventing degradation of protein content (Jarquin et al. 2020). In sorghum, the glossiness of leaves, which is governed by few recessive genes, is associated with drought tolerance and shoot fly resistance. These glossy genes need to be explored in pearl millet.

The sugars present in the green stalk determines the quality of the pearl millet fodder. Pearl millet genotypes with considerable variation in sweetness and juiciness are observed with Brix value from 3% to 16%. Lignin content present in the cell wall has an adverse effect on fodder digestibility. Brown midrib is a phenotypic marker linked with low lignin content. One bmr loci is reported in pearl millet, which improves the IVDMD of the forage by lowering down lignin content in stem and leaves (Ansari et al. 2020).

6 Genome Editing in Pearl Millet

Conventional breeding approaches such as hybridization, selection, and molecular marker-based selection (MAS) explore only the existing natural variations. Latest genome editing techniques have made plant breeders capable of integrating new and specific variations. Site-specific nucleases such as zinc finger nuclease (ZFNs), transcriptional activator-effector nucleases (TALENs), and CRISPR-related endonuclease Cas9 (CRISPR/Cas9) are used to create double-strand break (DSB) in DNA. The inherent repair mechanism of the cell repairs the DSB by homologous recombination (HR) or non-homologous end joining (NHEJ). Homologous recombination is a precise pathway for generating knock-in and gene replacement, whereas NHEJ is error prone due to random insertion or deletion which generates knockouts. At present, CRISPR/Cas9 is considered the most convenient genome editing tool. It consists of Cas9 (endonuclease) and gRNA (guide RNA), which has two components: trans-activating RNA (tracrRNA) and CRISPR-derived RNA (crRNA). Cas9-sgRNA (single-guide RNA containing crRNA and tracerRNA) complex reaches the target site in the genome homologous to sgRNA sequence and generates DSB. The CRISPR/Cas9 technology is highly utilized, as it is less time-consuming, target-specific, and cost-effective and has wider acceptability. The CRISPR/Cas technology is considered a substitute for transgenic technology especially for commercial release (Ansari et al. 2020; Singh et al. 2023a).

CRISPR/Cas9 is used in various cereal crops such as rice, wheat, barley, maize, and sorghum. To date, no report of exploitation of genome editing tools in pearl millet is available. The well-established protocols of CRISPR/cas9 in maize and sorghum can be used in pearl millet crop improvement with some modifications (Cong et al. 2013). An *Agrobacterium tumefacien* binary vector Y158 was constructed with Cas9 gene, a U6 promoter driven sgRNA, a GFP-NptII as marker gene, and red fluorescence protein gene (DsRED2). The GFP expression was observed after 14 days of transformation of sorghum embryo with Y158 to distinguish the transformed cells from non-transformed cells. The results indicated only 33.33% of stable transformed GFP-positive groups contain DsRED2 sections. The CRISPR/Cas9 system was also used to mutate two genes underlying quantitative trait loci for plant height (SbGA2ox5) and flowering time (SbFT) in sorghum. The transformation efficiency was 14.3% (3/21) for SbFT and 13.3% (2/15) for SbGA2ox5; however, editing efficiency was highly variable with 33.3% and 83.3% for SbFT and SbGA2ox5, respectively (Char et al. 2020).

Basically, before going for embryo transformation of any crop, the efficiency of the sgRNA is tested by protoplast assay or tobacco assay. A new protocol of transient transformation of intact leaves of sorghum by *Agrobacterium*-mediated CRISPR/Cas construct containing fluorescent protein as marker gene is a good alternative system for testing the efficiency of sgRNA construct (Sharma et al. 2020). These methods of transient transformation with high reproducibility, simplicity and high scale of testing can be adopted in pearl millet to boost up genome editing experiments. Genome editing has great potential to improve the genetic makeup of the pearl millet cultivar for variations that do not persist naturally, but its protocols need to be established well. The genome editing techniques in related crops are footprints to carry out these studies in pearl millet for developing the crop varieties with favorable traits.

Research related to forage quality is highly complex, costly, and laboratory dependent requiring a multi-institutional confederation and multidisciplinary approach. Significant research has been made on other crops that could be adapted to design forage pearl millet research (Ansari et al. 2020).

7 Genomic Selection in Pearl Millet

Traditional linkage and association mapping studies have been used in pearl millet to explore the genetic architecture of most of the agronomic, quality, biotic, and abiotic stress-related traits, but these techniques have not achieved much success due to the complex nature of quantitative traits (Sandhu et al. 2021a). Genomic selection (GS) offers the advantage of studying complex traits and has shown the potential to increase the genetic gains in crops. GS is a breeding approach that uses the whole genome-wide markers to predict the genomic estimated breeding values (GEBVs) of the plants. The best candidates can be then selected based on these GEBVs (Meuwissen et al. 2001). To make predictions, GS uses the previous year's phenotypic and genotypic datasets for a set of lines with statistical models to train the GS model, which is subsequently used to predict the performance of lines with only genotypic information. The model's performance is assessed by the Pearson correlation coefficient between the predicted GEBVs and actual phenotypic values. The predicted GEBVs aid the breeders either to use them as a parent in the hybrid development or next-generation advancement in the breeding program. GS has proven superior to traditional phenotypic selection and marker-assisted selection to improve quantitative traits and accelerate crop improvement in several crop species (Rutkoski 2019; Sandhu et al. 2021a). GS uses the linkage disequilibrium between the molecular marker and associated QTL for predicting its effect.

The biggest advantage of GS is that the selection decisions could be made during the off-season, which leads to an increase in genetic gain on an annual basis. Furthermore, the development of better statistical models accounting for additive, dominance, and epistasis variance components opens up avenues for predicting hybrid's performances and better resource allocation (Sandhu et al. 2021a). With the sequencing of the pearl millet genome in 2017, the first study was conducted using the GS to predict the hybrid performance (Varshney et al. 2017). GS was used to predict test cross performances using whole-genome resequencing data under four different scenarios: control, early stress, late stress, and across environments. High prediction accuracy of up to 0.6 was obtained for predicting across environments. The accuracy obtained in this study strengthens the evidence that adopting GS in Bajra breeding has substantial potential to improve gain per year (Varshney et al. 2017). The hybrid performance was predicted using the ridge regression best linear unbiased prediction (rrBLUP) model by training the model on grain yield from 64 hybrids grown from 2004 to 2013 under replicated conditions in 5 different environments. Their study identified 170 best hybrid combinations, of which 11 were already in use, and thus they showed new 159 hybrid combinations, which will be explored. A hierarchical clustering method was used for analyzing the 167,910 total single crosses, and 2 heterotic pools were developed, which resulted in an 8% higher hybrid performance from single crosses (Varshney et al. 2017).

In India, approximately 80–100 pearl millet hybrids are grown on more than 5 M hectare areas. For these hybrid developments, thousands of test crosses are evaluated each year by 30–40 seed companies. The resulting hybrid combinations were later evaluated for multiple years and multiple locations for releasing the best hybrid for cultivation by farmers and seed multiplication. The high time and investment for evaluating the performances of hybrids and inbred benefit GS's adoption. Liang et al. (2018) evaluated the performances of two genotyping strategies, i.e., tGBS and RAD-seq, for characterizing the inbred developed by ICRISAT and assessing GS's utility for hybrid development. Twenty random rounds with fivefold cross-validation utilizing SNPs from tGBS and RAD-seq were performed using four genomic selection schemes to predict four different traits' hybrid performances. The prediction accuracies obtained for four different traits with rrBLUP were grain yield (0.48–0.51), days to flowering (0.87–0.89), thousand-grain weight (0.73–0.74), and plant height (0.72–0.73). They observed no difference in prediction accuracies for traits having no or low heterosis using either hybrid/inbred or hybrid only prediction scheme. In the case of traits showing high heterosis, there was a significant reduction in prediction accuracy when phenotypic data from inbred parents were included in the models. However, it was concluded that the adjustment of phenotypic values for inbreds and hybrid trait values relative to the population mean results in a moderate improvement in prediction accuracy for the hybrid's performance. Hence, the conclusion was that we could not directly use the historical inbred data for hybrid prediction without accounting for heterosis (Liang et al. 2018).

Being a poor man's crop, pearl millet has not gotten significant benefits in comparison to rice, wheat, and maize from the various policymakers and government officials to develop the upstream science. This is particularly seen for GS and genomics, where funding plays an important role in studying many genotypes in a diverse set of environments. Furthermore, pearl millet, due to its heterozygous nature, residual heterozygosity, high outcrossing nature, and inbreeding depression, poses the bottleneck in breeding programs for parental/inbred developments, which will be ultimately used as a training population for the GS. Rapid decay of linkage disequilibrium warrants dense genome-wide markers, making the genotyping pipeline more costly. GS prediction accuracies are also affected by high rates of segregation distortion observed in the specific training population. Three-way and top cross hybrids are more commonly grown in Africa and single-cross hybrids in Asia. However, the GS model development for parental lines demonstrating different heterotic combinations for hybrids is quite challenging. In the coming years, this issue could be resolved with the adoption or development of GS models that can account for genotype by environment interactions and precisely estimate the general combining ability and specific combining ability for different heterotic pools and agroecological zones (Sandhu et al. 2022).

8 Prospects of High-Throughput Phenotyping in Pearl Millet

During the last decade, there has been rapid adoption of the ground and aerial platforms with multiple sensors in agriculture for phenotyping various traits throughout the plants' growth stage to alleviate these phenotyping bottlenecks. High-throughput phenotyping (HTP) involves the application of all these tools for phenotyping the plants, which vary from ground-based imaging to aerial phenotyping with unmanned aerial vehicles. These HTP uses technologies such as spectroscopy, noninvasive imaging, robotics, image analysis, and high-performance computing (Sandhu et al. 2021b). All these tools vary in their capacity and cost but provide dense phenotypic data. Different studies have shown that these technologies have a higher correlation with the manually corrected field data. HTP has opened the prospectus for nondestructive field phenotyping for many traits, including biotic, physiological, and abiotic stresses (White et al. 2012). Both aerial and ground-based HTP are being rapidly adopted in agriculture in countries like the United States, Canada, and Australia for measuring multiple plant traits at different growth stages accurately and precisely.

Ground-based HTP platforms were developed for controlled environments and field applications such as robotic-type and manually driven vehicle type. Most ground-based phenotyping platforms are used for indoor facilities like greenhouses and growth chambers with precise environmental control. Furthermore, several field-based HTP platforms such as tractor-mounted systems, aerial vehicles, and pushcarts have recently been adopted by plant scientists for phenotyping. These phenotyping platforms use a different number of sensors and height-adjustable sensors for measuring phenotypes over the various growth stages of the plants (Araus and Cairns 2014). Data collection from these sensors is stored in specific data loggers, general-purpose laptops to industrial computers. Different sensors used include multispectral, hyperspectral, thermal, fluorescence, and time-of-flight sensors. Multispectral and hyperspectral provide information about the various physiological processes like drought, heat, water, pigments, and insect pest information with the extraction of multiple reflectance indices and image processing pipelines (Sandhu et al. 2021d).

In the case of pearl millet, the adoption rate of these HTP tools is also almost negligible. Owing to the capacity of sensors to rapidly collect temporal and spatial data, it will ultimately help study the morphological and physiological changes occurring in the plant under any abiotic stress, for which pearl millet is mostly resistant. Sorghum and maize are highly studied fodder crops using these HTP tools. Hyperspectral and RGB cameras were used on UAV for multi-temporal and spatial data collection during the sorghum development stages, and ultimately different deep learning models were employed that resulted in better prediction of plant biomass (Zhang et al. 2017). In fodder crops, especially bajra, maize, and ryegrass, visual scoring and cutting biomass are tedious for estimating the total plant biomass. However, high-resolution multi- and hyperspectral sensors and UAVs have aided in estimating the aboveground plant biomass throughput with high resolution and reducing cost and time. A study in ryegrass showed that extraction of vegetation indices, namely, normalized difference vegetation index (NDVI) from the crop, aided in predicting the biomass efficiently compared to a visual rating in the field. Thus, this technique should be employed in other fodder crops (Wang et al. 2019). Furthermore, data collected from these HTP tools can be used in genomic selection models to increase the prediction accuracy for biomass, yield, and quality-related traits in fodder crops. In sorghum, UAV and GS were used to predict the plant height and obtained a high correlation value of ($r = 0.85$) for predicting the plant height by using whole genome-wide markers and training data from remote sensing (Watanabe et al. 2017).

9 Prospects of Speed Breeding in Pearl Millet

For most crops, the breeding of new improved varieties takes several years. In the case of pearl millet, hybrid varieties are generally preferred over open-pollinated varieties. Development of the hybrids requires inbred parents, which are often produced after four to six generations of selfing. This is especially time-consuming for field crops generally restricted to only one or two generations per year. Speed

Cong L, Ran FA, Cox D et al (2013) Multiplex genome engineering using CRISPR/Cas systems. Science 339(80):819–823. https://doi.org/10.1126/science.1231143

Diack O, Kanfany G, Gueye MC et al (2020) GWAS unveils features between early- and late-flowering pearl millets. BMC Genomics 21:777. https://doi.org/10.1186/s12864-020-07198-2

Dujardin M, Hanna WW (1989) Crossability of pearl millet with wild *Pennisetum* species. Crop Sci 29:77–80. https://doi.org/10.2135/cropsci1989.0011183X002900010019x

Elshire RJ, Glaubitz JC, Sun Q et al (2011) A robust, simple genotyping-by-sequencing (GBS) approach for high diversity species. PLoS One 6:1–10. https://doi.org/10.1371/journal.pone.0019379

Gulia SK, Wilson JP, Carter J, Singh BP (2007) Progress in grain pearl millet research and market development. ASHS Press

Gupta SK, Rai KN, Singh P et al (2015) Seed set variability under high temperatures during flowering period in pearl millet (Pennisetum glaucum L. (R.) Br.). F Crop Res 171:41–53. https://doi.org/10.1016/j.fcr.2014.11.005

Hanna WW, Dujardin M (1986) Cytogenetics of *Pennisetum schweinfurthii* Pilger and its hybrids with pearl millet [1]. Crop Sci 26:449–453. https://doi.org/10.2135/cropsci1986.0011183X002600030001x

Harinarayana G, Melkania NP, Reddy BVS et al (2005) Forage potential of sorghum and pearl millet

Harlan JR, Wet JMJ (1971) Toward a rational classification of cultivated plants. Taxon 20:509–517. https://doi.org/10.2307/1218252

Jähne F, Hahn V, Würschum T, Leiser WL (2020) Speed breeding short-day crops by LED-controlled light schemes. Theor Appl Genet 133:2335–2342. https://doi.org/10.1007/s00122-020-03601-4

Jarquin D, Howard R, Liang Z et al (2020) Enhancing hybrid prediction in pearl millet using genomic and/or multi-environment phenotypic information of Inbreds. Front Genet 10:1294. https://doi.org/10.3389/fgene.2019.01294

Jauhar PP (1981) Cytogenetics and breeding of pearl millet and related species. Cytogenet Breed pearl millet Relat species

Jones ES, Liu CJ, Gale MD et al (1995) Mapping quantitative trait loci for downy mildew resistance in pearl millet. Theor Appl Genet 91:448–456. https://doi.org/10.1007/BF00222972

Kover PX, Valdar W, Trakalo J et al (2009) A multiparent advanced generation inter-cross to fine-map quantitative traits in Arabidopsis thaliana. PLoS Genet 5:1000551. https://doi.org/10.1371/journal.pgen.1000551

Kumar A, Arya RK, Kumar S et al (2012) Advances in pearl millet fodder yield and quality improvement through breeding and management practices. Forage Res 38(1):1–14

Kumar S, Hash CT, Thirunavukkarasu N et al (2016) Mapping quantitative trait loci controlling high iron and zinc content in self and open pollinated grains of pearl millet [Pennisetum glaucum (L.) R. Br.]. Front Plant Sci 7:1636. https://doi.org/10.3389/fpls.2016.01636

Kumar S, Hash CT, Nepolean T et al (2017) Mapping QTLs controlling flowering time and important agronomic traits in pearl millet. Front Plant Sci 8:1731. https://doi.org/10.3389/fpls.2017.01731

Kumari BR, Kolesnikova-Allen MA, Hash CT et al (2014) Development of a set of chromosome segment substitution lines in Pearl Millet [*Pennisetum glaucum* (L.) R. Br.]. Crop Sci 54:2175–2182. https://doi.org/10.2135/cropsci2013.09.0589

Kumari P, Godika S, Geat N, Pandia S (2020) Ergot of bajra and its management: review. J Pharmacogn Phytochem 9:274–277

Li H, Yan S, Zhao L et al (2014) Histone acetylation associated up-regulation of the cell wall related genes is involved in salt stress induced maize root swelling. BMC Plant Biol 14:1–14. https://doi.org/10.1186/1471-2229-14-105

Liang Z, Gupta SK, Yeh CT, Zhang Y, Ngu DW, Kumar R, Patil HT, Mungra KD, Yadav DV, Rathore A, Srivastava RK (2018) Phenotypic data from inbred parents can improve genomic prediction in pearl millet hybrids. G3 Genes Genomes Genetics 8(7):2513–2522

QTLomics Approach for Improvement of Finger Millet

Narkhede Gopal Wasudeo, G. Harish Kumar, Manchikatla Arun Kumar, D. P. Divyavani, Shabir Hussain Wani, and Yaswant Kumar Pankaj

1 Introduction

Finger millet, also known as *Eleusine coracana* L., is an important cereal crop, mostly grown in semi-arid tropical regions of Africa and Asia, for its drought resistance and nutritious nature. However, the average productivity of finger millet is low, often below 1.0 ton per hectares, because of many biotic and abiotic stresses, limited access to improved seed varieties, and less agronomic practices. This chapter, QTLomics for finger millet improvement, focuses on the application of quantitative trait locus (QTL) mapping and genomics-assisted breeding approaches to improving the agronomic performance of finger millet (Shamkuwar et al. 2024).

While QTL mapping works by associating the specific trait with genomic data, this method becomes all the more important for finger millet improvement by allowing the dissection between trait such as yield, disease resistance, and nutrient content, through studying the genetics behind them. In recent investigations relating to the agronomic traits of finger millet, one has found QTLs that are linked to the leaf blast resistance, tiller number, root length, and seed yield. For example, unambiguous QTLs with regard to their association with these traits were evidenced through association mapping analysis involving diverse genotypes (Tiwari et al. 2024; Ramakrishnan et al. 2016).

N. G. Wasudeo · G. H. Kumar · M. A. Kumar · D. P. Divyavani
Department of Genetics and Plant Breeding, School of Agriculture, SR University, Warangal, Telangana, India

S. H. Wani
National Seed Project (NSP), Sher-e-Kashmir University of Agricultural Sciences and Technology – Kashmir (SKUAST-K), Srinagar, Jammu and Kashmir, India

Y. K. Pankaj (✉)
Department of Soil and Crop Sciences, Texas A&M University, College Station, TX, USA

S. H. Wani, S. S. Gosal (eds.), *Accelerated Plant Breeding, Volume 5*,
https://doi.org/10.1007/978-3-032-04527-0_5

Finger millet (*Eleusine coracana* L.) is an important cereal crop growing primarily in the semi-arid tropics of Africa and Asia recognized for drought tolerance and net nutritional advantage. Despite the potential it carries, finger millet remains an underperforming average yield of generally below 1.0 tons ha^{-1} due to several biotic and abiotic stress factors, limited access to superior seed varieties, and improper agronomic practices. This chapter, QTLomics for finger millet improvement, will discuss all the uses of quantitative trait locus (QTL) mapping and genomics improvement breeding strategies for improving agronomic performance traits in finger millet (Wright and Devos 2024).

QTLomics does a very crucial job for integration of QTL mapping with genomic data, more particularly, for the detection of genetic markers related with specific traits. It is the platform on which future research for finger millet improvement bases the identification of the genetic base for traits such as yield, disease resistance, and nutrient content. Some of these studies have been able to provide multiple QTLs assigned to agronomic traits of finger millet such as resistance to leaf blast, tiller number, root length, and seed yield. An instance of this case is the research that unambiguously identified QTLs in these traits through association mapping study involving diverse genotypes (Antony Ceasar et al. 2018; Maharajan et al. 2023) .

The advent of genomic technologies has revolutionized plant breeding by facilitating the identification of candidate genes linked to desirable traits. High-throughput sequencing and comparative genomics have been used to explore and define genetic diversity in finger millet and in identifying novel QTLs. These technologies bring to breeders that much more effective use of existing genetic resources, enabling them to integrate desirable traits into new cultivars through marker-assisted selection. The chapter will talk much about how these advancements can be deployed to develop high-yielding and resilient finger millet varieties to the dimensions of increasing food security demand (Mane et al. 2024a, b; Millet et al. 2023).

In spite of the great strides gained in QTL mapping and genomic studies, further effort is much needed in translating this knowledge for practical breeding applications in crop plants (Choudhary et al. 2019; Lin et al. 2020; Wani et al. 2022; Gao et al. 2025). The tetraploid genome of finger millet makes it extremely difficult to develop markers for association studies on trait identification. Moreover, there is also an urgent requirement for better cooperation between researchers in the context of shared genetic resources and data. This chapter will discuss these issues and also recommend future avenues for research to improve efficiency in finger millet breeding programs.

QTLomics for finger millet improvement seeks to extend the extensive account of research status thus far in this subject over great depth by defining the need for integrating QTL mapping with modern genomic approaches. This is expected to lay the groundwork for enhanced varieties of finger millet qualified to thrive even in adverse environments and, at the same time toward food security, at national and global level (Kumawat et al. 2016).

2 Understanding QTLomics

2.1 *Quantitative Trait Loci*

Specific regions of the genome are directly involved in the variation of quantitative traits, which are phenotypic characters showing continuous variation due to multiple effects by many different genes acting in interaction with the environment. Thus, QTLs are useful for plant breeding and genetic studies concerned with most traits of interest such as yield, drought resistance, disease resistance, and other complex traits.

2.2 *Importance of QTLs in Plant Breeding*

Quantitative trait loci (QTLs) are very effective and important to plant breeding since they are some portions of genome that control the several quantitative traits, e.g., yield, disease resistance, drought tolerance, etc., which are important agronomic traits that are controlled by multiple genes. The study of QTLs gives the plant breeder a better understanding of the genetic basis for these traits, thus allowing improved varieties to be developed through more efficient and rapid breeding scheme (Kumawat et al. 2016; Pendergast IV et al. 2022). Here is a breakdown of their importance in plant breeding:

2.2.1 Understanding Complex Traits

Most of the traits, whether it is crop yield, plant height, resistance to environment stress, or any other trait, are controlled by multiple genes. These complex traits can be established at the genomic level through QTL mapping, which indicates the specific regions of the genome involved and their influence on the trait to understand how these complex traits are inherited and expressed.

2.2.2 Marker-Assisted Selection (MAS)

- *Enhanced Selection Efficiency*: The possible linking of QTLs with molecular markers (DNA sequences) makes them relatively easier to identify than phenotypic traits and makes it possible for the breeder to select for traits desired in plants at the molecular level during the early generations. As a consequence, this speeds up the breeding process.
- *Reduced Need for Phenotypic Screening*: Screening plants for specific traits may consume time and cost in conventional breeding. QTLs serve in that they can be marked by those conditions with associated markers, thereby making it possible

to select plants that can be expected to express the desired trait without waiting for the manifestation of the trait or the maturation of the plants.

2.2.3 Pyramiding of Traits

- *Combining Multiple Traits*: QTL mapping allows breeders to breed various advantages through different parent plants into one superior variety through pyramiding. This is pyramiding and allows breeders to create a single crop with greater resistance to diseases, improved drought tolerance, and higher yielding capacity (Ingle et al. 2023).

2.2.4 Faster Breeding Cycles

- *Precision and Speed*: Plant breeders, by identifying specific genetic loci linked with particular traits, could also develop superior plant varieties faster and more accurately. This allows the rapid release of new crop varieties, which is crucial for solving global challenges such as food security and climate change (Maharajan et al. 2023).

2.2.5 Genomic Selection

- *Predictive Breeding*: With recent developments in high-throughput sequencing technologies, breeders are incorporating QTLs with genomic selection (GS) to predict the genetic potential of an individual for complex traits. This approach uses genomic modeling to estimate breeding values from a large number of QTLs, enabling better selection of superior individuals without much extensive field testing.

2.2.6 Identifying Candidate Genes for Breeding

- *Gene Discovery*: Mapping QTLs can facilitate the identification of genomic regions that influence certain traits. Such regions would allow breeders to identify likely genes that participate directly in expressed phenotypes. This information would be important for generating specific genetic changes (e.g., adding or deleting genes using CRISPR) or possibly marker development for future breeding programs (Bančič et al. 2024; Puranik et al. 2020).

2.2.7 Improving Crop Resilience

- *Climate Change Adaptation*: QTLs can help map out the genetic regions that are associated with stress tolerance such as drought, heat, salinity, and pest resistance. This, in turn, helps in breeding for developing varieties that can be resilient under varying environmental changes due to climate change, making them more resistant (Devos et al. 2023; Gebreyohannes et al. 2024).

2.2.8 Reducing Genetic Load

- *Minimizing Unwanted Traits*: QTL analysis can prevent the introduction of undesirable characteristics through hybridization and ensure that only the desirable genetic material introgression into the breeding population (Devos et al. 2023; Guo et al. 2023; Jacob et al. 2024).

2.2.9 Genetic Diversity Conservation

- *Conserving Useful Variability*: QTLs studying the genetic basis of traits allow breeders to conserve and utilize genetic diversity from either wild relatives or landraces, which may contain valuable alleles that could otherwise go unnoticed (Lee et al. 2017; Shamkuwar et al. 2024; Backiyalakshmi et al. 2021).

2.2.10 Enhanced Precision in Crop Improvement

- *Targeted Breeding*: Identifying QTLs has made breeding much more precise and targeted, where it is possible to improve individual traits without simultaneously modifying others. This becomes very critical when working with genetically modified crops or when maintaining integrity over a certain traits.

2.3 *QTL Mapping*

Quantitative trait locus (QTL) mapping is a powerful tool used in genetics to identify regions of the genome that are associated with quantitative traits, such as yield, disease resistance, or drought tolerance, which are crucial for agronomic performance.

2.3.1 Identification of Genetic Regions (QTLs) Linked to Traits

- QTL mapping helps in identifying loci associated with quantitative trait variations across chromosomes.
- Such traits are identified by their multiple genes; each gene, however, displays small contributions.
- QTL mapping is a very useful tool to identify such loci and understand the genetic architecture behind complex agronomic traits.
- For example, in rice or maize, QTL mapping would identify loci in the genome that function in yield, seed size, and disease resistance or stress tolerance (Chandra et al. 2024; Devos et al. 2023).

2.3.2 Marker-Assisted Selection (MAS)

- Once QTLs are discovered, molecular markers such as SNPs or SSRs close to the QTLs may be developed.
- These markers may be used in breeding programs for the selection of plants harboring desirable alleles for agronomically important traits, which may be linked to higher performances before the trait is expressed.
- MAS accelerates the breeding process, increasing efficiency and precision in developing new varieties with improved traits (Ganesh et al. 2024; Gebreyohannes et al. 2024).

2.3.3 Understanding Genetic Interactions

- QTL mapping not only identifies individual loci but also provides insights into gene interactions, epistasis, and pleiotropy.
- This is essential for complex traits, where several genes are interacting to influence the agronomic performance.
- For example, some QTLs may interact synergistically with each other to provide higher disease resistance or drought tolerance, while others may act antagonistically. Such knowledge will guide the development of better breeding strategies (Guo et al. 2023; Hatakeyama et al. 2018).

2.3.4 Dissecting Complex Traits

- Many agronomic traits, such as yield, are polygenic, that is, they are determined by many genes. The QTL mapping technique can now dissect such complex traits into simpler components, making the targets of improvement better targeted.
- For example, instead of trying to improve overall yield directly, QTL mapping might identify specific components of yield, such as root depth, plant height, or

flowering time that can be optimized for improved performance (Jacob et al. 2024; Joshi et al. 2023).

2.3.5 Improving Crop Resilience

- QTL mapping is crucial in the identification of traits associated with crop resilience, such as resistance to disease, pests, and environmental stresses such as drought, heat, or salinity.
- Identification of the loci for the improvement of these traits of resilience allows breeders to cultivate crops that would better perform in changing environments and ensure a sustainable form of agriculture.

2.3.6 Cross-Species Insights

- QTL mapping is not limited to a single species. Information from QTL studies in one crop can be applied to related species, improving cross-species breeding and enhancing the genetic improvement of agronomically important crops.
- For instance, QTLs found in one grass species can be moved to other cereals such as wheat or barley and thereby enhance other traits such as grain quality or disease resistance (Kumar et al. 2021; Maharajan et al. 2023).

2.3.7 Fine-Mapping and Functional Genomics

- QTL mapping usually marks the point of initiation for a broader research process. Once identified, QTLs can be fine-mapped to locate more precisely the causal gene(s) responsible for the trait.
- The process of fine-mapping is followed by identifying the specific alleles for which such QTL regions exist and thus linking these to the functional genes, advancing the understanding of biological mechanisms responsible for important agronomic traits (Mane et al. 2024a, b; Pandian and Ramesh 2019).

2.4 Technological Advances in QTLomics

Quantitative trait locus (QTL) mapping continues to serve as an important tool in plant breeding for locating certain traits such as yield, disease resistance, and drought resistance. Recent developments in QTLomics have remarkably improved the precision, speed, and scale of plant breeding programs.

2.4.1 High-Throughput Genotyping Technologies

- The huge potential of next-generation sequencing technology has allowed researchers to identify and genotype a large number of markers across the genome. New generation sequencing techniques such as whole-genome sequencing (WGS) and genotyping-by-sequencing (GBS) have led to much wider access at a lower cost to the huge genetic data researchers can obtain from high-throughput sequencing methods. It offers far more high-resolution QTL mapping in a wider range of plant species, especially crops with complex genomes.
- Single nucleotide polymorphism (SNP) markers and microsatellites, which are commonly assessed in QTL studies, have made dashing strides over the last decade, owing to advances in genotyping platforms, allowing thousands of such markers from across the genome to be discovered in a single experiment (Pendergast IV et al. 2022; Ramakrishnan et al. 2016, 2017).

2.4.2 Advances in Phenotyping

Phenotyping was a bottleneck in QTL mapping mostly due to the labor- and time-consuming collection of data. However, recent technological advancements have significantly improved phenotyping in plant breeding:

- *High-throughput phenotyping (HTP)*: Ingenious tools for studying phenotypes include drones, remote sensing, and unmanned aerial vehicles (UAVs) equipped with imaging sensors (e.g., RGB cameras, multispectral, and hyperspectral sensors), all of which have allowed data collection of very detailed phenotypic characteristics across thousands of individuals in field environments.
- *Digital phenotyping*: With imaging technologies, such as 3D phenotyping, more accurate and repeatable measurements of traits can be done (e.g., plant height, leaf area, root architecture). Machine learning algorithms are also being used to process and analyze large phenotypic datasets (Tiwari et al. 2024; Wei et al. 2023).

2.4.3 Genome-Wide Association Studies (GWAS)

It was really GWAS that made an impact in detecting QTLs through identifying genetic variants influencing traits in natural populations. In contrast to controlled crosses commonly used in traditional QTL mapping, GWAS can be performed on a vast number of diverse, unstructured populations. The development of modern statistical methods and computational tools enables handling the huge amount of data produced in studies (Backiyalakshmi et al. 2021; Blume et al. 2022; Brhane et al. 2022a, b, 2024; Devos et al. 2023).

- *Bayesian methods* and *mixed linear models (MLMs)* are widely used to account for population structure and genetic relatedness, leading to more accurate and reproducible results.

- *Integration of GWAS with functional genomics*: Combining GWAS with transcriptomics, proteomics, and metabolomics data helps to uncover the underlying biological mechanisms of the identified QTLs (Dida et al. 2021; Gimode et al. 2016; Guo et al. 2023; Hatakeyama et al. 2018; Jacob et al. 2024; Joshi et al. 2023; Kasule et al. 2024; Mane et al. 2024a, b; Puranik et al. 2020; Sharma et al. 2019; Tiwari et al. 2020).

2.4.4 CRISPR and Genome Editing Technologies

With the advent of CRISPR/Cas9 and other genome editing technologies, new opportunities have been opened in QTLomics for precise modification of genes within the identified QTL region. Breeders can edit specific genes associated with QTLs and enable the functional validation in trait improvement by using these modern approaches.

- *Gene stacking*: Using CRISPR techniques, breeders can introgress one or several beneficial alleles from different QTL location into a single plant, improving various characteristics such as disease resistance, yield, or tolerance to stress.
- *Precision breeding*: Genome editing has proven to be valuable for precise breeding, meaning specific alterations to an organism's genome as determined by QTL information that will enhance selected traits without additional noise in the unwanted genetic variation.

2.4.5 Linkage Disequilibrium (LD) Mapping and QTL Fine-Mapping

Technological advances on both sequencing and bioinformatics sides have enhanced the research concerning LD patterns, highly valuable for fine-mapping the studied QTLs. This study allowed the researchers to increase a specific number of recombination events using high-density markers, narrowing a broader QTL region toward specific candidate genes.

- *Dense marker arrays* help increase the resolution of QTL maps, leading to more accurate identification of the underlying genes controlling complex traits.
- *Fine mapping tools* such as *bulk segregant analysis (BSA)* and *next-generation sequencing (NGS)* allow for better resolution in pinpointing the exact genomic regions controlling key traits (Gebreyohannes et al. 2024).

2.4.6 Big Data and Artificial Intelligence (AI)

The involvement of big data approaches and artificial intelligence in QTLomics has intensified the data analysis, interpretation, and decision-making aspects of plant breeding. With the help of AI and machine learning, the large-sized datasets on genomics, phenomics, and environmental factors are used for the more accurate prediction of QTL effects. Some specific areas where AI is used include following:

- *Predictive modeling*: The prediction made by deep learning and random forest algorithms for different environmental conditions becomes highly meaningful in terms of QTL interactions, thus paving way for optimal selection strategies among breeders.
- *Data mining*: Methods developed by AI have uncovered hidden patterns and relationships in the large volume genomic and phenotypic data that improved the discovery of QTLs (Pazhamala et al. 2021; Zhu et al. 2024).

2.4.7 Marker-Assisted Selection (MAS) and Genomic Selection (GS)

Marker-assisted selection (MAS) has been employed over decades in hastening the breeding process by selection of plant materials using specific QTL markers. However, recently, more powerful tool emerged called genomic selection (GS); it employs genome-wide marker data to make predictions of the genetic value of individuals rather than their whole-genome profiles (Odeny et al. 2020).

- *Accelerated breeding cycles*: Genomic selection quickly predicts breeding values, making it not necessary to obtain the complete phenotypic information from all individuals. Its value lies in difficult and costly measurements of the trait.
- *Integrated with QTL mapping*: The complex traits can be influenced by GS when used with QTL mapping to locate the genetic regions that affect the trait and consequently helps to improve the breeding process (Yadav et al. 2020).

2.4.8 Multi-Omics Approaches

The integration of genomics, transcriptomics, proteomics, and metabolomics in QTLomics is making it more comprehensive toward an in-depth understanding of genetic and molecular basis of complex traits. It enables researchers to learn the corresponding regulatory networks and molecular interactions that are responsible for altering traits, in addition to being able to identify QTLs.

- *Functional genomics*: Multi-omics strategies will enlighten how genes within the QTL regions behave at the transcriptional, proteomic, and metabolomic levels, thus revealing the pathway from genetic variation to traits (Kayastha et al. 2024; Mane et al. 2024a, b; Panda et al. 2022; Rakkammal et al. 2022; Singh et al. 2022; Sood et al. 2019; Yang et al. 2021).

3 Application of QTLomics in Finger Millet

Inclusion of QTLomics improves the genetic groundworks of traits like drought avoidance, disease tolerance, and better grain quality and enables the breeder to develop yielding resourced improved resilience and better nutritious quality finger millet breeds.

- *Identification of QTLs for Key Traits:* To ascertain particular genomic regions linked to the desirable agronomic traits, it is possible to employ either bi-parental mapping populations (e.g., F_2s, recombinant inbred lines or RILs) or the more sophisticated genome-wide association studies (GWAS). These important traits include following:
- *Drought tolerance:* Identify QTLs associated with traits such as stomatal conductance, leaf area, root depth, and osmotic adjustment.
- *Disease resistance:* QTL identification for the resistance to pathogens like blast disease caused by *Pyricularia grisea* or rust caused by *Puccinia* species.
- *Nutritional quality:* Identifying QTLs for basic micronutrients, namely, calcium, iron, and zinc, including amino acid profiles that ultimately end up in the value of nutrient content of the grain.
- *Yield-related traits:* QTLs for components like grain size, tillering capacity, panicle length, and harvest index (Chandra et al. 2024; Ganesh et al. 2024; Jacob et al. 2024; Mane et al. 2024a, b; Tiwari et al. 2024).
- *Applications:*
  - *Marker-Assisted Selection (MAS):* The development of molecular markers like SNPs, SSRs, and AFLPs can then be applied in marker-assisted selection (MAS) when QTLs of the desired traits are identified. This provides an opportunity for a breeder to select individuals possessing favorable QTLs without the necessary assessment of traits phenotypically. MAS speeds up the whole breeding process while making it even more efficient in terms of precision, particularly for traits so complexly governed by several genes.
  - *Genomic Selection (GS)*: Tailored breeding techniques using whole-genome data for the prediction of individual breeding values constitute genomic selection. In finger millet, GS can be applied to traits complex in nature and lowly heritable like drought tolerance and yield stability across a spectrum of environments. Using genome-wide markers, GS will let one identify better genotypes even before they enter a field-testing phase.
  - *Gene Editing and Functional Genomics*: QTLomics reveals the genes and regulatory pathways controlling key traits. With this information, gene editing technologies like CRISPR/Cas9 are well-suited to direct modification of genes of interest toward improving traits such as disease resistance or nutritional quality. Functional genomics tools like RNA-seq can also be used to study gene expression with environmental stresses to yield more information about mechanisms of stress tolerance (Ingle et al. 2023; Joshi et al. 2023; Kumawat et al. 2016; Pandian and Ramesh 2019).

- *Genetic Diversity Assessment*: It is important to understand the genetic diversity within the finger millet germplasm for effective breeding purposes. QTLomics approaches have come in handy in mapping the genetic variation present in different varieties, landraces, and wild relatives of finger millet. Such understanding would help with the introgression of desirable traits from diverse sources into elite breeding lines, which would result in an overall improvement of the genetic pool as well as the resilience to environmental stresses (Shamkuwar et al. 2024).

4 Integration of Genomic Resources

4.1 Availability of Genomic Data

4.1.1 Sequencing Efforts in Finger Millet

Sequencing the genome of finger millet has become a widely achieved accession in the field of research on this crop. Different sequencing initiatives to unravel genome structure, genetic diversity, and develop molecular tools for improved crop productivity have been carried out.

(a) *First Sequencing Project (2017):* Finger millet genomic research made its first giant leap in 2017 with the announcement of draft genome sequence for *Eleusine coracana*. The genome sequencing project was undertaken as part of an international program in conjunction with researchers from India and other countries. This draft genome was based on the genome of the finger millet variety frequently cultivated in India, referred to as "ICMR 1101." The size of this draft genome of about 1.2 Gb marks the beginning of an important stride into further knowledge of the genetic structure of the crop and the basis for traits like drought tolerance, pest resistance, and grain quality. It would also help engage in the identification of key genes and pathways that could be targeted in breeding programs.

(b) *Advancements and Improvements in Sequencing (2020–2024):* Initial sequencing followed by advances in sequencing technologies like the PacBio long-read sequencing, Oxford Nanopore, and Hi-C scaffolding have made possible the assembly of complete and accurate genomes for finger millet. This sets the groundwork for improved gene annotations especially from stress resistance to seed development and nutrition-related genes (Tiwari et al. 2024; Gimode et al. 2016; Pathak et al. 2022).

4.1.2 Genomic Resources for Finger Millet

The genomic resources essential for finger millet have increased significantly and have ultimately provided a platform for both basic and applied types of research. These include genome sequences, gene annotations, and genetic diversity data.

Reference Genomes and Databases Various draft genome sequences of finger millet, including refined ones, are publicly available in databases like NCBI, Ensembl Plants, and the Finger Millet Genome Database (FMDB). The models of the gene and its function annotations can also be explored using these platforms.

(a) *Genomic Markers and QTLs:* Single nucleotide polymorphisms (SNPs), simple sequence repeats (SSRs), and insertion-deletion markers (InDels) are multi-genomic tools discovered for finger millet. Now, these genomic tools are used to assist in marker-assisted selection (MAS). Important quantitative trait loci (QTLs) related to valuable traits such as drought tolerance, resistance to major diseases, and grain quality have been mapped and made available to breeding programs, which will help them improve their breeding strategies in a more specific way.

(b) *Transcriptomics and Functional Genomics:* Extensive RNA-seq studies performed across tissues of finger millet at different stress conditions have revealed genes responsible for their biotic and abiotic stress responses. This study will prove extremely useful for identifying potential new genes valuable in improving the resilience of this crop. Functional genomics approaches like CRISPR-Cas9 target gene editing further unlock the potential for intervention on traits such as disease resistance and nutrient content (Brhane et al. 2022a, b; Kasule et al. 2024; Millet et al. 2023).

(c) *Genetic Diversity and Breeding Resources:* The sequencing of various accessions of finger millet has made it possible to study genetic diversity in different geographic regions and cultivars. Such information becomes very crucial for breeding programs that aim to enhance yield, pest control, and diseases resistance, as well as value addition through nutritional improvement. Genomic selection (GS), which is an approach employing genomic data to estimate the breeding value of individuals, is also being considered as a tool for speeding up the breeding of improved finger millet varieties (Gimode et al. 2016; Pathak et al. 2022).

4.1.3 Challenges and Future Directions

Despite these advancements, several challenges remain in the genomic research of finger millet:

- *Incomplete Genomes:* Though sequencing advancement has been high, sequencing of finger millet genomes has not yet been completed, leaving many gaps, particularly in the more complex regions of the genome. This will impede any in-depth functional annotation of the genome and the understanding of specific agronomic traits.

- *Lack of Comprehensive Genomic Resources:* It lacks, although some genomic data are available, extensive, top-quality resources such as gene expression data for various environmental conditions, complete genomic maps, and a functional genome. This hampers the capacity for high-throughput studies and genomic selection on a much broader level.
- *Integration with Breeding Programs:* Merging genomic knowledge with breeding activities is another major hindrance. The identification of several genetic tools and QTLs has yet led to their much application in breeding programs, as there is no proper integration with traditional breeding practices. Besides, finger millet genetic improvement is further hampered by its being a minor crop and the inadequate advanced breeding infrastructure (Joshi et al. 2023; Kumawat et al. 2016; Mane et al. 2024a, b; Pandian and Ramesh 2019; Puranik et al. 2020; Sharma et al. 2019; Sood et al. 2019).

4.1.4 Opportunities for Future Research

The future of finger millet genomics looks promising, with several opportunities for further advancement:

(a) *Advanced Sequencing Technologies:* The development of novel sequencing and genotyping technologies such as long-read sequencing and high-throughput sequencing will further enhance the quality and completeness of the finger millet genome, thus providing better functional annotation and greater detail about genetic diversity.

(b) *Genomic Selection and CRISPR-Cas9:* Genomic selection and genome editing technologies such as CRISPR-Cas9 would greatly benefit finger millet breeding by enhancing agronomic characters such as yield, disease resistance, and drought tolerance thereby accelerating the time needed to develop improved varieties.

(c) *Pan-Genomics and Diversity Studies:* Expanding the collection of diverse finger millet germplasm and sequencing a wide range of accessions will provide thorough insights into its genetic diversity, enabling the development of crops that are more resilient and nutritious.

(d) *International Collaborations:* It is very evident that global collaborations involving researchers and governmental organizations will become very crucial in addressing issues of food security and climate change by utilizing the finger millet genomics technologies in these dry regions of Africa and Asia (Tiwari et al. 2020; Brhane et al. 2022a, b, 2024; Gimode et al. 2016; Jacob et al. 2024; Joshi et al. 2023; Sood et al. 2019).

4.2 Comparative Genomics

In finger millet, the functions of genes can be investigated and validated from related species through comparative genomics and functional validations.

4.2.1 Genomic Resources from Closely Related Species

Finger millet originates from the Poaceae family (grass family) and the *Eleusine* genus. This genus has various species related to finger millet that could act as significant models for understanding gene functions. Some key species include the following:

(a) *Brown top millet (E. indica):* This particular species is closely associated with finger millet and shares many agronomic traits.
(b) *Maize (Zea mays):* Along with many other cereals, though not closely related, maize is among the well-studied model organisms within the Poaceae, and with extensive genomic and functional data.
(c) *Sorghum (Sorghum bicolor):* Sorghum is another member of the Poaceae family and possesses high degree of annotations on its genome, as well as being actively used in comparative genomics studies with the use of finger millet.
(d) *Rice (Oryza sativa):* Although very distantly related, rice is another grass species in which the genome has been extensively studied. These genetic pathways can be possibilities for inferring functions of finger millet genes (Joshi et al. 2023; Kumawat et al. 2016; Mane et al. 2024a, b; Pandian and Ramesh 2019; Puranik et al. 2020; Sharma et al. 2019).

Comparing one or more DNA sequences of different organisms could help associate functional assignments with predicted gene families and genomic sequences in finger millet. If a gene is conserved in other species such as maize, sorghum, or rice, it could indicate that the gene has important functionalities for example in stress, growth, or grain development in finger millet.

4.2.2 Comparative Genomics and Synteny Analysis

By means of synteny analysis between finger millet and its closely related species, one can identify such orthologous genes useful for predicting gene function. The process generally involves:

(a) *Identifying conserved gene regions*: These are the regions located in related species' genomes where genes are situated in the same order.
(b) *Gene function prediction*: Conserved regions mapped into other species harbor possible genes that might have functions resembling those of finger millet. It could be that genes that confer stress resistance or drought tolerance in maize or sorghum are supposed to play similar roles in finger millet. Most of the

analyses involve databases such as *Ensembl Plants* or *Gramene*, containing gene annotations and comparative data for grass taxa (Ramakrishnan et al. 2017).

4.2.3 Functional Validation Using Model Organisms

Completion of finger millet genome sequencing and annotation is tedious and time consuming; therefore, functional validation for genes will often use model organisms such as *Arabidopsis*, rice, or maize. Techniques include the following:

(a) *Gene expression analysis*: Use an RNA sequencing (RNA-seq) or quantify using PCR to analyze gene expressions under diverse environments, e.g., drought, salinity.
(b) *Gene knockout or overexpression*: Molecular techniques can be specified in inserting potential finger millet genes into model organisms like *Arabidopsis* or maize, according to genetic manipulation innovations, and the subsequent phenotypic characterization can involve overexpressing or silencing the genes in these models.
(c) *CRISPR/Cas9 genome editing*: This methodology can also be utilized for gene knockout studies in closely related species. For example, maize or sorghum could be used to validate the functioning of finger millet genes relating to drought or disease resistance (Tiwari et al. 2024; Gimode et al. 2016; Pathak et al. 2022).

4.2.4 Transcriptomic and Proteomic Data

(a) *Gene expression comparison*: Comparative analysis of gene expression data between finger millet and model species like maize or sorghum under different stress conditions can assist in identifying candidate genes for important traits.
(b) *Proteomics*: Functional analysis of finger millet genes could also be performed using techniques such as mass spectrometry or by studying protein-protein interaction networks, supplemented by data from model species like rice or maize (Kayastha et al. 2024; Mane et al. 2024a; b, Panda et al. 2022; Rakkammal et al. 2022; Singh et al. 2022; Sood et al. 2019; Yang et al. 2021).

4.2.5 Trait-Based Comparisons with Related Species

Phenotypic studies are beneficial for the comparison of closely related finger millet species under varying environmental stresses such as drought, high salinity. For example:

(a) *Drought tolerance*: Maize, sorghum, and rice are used in studying drought tolerance. Root development and osmotic regulation are inferred to have similar pathways in finger millets.

(b) *Disease resistance*: The genes that give resistance to pathogens in maize and sorghum also function in finger millet. Identification of candidate resistance genes for breeding programs can be done by comparative study (Ramakrishnan et al. 2017).

5 Breeding Strategies Under QTLomics

5.1 Marker-Assisted Selection (MAS)

QTLs are an important aspect for traits such as disease resilience, drought tolerance, yield, or even fruit quality. In breeding programs, molecular markers associated with these QTLs, for instance, SNPs, microsatellite markers, or RFLPs, can be used in marker-assisted selection. Rather than having to wait for the trait to express phenotypically, plants or animals can be screened for the presence of genetic markers, thereby accelerating the selection process while improving the accuracy of trait selection (Sharma et al. 2019).

5.1.1 Introgression of Desirable Traits

However, once a QTL is identified, breeders can start using it to introduce desirable traits into elite breeding lines. This is especially useful for complicated traits that are controlled by many genes. It bridges the beneficial alleles from the donor parent into the target population through QTL-guided improvement for traits such as high yield, stress tolerance, or disease resistance without the loss of other essential characteristics.

5.1.2 Genomic Selection (GS)

With identification processes of QTLs on a certain trait, it is quite possible to organize genomic selection because it would include genome-wide data from many markers (linked to QTLs) in estimating individual breeding values. Marker information is then compiled into a prediction model used for the selection of individuals occupying a favorable position in the predicted performance for the trait.

5.1.3 Increasing Genetic Gain

This sustains a genetic gain by increasing the efficiency of selection, since it enables screening with much-desired phenotypes, saving time and money spent on standard selection approaches in breeding programs. Breeders can also use QTL information to target particular genes or regions in the genome affecting important traits.

5.1.4 Stacking Beneficial QTLs

To strengthen a trait expression, sometimes, it is necessary to pyramid yield-influencing QTLs conforming to different phenotypes. It is possible, for instance, to consider pyramiding QTLs for yield improvement, drought resistance, and disease tolerance into a population to develop plants or animals with enhanced performance.

5.1.5 Validation of QTLs

QTLs are to be validated in various environments or populations for effective applications in breeding. Breeders often conduct validation studies after QTL mapping to check whether the QTLs affect the trait in other environments or populations. This is the process of gaining information about the most reliable, and thus stable, QTLs to be used in breeding programs.

5.1.6 Reducing Genetic Disequilibrium

The use of QTLs allows for more informed and controlled crosses between individuals. Breeders can select parents with complementary genetic backgrounds for the desired traits, which help to reduce genetic disequilibrium and improve the overall genetic quality of the breeding population.

5.1.7 Accelerating Breeding Cycles

By focusing on genetic markers tied to important QTLs, breeders speed the breeding process. This is perhaps most relevant to crops or livestock with protracted breeding cycles. One example is marker-assisted backcrossing, where the time for conjoining a new trait into a breeding line is shortened (Ganesh et al. 2024; Joshi et al. 2023; Kayastha et al. 2024; Kumawat et al. 2016; Mbinda and Masaki 2021; Nair and Devi 2024; Ramesh et al. 2024).

6 Challenges in Implementation

6.1 Genetic Diversity in Finger Millet Breeding

Genetic diversity is the broad base in increasing the traits of finger millet populations within a specific population to improve yield, resistance to diseases, and environmental tolerance.

6.1.1 Challenges

- *Narrow Genetic Base*: The fact that a number of parent lines, which are limited in their availability, are the bases for many finger millet varieties used in breeding programs reduces the genetic diversity in these varieties.
- *Loss of Landraces*: The introduction of high-yielding varieties completely replaces traditional forms or landraces of finger millet which may possess very valuable traits such as drought resistance, pest tolerance, etc. This has inevitably resulted in loss of genetic resources.
- *Inbreeding Depression*: Breeds are most likely to be reduced in population fitness in general, when the breeding programs concentrate overly on few high-performing individuals.

6.1.2 Strategies

- *Conservation of Landraces*: Efforts should strive to conserve and integrate the traditional landraces into the continuing breeding programs. This would lead to greater genetic diversity as well as the maintenance of desirable forms such as drought resistance or pest tolerance, among others.
- *Hybridization*: It will introduce new genetic combination to the well-known hybrids that are generated by crossing genetically diverse varieties (including landraces). The objective of this hybridization is to bring positive change in traits like yield, disease resistance, and adaptation to different climatic conditions.
- *Marker-Assisted Selection (MAS)*: With this molecular means, breeders identify with desirable genes from heterogeneous germplasm. This results in improved trait selection while maintaining genetic diversity and most preferably specific traits (Ganesh et al. 2024; Joshi et al. 2023; Kayastha et al. 2024; Kumawat et al. 2016; Mbinda and Masaki 2021; Nair and Devi 2024; Ramesh et al. 2024).

Selection Pressure in Finger Millet Breeding

In breeding programs, selection pressure indicates the strength or intensity with which specific traits are selected. Very high selection pressure for certain traits can make those traits improve rapidly, but this may also result in some unexpected results.

6.1.3 Challenges

- *Reduced Genetic Variation*: Evolving in terms of time or traits such as high yield or early maturity, continuous genetic selection increases vulnerability to pests, diseases, or environmental stresses.

- *Compromised Resistance*: Focusing too much on improving yield might compromise resistance to biotic and abiotic stresses, like drought or disease, which are critical for finger millet's survival in marginal conditions.
- *Pest and Disease Pressure*: Adapt continuous selection for high-yield varieties without considering the resistance of strains toward pest and pathogen attack toward the formation of pest populations that are better adapted to these kinds of selected crops, which will later increase the use of chemical interventions.

6.1.4 Strategies

- *Balanced Selection*: Breeding programs for crops should balance between the selection for high yield and for other traits associated with disease, drought, and adaptation to various agro-ecologies. It is this aspect that guarantees long-term productivity and sustainability.
- *Broad-Spectrum Disease Resistance*: Apart from yield selection, traits such as resistance to fungal diseases, insect pests, and other biotic stresses should also be included in the selection. This will make the varieties less reliant on chemical control and develop their resilience.
- *Sustainable Agricultural Practices*: Incorporation of crop rotation, intercropping, and different agro-ecological practices would be accompanied by breeding efforts to reduce the hazards associated with selection pressure on the environment and increase overall sustainability.
- *Use of Genomic Selection*: One of the advantages of genomic selection techniques is that they speed up breeding via direct selection of genetic phenotypes of desirable traits even before such traits have expressed phenotypically. This would ensure a good selection for diverse and sustainable varieties (Ganesh et al. 2024; Joshi et al. 2023; Kayastha et al. 2024; Kumawat et al. 2016; Mbinda and Masaki 2021; Nair and Devi 2024; Ramesh et al. 2024).

Strategies for Sustainable Breeding

The approaches that can be termed as being sustainable in breeding are such ones that safeguard genetic diversity in the crop, reduce the negative selection pressure, and ensure the long-term adaptability of the finger millet crops.

Strategies

- *Participatory Breeding*: Involving farmers in the breeding process is to select traits that have been tested in fluctuating conditions, such as drought, pest resistance, and local adaptation for varietal selection. It also conserves natural diversity and prevents using only market-based traits as selection criteria.

- *Ex Situ and In Situ Conservation*: By establishing gene banks and keeping conservation areas for various germplasm of finger millet, wide ranges of genetic materials can be preserved for use in future breeding.
- *Climate Change Adaptation*: Finger millet breeding programs need to integrate traits, such as heat tolerance and water-use efficiency, that will enhance their performance under changing climatic conditions. These features will help to minimize the adverse effects of climate change on production systems (Mbinda and Masaki 2021; Nair and Devi 2024; Ramesh et al. 2024).

7 Future Directions

Modern genomics has great potential to compliment traditional breeding efforts in making strides toward developing better finger millet (*Eleusine coracana* L.) varieties. Such integrated approaches would be able to address several challenges such as low yield, poor disease resistance, drought tolerance, and other nutritional quality.

7.1 Traditional Breeding

It entails the selection of desirable parent plants to formulate new improved varieties. In finger millet, this could relate to the selection of plants with better disease resistance, grain quality, or yield potential.

(a) *Hybrid breeding* for superior traits like larger grain size, higher resistance to pests, and drought tolerance.
(b) *Phenotypic selection* where traits such as plant height, tillering ability, and disease resistance are visually assessed.

Conventional breeding methods take time, often involving numerous generations of plants before arriving at the final result. Furthermore, the most complex genetic background with different traits has polygenic nature, such as finger millet; hence, it becomes highly challenging to effect specific improvements in it.

7.1.1 Modern Genomics

Modern genomics provides power to boost the traditional breeding like accelerating trait identification and increase in efficiency of breeding. Some key genomic techniques include the following:

(a) *Genetic Mapping and QTL Identification*: Through the identification of quantitative trait loci (QTLs) associated with important traits such as drought tolerance, disease resistance, or yield, breeders are able to choose more efficiently for these traits in their breeding programs (Table 1).

Table 1 Successful QTL mapping projects

Project title	Traits mapped	Methodology	Key findings	References
QTL mapping for leaf blast resistance	Leaf blast resistance, agronomic traits	Association mapping	Identified 7 QTLs associated with disease resistance and agronomic traits	Ramkrishnan et al. (2016)
QTLs for seedling stage phosphorus deficiency tolerance	Phosphorus deficiency tolerance	Linkage mapping	Revealed QTLs related to phosphorus uptake under low availability conditions	Ramkrishnan et al. (2016)
QTLs for drought tolerance and agronomic traits	Drought tolerance, biomass	Association mapping	Mapped QTLs associated with growth traits under drought stress conditions	Maharajan et al. (2023)
High-density linkage map for blast resistance	Blast resistance	High-density linkage mapping	Provided a detailed map with multiple QTLs linked to blast resistance	Pendergast et al. (2022)
Genetic mapping for protein and tryptophan content in seeds	Protein content, tryptophan	Linkage mapping	Identified QTLs associated with nutritional quality traits in seeds	Antony Ceasar et al. (2018)
QTL mapping for morphological traits under stress conditions	Morphological traits	Association mapping	Mapped several QTLs related to morphological adaptations under stress conditions	Ramkrishnan et al. (2017)
Comparative genomics analysis of QTLs in finger Millet	Various agronomic traits	Comparative genomics	Identified candidate genes linked to important agronomic traits through comparative analysis	Ramkrishnan et al. (2016)
Identification of QTLs for yield and related traits	Yield, tiller number	Linkage mapping	Discovered significant associations between yield components and specific markers	Antony Ceasar et al. (2018)
QTL mapping for grain quality traits	Grain quality	Association mapping	Mapped QTLs associated with grain quality parameters, enhancing breeding strategies for quality improvement	Antony Ceasar et al. (2018)
Genomic analysis of finger Millet genotypes for trait mapping	Multiple agronomic traits	Genomic analysis	Utilized genomic tools to identify traits linked to agronomic performance across different genotypes	Antony Ceasar et al. (2018)
Marker-assisted selection for agronomic traits	Agronomic performance	Marker-assisted selection	Developed markers linked to key agronomic traits for use in breeding programs	Ramkrishnan et al. (2016)

(b) *Marker-assisted Selection (MAS)*: Molecular profiling using MAS will probably accelerate the whole breeding process because plants are selected based on alleles of a particular desired marker trait such as pest resistance or drought tolerance, as molecular markers for many desirable traits are available.
(c) *Genome-wide Association Studies (GWAS)*: Using GWAS, different genetic variants may be identified for the traits of interest by the breeders that possibly act as more accurate targets for the selection.
(d) *CRISPR-Cas9 and Gene Editing*: The prospect that CRISPR technology would offer a facility to directly edit the genome of finger millet by introducing or modifying genes to develop beneficial traits like nutritional quality or pest resistance is quite promising (Joshi et al. 2023; Kumawat et al. 2016; Mane et al. 2024a, b; Pandian and Ramesh 2019).

7.1.2 Advantages of Combining Both Approaches

By integrating traditional breeding with genomics, the potential benefits include the following:

- *Accelerated Breeding Cycles*: Genomics speedily identifies desirable traits and eliminates the long intervals of classical selection methods.
- *Improved Precision in Selection*: Genetic tools allow for selection of plants with the precise combinations of traits and thereby make breeding more efficient and effective.
- *Enhanced Genetic Diversity*: Genomic technologies can be used for evaluating genetic diversity and directing breeding strategies that will be useful in maintaining or improving genetic diversity for sustainability of new varieties' resilience.
- *Increased Stress Tolerance and Resilience*: Genomic insights could assist in the identification of genes responsible for traits such as drought tolerance or pest resistance, thereby enhancing the adaptability of finger millet to climate change and other environmental stresses.
- *Better Nutritional Profiles*: This enhances the nutritional quality of finger millet through genomic selection and gene editing, for example, its mineral content, especially with respect to iron, zinc, and calcium (Mbinda and Masaki 2021; Nair and Devi 2024; Ramesh et al. 2024; Sharma et al. 2019).

7.1.3 Challenges and Considerations

While the integration of traditional breeding and modern genomics holds great promise, there are challenges:

- *High Initial Costs*: Availability of genomic tools and technologies is so costly that they are mostly inaccessible to small-scale farmers, particularly in developing countries where finger millet is grown.

- *Complexity of Finger Millet Genomics*: Finger millet is endowed with a large and very complex genome wherein many of the traits with economic significance are regulated by multiple loci. This makes it difficult to apply genomics in an efficient manner.
- *Farmer Adoption*: Farmers do not always adopt new varieties because they are not used to or confident in the technology, so strong extension services and education programs will be required (Brhane et al. 2022a, b, 2024; Dida et al. 2021).

Case Studies and Success Stories

7.1.4 Impact of QTLomics on Agronomic Traits

Genetic Improvements and Yield Enhancement

Genetic improvement has been a primary factor in increasing the yield potential of finger millet. It includes breeding for higher grain size, improved tillering, drought tolerance, and disease resistance (Table 2).

- *High-yielding varieties*: The traditional finger millet varieties generally were less productive. New high-yielding varieties have been developed through conventional breeding and current modern genetic techniques. These varieties have been selected on the basis of greater panicle size and grain filling. Varieties ML365 and ICTP 8203 improved yields dramatically, illustrating this development.
- *Improved plant architecture*: Droughts have proved beneficial for breeding purposes, as it has paved the way for improved plant architecture by making use of compactness measurements, improved tillering, and resistance to lodging, which, in turn, has ensured that yield per unit area increases as a consequence.

7.1.5 Drought Resilience

Drought has been one of the major constraints in finger millet cultivation. Even though it has deep roots, which make it more drought resistant than several other cereal crops, there has been further improvement in breeding drought-resilient varieties focusing on:

- *Root architecture and water-use efficiency*: Improved root systems were selected for deeper and more efficient extraction of moisture under drought conditions. Such varieties are meant for consideration under conditions that most times lead to scarcity of water which is quite vital in yield maintenance in arid and semi-arid countries.

Table 2 QTLomics for finger millet improvement

QTLomics study	Traits studied	Key findings	References
Leaf blast resistance	Plant height, root length, tiller number, seed yield	Identified 7 QTLs associated with leaf blast resistance and agronomic traits; markers UGEP101 and UGEP95 linked to blast resistance	Ramkrishnan et al. (2016)
Aluminum tolerance	Al-tolerance vs. Al-susceptibility	RNA-seq identified 198,546 unigenes; significant differences in gene expression between tolerant and susceptible genotypes	Brhane et al. (2022a, b)
Agro-morphological traits	Biomass, root traits under low phosphorus	QTL mapping revealed associations with biomass and root traits; utilized SSR markers for genetic analysis	Maharajan et al. (2023)
Drought resistance	Drought tolerance traits	Identified QTLs associated with drought resistance; markers linked to improved water-use efficiency	Gupta et al. (2017)
Seed quality traits	Protein and amino acid content	QTLs identified for protein content and amino acid profiles in seeds; implications for nutritional enhancement	Joshi et al. (2023)
Disease resistance	Finger and neck blast resistance	Mapping revealed QTLs associated with resistance to fungal diseases; potential for marker-assisted selection in breeding programs	Ramkrishnan et al. (2016)
Morphological traits	Plant height, leaf length, tiller number	Several QTLs identified influencing key morphological traits; important for agronomic performance improvements	Brhane et al. (2022a, b), Maharajan et al. (2023)
Yield components	Grain yield, harvest index	QTLs associated with yield components were mapped; findings can assist in developing high-yielding varieties	Maharajan et al. (2023), Joshi et al. (2023)
Biotic stress resistance	Resistance to pests and diseases	Identification of multiple QTLs linked to biotic stress resistance; potential applications in breeding for resilient varieties	Ramkrishnan et al. (2016)

- *Drought-tolerant traits*: Planning and breeding have been done for traits like leaf rolling, reduced transpiration, and better moisture retention, which help the finger millet plant survive under water-limited growing conditions. The traits developed have also given birth to improved varieties that are more tolerant to erratic rainfall patterns and droughts (Maharajan et al. 2023; Mane et al. 2024a, b; Chandra et al. 2024; Ganesh et al. 2024).

7.1.6 Disease Resistance

Finger millet's disease infestation has a high impact on its productive potential due to high susceptability to wide range of diseases. Among the different diseases are blast (*Magnaporthe grisea*), smut (*Ustilago exitalis*), and bacterial wilt (*Pseudomonas solanacearum*). Genetic improvement has led to the development of disease-resistant varieties to overcome these challenges.

- *Blast resistance*: The major research work has focused on developing blast-resistant varieties of finger millet as this can cause a severe loss to the crop. Many varieties such as Rashmi, Pramila have been reported to resist blast, thus reducing yield losses in such locations where the disease is prevalent.
- *Smut and wilt resistance*: Research on breeding against smut and bacterial wilt has led to promising disease-resistant varieties acting as a buffer toward chemical control measures toward more sustainable agriculture.
- *Resistant cultivars*: Modern breeding techniques like marker-assisted selection advancement now allow disease-resistant gene identification in a much more precise and efficient way, enabling breeders to develop multi-disease-resistant varieties of crops soon (Jacob et al. 2024; Joshi et al. 2023).

7.1.7 Nutritional Quality and Agronomic Benefits

Along with yield and resilience, enhancement of the nutritional profile of finger millet is now receiving attention. Finger millet is a good source of calcium, iron, and dietary fiber, thus making it as an important food crop in many developing areas.

- *Improved nutrient content*: Specific types of varieties developed with improved nutritional qualities, like increased minerals, are created through the method of selective breeding. This is usually beneficial to human beings, especially in places where malnutrition becomes dominant along with other micronutrient deficiencies.
- *Better grain quality*: The grain quality enhancement is another genetic improvement that makes the grains better suited for processing and consumption. Improvement has also included development in color, texture, and milling quality of grains, often resulting in better marketability (Kayastha et al. 2024; Puranik et al. 2020; Sharma et al. 2019).

7.1.8 Climate Resilience and Adaptation

Finger millet is a tough crop, but climate change is dictating that it be improved for better resilience. The major focus has been to develop varieties that withstand high temperature, erratic rainfall, and other changing environmental conditions.

- *Heat tolerance*: Harnessing heat-tolerant varieties has become critical in regions suffering increased temperatures because of climate change. These varieties yield even under high temperatures.
- *Shorter growing season*: Farmers have been able to harvest crops ahead of a severe drought or heat spell because of the breeding of early maturing varieties which, in the long run, will ensure reliability in yield.

7.1.9 Agro-Ecological Adaptation

All agronomic improvements have been directed toward better adaptation of finger millet under various agro-ecological conditions.

- There is a potential for improved soil fertility use. Varieties developed to be better able to use available nutrients on soils of low fertility have been an important consideration for farmers with minimal input. Particularly in those places where chemical fertilizers will not be available or will not be affordable, that will add heavy weight on this aspect.
- Intercropping finger millet is mostly done along with legumes or pulses. Through breeding for better husbandry in intercrop systems, finger millet has also been improved. Using hand mixing in involved mixed cropping systems enhances soil health and sources of income.

8 Conclusion

QTLomics has brought a revolution in genetic improvement of finger millet providing an excellent opportunity to get a closer grasp of complex architectures for critical agronomic traits. By identifying and mapping quantitative trait loci for traits such as yield, disease resistance, and stress tolerance, QTLomics offers solid ground for marker-assisted breeding, which can indeed speed up the development of better-performing finger millet varieties. Integrated breeding with genomics approaches has much to offer in tackling the challenges of climate change, food security, and sustainable agriculture. However, a lot more will be required to ensure QTLomics realizes its true potential in finger millet improvement through fine-tuning of the identified QTLs, understanding gene interaction, and connecting genomic discoveries with real-world breeding applications. Such research and breeding will benefit the small farmers, especially those who depend on finger millet for their food and nutritional security. Collaborative efforts among researchers, breeders, and farmers will be vital in translating QTLomics discoveries into tangible benefits for smallholder farmers, particularly in regions where finger millet plays a vital role in food and nutrition security.

References

Antony Ceasar S, Maharajan T, Ajeesh Krishna TP, Ramakrishnan M, Victor Roch G, Satish L, Ignacimuthu S (2018) Finger millet [*Eleusine coracana* (L.) Gaertn.] improvement: current status and future interventions of whole genome sequence. Front Plant Sci 9:1054

Backiyalakshmi C, Vetriventhan M, Deshpande S, Babu C, Allan V, Naresh D, Gupta R, Azevedo VC (2021) Genome-wide assessment of population structure and genetic diversity of the global finger millet germplasm panel conserved at the ICRISAT Genebank. Front Plant Sci 12:692463

Bančič J, Odeny DA, Ojulong HF, Josiah SM, Buntjer J, Gaynor RC, Hoad SP, Gorjanc G, Dawson IK (2024) Genomic and phenotypic characterization of finger millet indicates a complex diversification history. The Plant Genome 17(1):e20392

Blume R, Yemets A, Korkhovyi V, Radchuk V, Rakhmetov D, Blume Y (2022) Genome-wide identification and analysis of the cytokinin oxidase/dehydrogenase (ckx) gene family in finger millet (*Eleusine coracana* L.). Front Genet 13:963789

Brhane H, Haileselassie T, Tesfaye K, Ortiz R, Hammenhag C, Abreha KB, Geleta M (2022a) Novel GBS-based SNP markers for finger millet and their use in genetic diversity analyses. Front Genet 13:848627

Brhane H, Haileselassie T, Tesfaye K, Ortiz R, Hammenhag C, Abreha KB, Vetukuri RR, Geleta M (2022b) Finger millet RNA-seq reveals differential gene expression associated with tolerance to aluminum toxicity and provides novel genomic resources. Front Plant Sci 13:1068383

Brhane H, Haileselassie T, Tesfaye K, Lule D, Abreha KB, Hammenhag C, Ortiz R, Geleta M (2024) Genome-wide association analysis revealed genetic markers linked to grain yield and yield related traits in finger millet grown in acidic soils. Plant Soil:1–16

Chandra AK, Pandey D, Sood S, Joshi DC, Tiwari A, Sharma D, Gururani K, Kumar A (2024) Uncovering the genomic regions underlying grain iron and zinc content using genome-wide association mapping in finger millet. 3 Biotech 14(2):47

Choudhary M, Wani SH, Kumar P, Bagaria PK, Rakshit S, Roorkiwal M, Varshney RK (2019) QTLian breeding for climate resilience in cereals: progress and prospects. Funct Integr Genomics 19:685–701

Devos KM, Qi P, Bahri BA, Gimode DM, Jenike K, Manthi SJ et al (2023) Genome analyses reveal population structure and a purple stigma color gene candidate in finger millet. Nat Commun 14(1):3694

Dida MM, Oduori CA, Manthi SJ, Avosa MO, Mikwa EO, Ojulong HF, Odeny DA (2021) Novel sources of resistance to blast disease in finger millet. Crop Sci 61(1):250–262

Ganesh BT, Dilipbhai KA, Ragi S, Katral A (2024) Breeding finger millet (*Eleusine coracana* L. Gaertn) for improvement of quality characters and yield. In: Genetic improvement of small millets. Springer Nature Singapore, Singapore, pp 213–224

Gao L, Zhu Q, Li H, Wang S, Fan J, Wang T, Yang L, Zhao Y, Ma Y, Chen L, Li X (2025) Construction of a genetic linkage map and QTL mapping of the agronomic traits in foxtail millet (Setaria italica). BMC Genomics 26(1):152

Gebreyohannes A, Shimelis H, Mashilo J, Odeny DA, Tadesse T, Ojiewo CO (2024) Finger millet (*Eleusine coracana*) improvement: challenges and prospects – a review. Plant Breed 143(3):350–374

Gimode D, Odeny DA, de Villiers EP, Wanyonyi S, Dida MM, Mneney EE, Muchugi A, Machuka J, De Villiers SM (2016) Identification of SNP and SSR markers in finger millet using next generation sequencing technologies. PLoS One 11(7):e0159437

Guo S, Chai S, Guo Y, Shi X, Han F, Qu T, Xing L, Yang Q, Gao J, Gao X, Feng B, Song H, Yang P (2023) Mapping of major QTL and candidate gene analysis for hull colour in foxtail millet (*Setaria italica* (L.) P Beauv.). BMC Genomics 24(1):458

Gupta SM, Arora S, Mirza N, Pande A, Lata C, Puranik S, Kumar J, Kumar A (2017) Finger millet: a "certain" crop for an "uncertain" future and a solution to food insecurity and hidden hunger under stressful environments. Front Plant Sci 8:643

Hatakeyama M, Aluri S, Balachadran MT, Sivarajan SR, Patrignani A, Grüter S et al (2018) Multiple hybrid de novo genome assembly of finger millet, an orphan allotetraploid crop. DNA Res 25(1):39–47

Ingle K, Thakur N, Moharil MP, Suprasanna P, Awio B, Narkhede G, Kumar P, Ceasar SA, Abdi G (2023) Current status and future prospects of molecular marker assisted selection (MAS) in millets. In: Pudake RN, Solanke AU, Kole C (eds) Nutriomics of Millet crops, 1st edn. CRC Press

Jacob J, Visarada KBRS, Malathi VM, Venkateswarlu R, Karunakaran B, Kannababu N (2024) Advanced biotechnological tools for genetic improvement of finger Millet. In: Genetic improvement of small millets. Springer Nature Singapore, Singapore, pp 329–349

Joshi P, Gupta SK, Ojulong H, Sharma R, Vetriventhan M, Kudapa H, Choudhary S, Naresh D, Kholova J, Sajja S (2023) Finger millet improvement in post-genomic era: hundred years of breeding and moving forward. In: Smart plant breeding for field crops in post-genomics era. Springer Nature Singapore, Singapore, pp 221–253

Kasule F, Diack O, Mbaye M, Kakeeto R, Econopouly BF (2024) Genomic resources, opportunities, and prospects for accelerated improvement of millets. Theor Appl Genet 137(12):1–32

Kayastha S, Sahoo JP, Mahapatra M, Panda N (2024) Understanding the molecular breeding and omics approaches for finger millet (*Eleusine coracana* L.) improvement towards a global sustainable nutritional security. Crop Design 3(1):100049

Kumar SM, Madhura JN, Sheshashayee MS, Hittalamani S, Udayakumar M (2021) Development of microsatellite markers in finger millet. Indian J Genetics Plant Breed 81(03):462–465

Kumawat G, Gupta S, Ratnaparkhe MB, Maranna S, Satpute GK (2016) QTLomics in soybean: a way forward for translational genomics and breeding. Front Plant Sci 7:1852

Lee KJ, Yoon MS, Shin MJ, Lee JR, Cho YH, Lee HS, Ma KH, Lee GA (2017) Development of SSR markers and their use in studying genetic diversity and population of finger millet (*Eleusine coracana* L Gaertn.). Plant Breed Biotechnol 5(3):183–191

Lin F, Wani SH, Collins PJ, Wen Z, Li W, Zhang N, McCoy AG, Bi Y, Tan R, Zhang S, Gu C (2020) QTL mapping and GWAS for identification of loci conferring partial resistance to *Pythium sylvaticum* in soybean (Glycine max (L.) Merr). Mol Breed 40:1–1

Maharajan T, Ajeesh Krishna TP, Rakkammal K, Ramakrishnan M, Ceasar SA, Ramesh M, Ignacimuthu S (2023) Identification of QTL associated with agro-morphological and phosphorus content traits in finger millet under differential phosphorus supply *via* linkage mapping. Agriculture 13(2):262

Mane RS, Prasad BD, Sahni S, Quaiyum Z (2024a) Biotechnological approaches toward blast disease resistance in finger Millet. Int J Plant Soil Sci 36(7):422–428

Mane RS, Prasad BD, Sahni S, Quaiyum Z, Sharma VK (2024b) Biotechnological studies towards improvement of finger millet using multi-omics approaches. Funct Integr Genomics 24(5):148

Mbinda W, Masaki H (2021) Breeding strategies and challenges in the improvement of blast disease resistance in finger millet: a current review. Front Plant Sci 11:602882

Millet F, Ceasar SA, Babu BK (2023) Genetic and genomic resources for harnessing the health-related genes in finger millet. In: Compendium of crop genome designing for nutraceuticals. Springer Nature Singapore, Singapore, pp 267–282

Nair VD, Devi R (2024) Breeding finger millet for abiotic stress tolerance: strategies and challenges. In: Genetic improvement of small millets. Springer Nature Singapore, Singapore, pp 225–277

Odeny DA, Niazi A, Tesfaye K, Lule D, Wanyonyi S, Kunguni JS (2020) genomic designing for climate smart finger millet. In: Genomic designing of climate-smart cereal crops, pp 287–307

Panda D, Behera PK, Panda A, Nayak JK (2022) Advancement in omics technologies for enhancing abiotic stress tolerance in finger millet. In: Omics approach to manage abiotic stress in cereals. Springer Nature Singapore, Singapore, pp 559–574

Pandian S, Ramesh M (2019) Decoding of finger millet genome: a milestone of millet genomics. Signal Transduction Insights 8:1178643418820541

Pathak RK, Singh DB, Pandey D, Gaur VS, Kumar A (2022) Finger millet transcriptome analysis using high throughput sequencing technologies. In: In the finger Millet genome. Springer International Publishing, Cham, pp 123–134

Pazhamala LT, Kudapa H, Weckwerth W, Millar AH, Varshney RK (2021) Systems biology for crop improvement. Plant Genome 14(2):e20098

Pendergast TH IV, Qi P, Odeny DA, Dida MM, Devos KM (2022) A high-density linkage map of finger millet provides QTL for blast resistance and other agronomic traits. Plant Genome 15(1):e20175

Puranik S, Sahu PP, Beynon S, Srivastava RK, Sehgal D, Ojulong H, Yadav R (2020) Genome-wide association mapping and comparative genomics identifies genomic regions governing grain nutritional traits in finger millet (*Eleusine coracana* L Gaertn.). Plants People Planet 2(6):649–662

Rakkammal K, Priya A, Pandian S, Maharajan T, Rathinapriya P, Satish L et al (2022) Conventional and omics approaches for understanding the abiotic stress response in cereal crops - an updated overview. Plants 11(21):2852

Ramakrishnan M, Antony Ceasar S, Duraipandiyan V, Vinod KK, Kalpana K, Al-Dhabi NA, Ignacimuthu S (2016) Tracing QTLs for leaf blast resistance and agronomic performance of finger millet (*Eleusine coracana* (L.) Gaertn.) genotypes through association mapping and in silico comparative genomics analyses. PLoS One 11(7):e0159264

Ramakrishnan M, Ceasar SA, Vinod KK, Duraipandiyan V, Ajeesh Krishna TP, Upadhyaya HD, Al-Dhabi NA, Ignacimuthu S (2017) Identification of putative QTLs for seedling stage phosphorus starvation response in finger millet (*Eleusine coracana* L. Gaertn.) by association mapping and cross species synteny analysis. PLoS One 12(8):e0183261

Ramesh GV, Gudi S, Singh N, Bhandhari D (2024) Breeding finger millet for biotic stress resistance. In: Genetic improvement of small millets. Springer Nature Singapore, Singapore, pp 279–295

Shamkuwar S, Srivastava K, Tirkey AE, Prakash D, Madankar K, Saha S (2024) Reproductive biology, genetics, evolution, and diversity in finger millet (*Eleusine coracana* (L.) Gaertn.). In: Genetic improvement of small millets. Springer Nature Singapore, Singapore, pp 175–211

Sharma D, Sood S, Kumar A (2019) Paradigm shift from marker-assisted breeding to genomics-assisted breeding for calcium nutrition in finger Millet. In: Jaiwal P, Chhillar A, Chaudhary D, Jaiwal R (eds) Nutritional quality improvement in plants. Concepts and strategies in plant sciences. Springer, Cham, pp 55–77

Singh S, Mahalle MD, Mukhtar M, Jiwani G, Sevanthi AM, Solanke AU (2022) Advances in omics for enhancing abiotic stress tolerance in finger millets. In: Omics of climate resilient small millets. Springer Nature Singapore, Singapore, pp 235–257

Sood S, Joshi DC, Chandra AK, Kumar A (2019) Phenomics and genomics of finger millet: current status and future prospects. Planta 250:731–751

Tiwari A, Sharma D, Sood S, Jaiswal JP, Pachauri SP, Ramteke PW, Kumar A (2020) Genome-wide association mapping for seed protein content in finger millet (*Eleusine coracana*) global collection through genotyping by sequencing. J Cereal Sci 91:102888

Tiwari K, Tiwari S, Kumar N, Sinha S, Krishnamurthy SL, Singh R, Kalia S, Singh NK, Rai V (2024) QTLs and genes for salt stress tolerance: a journey from seed to seed continued. Plants 13(8):1099

Wani SH, Wang D, Singh GP (eds) (2022) QTL mapping in crop improvement: present progress and future perspectives. Academic, p 350

Wei W, Li S, Li P, Yu K, Fan G, Wang Y, Zhao F, Zhang X, Feng X, Shi G, Zhang W, Song G, Dan W, Wang F, Zhang Y, Li X, Wang D, Zhang W, Pei J, Wang X, Zhao Z (2023) QTL analysis of important agronomic traits and metabolites in foxtail millet (*Setaria italica*) by RIL population and widely targeted metabolome. Front Plant Sci 13:1035906

Wright H, Devos KM (2024) Finger millet: a hero in the making to combat food insecurity. Theor Appl Genet 137(6):139

Yadav S, Kumar A, Sood S (2020) Unravelling the genetics of calcium content in finger millet grains through association mapping. Indian J Genetics Plant Breed 80(04):432–440
Yang Y, Saand MA, Huang L, Abdelaal WB, Zhang J, Wu Y, Li J, Sirohi MH, Wang F (2021) Applications of multi-omics technologies for crop improvement. Front Plant Sci 12:563953
Zhu W, Li W, Zhang H, Li L (2024) Big data and artificial intelligence-aided crop breeding: Progress and prospects. J Integr Plant Biol:1–18

Current Status and Prospects of Genomics in Guar Breeding Program

Anju Biswas, Lovepreet Singh, Ishveen Kaur, Sangram Singh Sandhu, Japneet Kaur, Shakil Hosain, and Karansher Sandhu

1 Introduction

Guar (*Cyamopsis tetragonoloba*) is a summer legume grown in tropical and subtropical regions. The name guar is derived from the Sanskrit word go, or gav, which means cow (Monier-Williams 1964). It is a self-pollinated diploid (2n = 2× = 14) species belonging to the Fabaceae family with a genome size of 560 Mb (Ravelombola et al. 2021). Guar is also familiar with kutti, guari cluster bean, chavlikayi, etc. (Kumar and Rodge 2012). This summer legume is cultivated as a field crop in the dry regions of India and west of Pakistan. In some places in South India, it grows as a kitchen garden (Hymowitz 1972). It is anticipated that this crop was first domesticated in Pakistan, mainly the northwestern part of Pakistan (Gopala

A. Biswas
Department of Agronomy, University of Florida, Gainesville, FL, USA

L. Singh
Department of Plant and Soil Sciences, Mississippi State University, Mississippi State, MS, USA

I. Kaur
Agriculture, Environmental and Sustainability Sciences, College of Sciences, University of Texas Rio Grande Valley, Edinburg, TX, USA

S. S. Sandhu
College of Agriculture, Punjab Agricultural University, Ludhiana, Punjab, India

J. Kaur
College of Sciences, Panjab University, Chandigarh, Punjab, India

S. Hosain
Department of Plant Sciences, North Dakota State University, Fargo, ND, USA

K. Sandhu (✉)
Department of Crop and Soil Sciences, Washington State University, Pullman, WA, USA
e-mail: k.sandhu@wsu.edu

S. H. Wani, S. S. Gosal (eds.), *Accelerated Plant Breeding, Volume 5*,
https://doi.org/10.1007/978-3-032-04527-0_6

Krishnan et al. 2011). It is growing in India, Pakistan, and part of Texas, USA (Morris 2010) during the summer season. This crop is mainly used as a vegetable and a source of fiber and in the gum industry. In general, guar is used in broader aspects such as human consumption, cattle feed, medicinal purposes, laxatives, religious purposes, industrial, and other uses. Guar is popular because it contains galactomannan, also known as guar gum. This gum is used as a stabilizer in frozen desserts such as ice cream, salad dressing, and other bakery products, as well as in the pharmaceutical mining; paper, oil, food, and meat industries; and even in the beverage industry (Morris 2010; Mudgil et al. 2014; Zanoni et al. 1980). Guar is considered a good nutrient source for human beings. Green guar pods and seeds are sources of fat, protein, fiber, carbohydrate, and minerals (Ravelombola et al. 2021). It can also be used for green manure for increasing soil fertility (Dennis and Johnson 1981). Guar can fix nitrogen in the soil with the help of bacteria, improve soil fertility, and reduce nitrogen fertilizer application (Chaudhary and Sindhu 2016).

India, being a lead producer of guar, can generate approximately 3.4 million metric tons of guar seeds annually. India provides more than 80% of guar seed production, a net worth of about $2 billion (https://www.fas.usda.gov/). Pakistan is the second-largest (15%) guar producer (Ravelombola et al. 2021). Meanwhile, the USA is the largest Guar importer of guar gum (Chudzikowski 1971), with a net value of $1.1 billion per year. Despite the demand, guar cultivation is not common in the USA. Approximately nine to ten cultivars have been released in the USA since 1964. The breeding effort is underway to maintain the huge demand for this crop. Guar was introduced in the USA in 1903 (Hymowitz 1972). The Agricultural Research Service (ARS) of United State Department of Agriculture (USDA) is maintaining the accessions in Griffin, Georgia, at the plant genetic resources conservation unit. The USDA's ARS has 1298 accessions obtained from India, Pakistan, and some breeding lines from here (Morris 2010).

Despite the significant potential, the breeding efforts are insufficient due to the limited information on genetics and genomics (Ravelombola et al. 2021). Negligible works have been done on guar quality, disease resistance and genome editing, genomic selection, high-throughput phenotyping, and data science. In this modern breeding era, breeders are robustly using modern techniques (genomics, phenomics, crop growth model, and combining all these) to reduce the breeding cycle and enhance genetic gain. Some resources such as molecular markers (SSR, SNPs) have been developed for guar, which is not enough yet for this highly demanding crop. Still, molecular markers are scant for this crop. The development of genomic resources can accelerate guar breeding (Verma et al. 2025). Scarcer genome sequence information available through the current technologies are way cheaper than before. Molecular markers can help select important traits that are expensive to phenotype. Genomic selection is a process that estimates genomic-estimated breeding value (GEBV) from the seedling and predicts the performance of that crop and reduces the timeframe for releasing a new cultivar sooner than using conventional breeding. Also, for phenotyping nowadays, unmanned aerial vehicles (UAVs) are extensively used for precision phenotyping which is less laborious and time-consuming. Phenotyping a larger population can help capture rare alleles that affect the important traits. Next-generation sequencing (NGS) allows high-throughput genotyping, while UAVS allows high-throughput phenotyping. Guar breeding can take all those advantages mentioned above (genomic selection and high-throughput phenotyping), make selection faster, and release new cultivars to meet the potential demand of guar.

2 Genetic and Genomics Resources in Guar

Guar (*Cyamopsis tetragonoloba* L.) is a diploid species with a chromosome number of ($2n = 2\times = 14$). This leguminous crop has a vast genetic diversity yet to be fully explored. Lack of information in the molecular and genomic resources in guar limits and slowed the further advancement and breeding of this important crop. Krishnan et al. (2011) believe guar was domesticated in the northwestern part of the Indo-Pakistan region; however, no strong evidence was provided, and further validation of domestication of guar is needed. The genome size of guar is intermediate in comparison to other legumes and was determined at 581 Mbp using the procedure of flow cytometry by Tyagi et al. (2019). With the same procedure, the genome size of the two wild relatives of Guar, *Cyamopsis serrata* at 980 Mbp and *Cyamopsis senegalensis* at 944 Mbp, were also determined by Tyagi et al. (2019). The first whole-genome assembly of guar accession "Vaviloskij 130" was established by Grigoreva et al. (2019) using the second and third generation sequencing methods by combining the Illumina short reads and Oxford Nanopore long reads. The study also reveals that the *C. tetragonoloba* genome is highly related to the genomes of the genera *Vigna*, *Abrus*, *Glycine*, and *Lupinus*. High-quality and complete reference genome are still in need to accelerate the molecular breeding program in guar. The de novo assembly of *C. tetragonoloba* genome by Grigoreva et al. (2019) still has a 50% gap.

Availability of genomic resources is important for crop improvement and developing new cultivars. Molecular markers such as AFLP, RAPD, SSR, and SNPs have been developed and used in guar research. In addition, allozyme-based genetic diversity analysis was also conducted in guar. However, very few genome-wide association study (GWAS) or quantitative trait loci (QTL) have been reported in guar (Ravelombola et al. 2021; Malani et al. 2025). Using amplified fragment length polymorphism (AFLP) marker, genetic diversity analysis of eight guar accessions was conducted by Gresta et al. (2016). The study used eight primer pairs to amplify the AFLP markers and reported successful amplification. Genetic diversity analysis of 32 guar accessions by Pathak et al. (2010) using random amplified polymorphic DNA (RAPD) marker reported nine of the accessions were highly distinct from the others. Sharma et al. (2014) used RAPD and inter-simple sequence repeat (ISSR) markers to analyze the genetic diversity of 35 guar accessions and found 90.3% and 97% average polymorphism from RAPD and ISSR, respectively. RAPD markers were also used for genetic diversity analysis of 54 guar accessions by Kumar et al. (2015). They used 25 RAPD primer pairs and found 9 of them were polymorphic.

Kuravadi et al. (2014) suggested 187 simple sequence repeat (SSR) markers that can be effectively used in guar research. Using ISSR and RAPD markers, genetic diversity was performed by Ansari et al. (2016) on a panel of 104 guar accessions. Nuclear microsatellites (nSSR) were also used for the diversity analysis in 46 guar accessions by Santonoceto et al. (2019). They used 157 nSSR markers and found 48 nSSRs were polymorphic. Thakur and Randhawa (2018) identified 5773 SSRs and 3594 SNPs (single nucleotide polymorphism) using transcriptomic reads from two guar accessions "RGC-1066" and "M-83." Allozyme-based genetic diversity of 55 guar accessions was investigated by Brahmi et al. (2004). The guar chloroplast genome size, about 153 Mb, was reported by Kaila et al. (2017). Despite the works that have already been done in guar genomics, more work in developing molecular markers is needed to conduct molecular breeding through marker-assisted selection or genomic selection to accelerate guar breeding.

3 Disease Resistance in Guar

Thakur and Prasad (2020) recently reported an annual loss of 60% in yield due to loss by pests, pathogens, and the diseases associated with them. Such tremendous losses have diverted the attention of scientists to concentrate more efforts in the improvement of guar/cluster bean crop through different conventional as well as non-conventional programs, which has resulted in the development of disease-resistant varieties (Table 1). Guar crop is infested with different types of diseases mainly caused by fungi, bacteria, virus, pests, and other parasitic and non-parasitic sources. Among the fungal diseases, *Alternaria* leaf spot, powdery mildew, dry root rot, *Cercospora*, *Myrothecium*, *Curvularia* leaf spot, anthracnose, *Ascochyta* leaf blight, *Fusarium*, and *Rhizoctonia* blight are the most common ones. Among the bacterial diseases, bacterial blight and bacterial leaf spot are found widely on guar crops. Cluster bean is also infested with mosaic and tobacco ringspot viral disease along with different insect pests such as whitefly, thrips, weevils, jassids, stink bugs, pea leaf miner, blossom midge, and nematodes that feed on the leaves and pods, thus hampering the growth and yield of crop.

4 Fungal Diseases

Alternaria leaf blight, a seedborne disease, has been reported to cause an annual loss of approximately 43–47% in yield when the crop is infested at seedling and reproductive stage. This disease, caused by *Alternaria cyamopsidis*, is characterized by the presence of round or irregular spots which later on turn brown–grayish in appearance. The presence of water-soaked lesions is also observed, which later merges together to cause chlorosis with an increase in disease severity (Sharma et al. 2019). Saharan and Saharan (2001) have also reported the reduction in crude

Table 1 Different varieties that are resistant to different diseases in guar crop

Sr. no.	Type	Name of disease	Causal organism	Name of genotype	Method of improvement	Advantages	References
1.	Fungus	*Alternaria* leaf spot	*Alternaria cyamopsidis*	HG 2-20	Silencing of $AcCHS_7$ gene with pSilent CHS_{7-2} and $pARTCHS_7$	70% more survival rate as compared to wild	Thakur and Prasad (2020)
2.	Fungus	*Alternaria* leaf spot	*Alternaria cyamopsidis*	Resistant cultivars: HG1182, CAZG9007, RGC1014	RGC1014- RAPD markers identified	Fastened identification of disease resistant genes	Thakur and Randhawa (2018)
3.	Fungus	*Alternaria* leaf spot	*Alternaria cyamopsidis*	Variety- GAUG 9008, GAUG 9406, GAUG 9407	RAPD markers	Reduction in yield losses due to the presence of disease-resistant gene	Kumar and Rodge (2012)
4.	Fungus	*Alternaria* leaf spot	*Alternaria cyamopsidis*	HG-75 HG-365	Resistant variety	Sources of leaf blight resistance	Saharan and Saharan (2004)
5.	Fungus	Powdery mildew	*Oidium taurica, Erysiphe polygoni, Sphaerotheca fuliginea*	GAUG 63	Resistant genotype	Moderately resistant to powdery mildew	Lodha et al. (2009)
6.	Fungus	Dry root rot	*Rhizoctonia bataticola*	HG-75 RCG-1030	Resistant genotype	Prevent the occurrence of root rot disease	Lodha and Burman (2000)
7.	Fungus	Dry root rot	*Rhizoctonia bataticola*	Variety- CAZRI 27-1	Resistant variety	High-yielding variety with tolerance to disease	Lodha and Burman (2000)

(continued)

Table 1 (continued)

Sr. no.	Type	Name of disease	Causal organism	Name of genotype	Method of improvement	Advantages	References
8.	Fungus	Charcoal rot disease	*Macrophomina phaseolina*	GG1	Resistant genotype	Prevent disease development	Ravelombole et al. (2021)
9.	Bacteria	Bacterial leaf blight	*Xanthomonas axonopodis* pv. *cyamopsidis*	Lines- GAUG 9406, GG1, RGC 1027 Sources of resistance- G85, G102, G225	Resistant genotype	Dominant resistant lines developed by screening 375 accessions	
10.	Bacteria	Bacterial leaf blight	*Xanthomonas axonopodis* pv. *cyamopsidis*	HG 75, HG 563	Contains two genes, one for disease resistance, second one for inhibiting the expression of gene for disease	Broadens the genetic base by increasing the chances of desirable recombinants having disease-resistant gene	Kumar and Rodge (2012)
11.	Bacteria	Bacterial leaf blight	*Xanthomonas axonopodis* pv. *cyamopsidis*	M-83, RGC-1066	Transcriptome (RNA- sequencing)	62,146 coding sequences, 5773 microsatellite, 3594 SNP identified	

protein and organic matter content due to the development of infection by fungus. Additionally, the levels of phenol and peroxidase enzymes were found to be elevated upon colonization of the fungus as part of plant defense strategy, which can be exploited to form resistant lines (Bhatt et al. 2017).

Most of the breeding work in guar is pinned down to the development of the resistant genotypes and varieties, as interspecific hybridization is relatively difficult on guar due to the presence of pre- and post-fertilization barriers. Sona, IC-09229, Naveen, Suvidha, RGC-471, and plg-85 are some of the resistant genotypes which are released as varieties of seeds, while Pusa Mausami, Pusa Sadabahar, Sharad Bahr, PLG-850, and Pusa Navbahr are released for vegetable production that confers resistance against *Alternaria* blight (Sharma et al. 2019). Some of the moderately resistant lines to leaf blight such as HG-75 and HG- 365 were also reported by Saharan and Saharan (2004), which helped in delaying disease progression. This present work is more focused on marker-assisted selection to form a large number of SNP. SSR and InDel markers (Thakur and Randhawa 2018) facilitate marker-assisted selection. For instance, Thakur and Prasad (2020) recently reported silencing of A*cCHS*$_7$ gene in HG 2-20 variety with pSilent CHS7-2 and PARTCHS7. This posttranscriptional silencing jeopardized mycelium and spore development and multiplication of *A. cyamopsidis*, thus reducing disease development and progression.

The second most devastating fungal disease on guar crop is powdery mildew. Three pathogens potentially responsible for this disease are *Oidium taurica*, *Erysiphe polygoni*, and *Sphaerotheca fuliginea*, wherein they deposit white powdery mass on the upper surface of the leaf (Deore and Sawant 2000). It has been observed that not much work in terms of breeding has been done to manage powdery mildew, as approximately 24 varieties have been found to be moderately resistant to this fungus, and GAUG- 63 is one of them (Lodha et al. 2009). This disease is primarily controlled by different cultural methods.

Dry root rot is the third prominent fungal disease on guar. The main pathogens responsible are *Rhizoctonia bataticola* and *Macrophomina phaseolina*, which turn the leaves yellow, followed by their drooping within 2–3 days. The pycnidiospores of this fungus remain viable for 1 year, hence carrying infection over subsequent generations. The genotypes found to be resistant to *R. bataticola* are HG-75 and RCG-1030. Moreover, the high-yielding variety CAZRI 27-1 is also found to be tolerant to powdery mildew (Lodha and Burman 2000). Other promising lines for root rot resistance are GG1, HGS-844, and GAUG-9406 (Pathak 2015).

5 Bacterial Disease

Different types of pathogens are constantly evolving, severely affecting yield and quality of the guar crop. A plethora of different bacterial, viral, and insect pests have been identified on cluster bean crops as mentioned earlier. However, a major chunk of these diseases are controlled by cultural and biological methods. For instance, breeding work is mostly concentrated to eradicate bacterial blight in case of guar. Bacterial blight or commonly known as leaf blight is caused by *Xanthomonas axonopodis* pv. *cyamopsidis*. It is observed to cause an annual loss of (47–50)% in yield at (69–72)% blight intensity under arid conditions (Kumhar et al. 2018) and even extending to (50–70)% loss on complete wipeout. It is a seedborne disease which starts by forming irregular, water-soaked brown spots, surrounded by yellow concentric rings on leaf, which advances to petioles and stems on further congregation of this disease. These spots later on merge to form large patches giving it a blight-type appearance.

Four promising genotypes, which act as important sources of leaf blight resistance, are HG75, G85, G102, and G225, and different resistant lines that impart resistance against this disease are GAUG 9406, GG1, and RGC 1027. Other resistant genotypes such as W522, 590, 647, 648, GP 380, GP 508, and GP 509B also impart disease resistance against blight (Kalaskar et al. 2014). Further, the genotypes GAUG- 0522, HG-75, HG-365, and GG-2 are determined to be potential sources that can be exploited for future breeding programs against bacterial blight and for high yield in guar crop (Kalaskar et al. 2014). It was further observed that resistance in two genotypes, i.e., HG-75 and HG-563, is controlled by two genes; one of them controls the infection, and the second one prevents disease development. Further, have also reported a dominant type of resistance with possible additive, dominance, and epistatic interactions in case of bacterial blight disease. Moreover, significant decreases in carbohydrates and in the activity of peroxidases and polyphenol oxidase enzymes have also been observed, which can also act as potential biomarkers.

Extensive work both in conventional and new molecular breeding has been done to control bacterial blight due to heavy losses incurred every year. Kalaskar et al. (2014) developed polyclonal antibodies against *X. campestris* to manage this disease, and enzyme-linked immunosorbent assays (ELISA) protocol has also been developed for identifying virulent and non-virulent strains of this bacteria. Additionally, 62,146 coding sequences, 5773 microsatellite (SSR) markers, and 3591 SNPS were characterized due to *X. cyamopsidis* interaction with host genotypes (see sources from Radchenko and Sokolova).

Since guar is a self-pollinated crop, distant hybridization is still a challenge due to incompatibility in species. For instance, a wide type of crossing between cluster bean and *C. serrata* proved out to be unsuccessful due to unreceptiveness of stigma to foreign pollen (Kumar et al. 2017). This disadvantage further narrows down the scope to broaden the genetic base by the formation of successful recombinants. In such a scenario, wild relatives act as one of the few sources for natural variation; hence, interspecific cross between cluster bean and *C. senegalensis* and *C. serrata* is exploited to impart disease resistance in the guar crop. Moreover, other conventional breeding techniques such as mutation breeding with physical and chemical mutagens and heterosis breeding are also extensively explored to produce new cultivars and genotypes or to improve various characteristics such as disease susceptibility, tolerance, or resistance, with mixed responses of success. DNA-based markers help to tap genetic diversity in different varieties for different traits. Hence, a considerable number of efforts are diverted to identify SSR and SNP markers from RNA-seq data of root tip culture (Sahu et al. 2020; Thakur and Prasad 2020). Also, random amplified polymorphic DNA (RAPD) technique is also used massively to identify DNA markers, germplasm selection, cultivar identification, etc., owing to their low cost and efficiency. However, till date, no male sterile line has been reported so far in guar crop (Jukanti et al. 2019), which further urges scientists to explore other unconventional sources to overcome such unprecedented challenges Thus, more efforts need to be extended to develop modern tools to impart disease resistance to guar.

6 Breeding for Forage Quality Traits in Guar

Guar has many traditional, edible, and industrial uses other than its usage as a fodder crop. But the oldest use of this crop as fodder is known from the past as it is emerging now as an industrial crop. It is even believed that its crop name is the result of two Sanskrit words *gau* and *ahaar*, which translate to cow and fodder, respectively. Guar is a leguminous fodder crop which is grown mainly in subtropical areas of India and Pakistan. Guar is mainly grown in arid and semi-arid regions of the world because of its less water requirements and heat-tolerant nature, and it is tolerant to frost (Whistler and Hymowitz 1979; Gurmani et al. 2007). It is also grown to a limited extent in other countries such as Australia, Bangladesh, Brazil, Myanmar, South Africa, southwestern USA, and Sri Lanka. There are two commonly available varieties in India: one is dwarf and other is tall. The hairy fodder type is a dwarf one (Kuravadi et al. 2013). Guar has been used for forage purposes in India and Pakistan, but it has a potential to be adopted as forage in other countries for ruminants especially when there is a fodder shortage. Recently, Baath et al. (2020) conducted a study on the evaluation of summer forage capabilities of guar in Southern Great plains in the USA and reported leaf/stem ratio of 1.3–1.5 kg kg^{-1} and guar cultivars generated high N (325 g kg^{-1}), digestible biomass. Leaves and beans of guar are highly nutritious, and it can be preserved as silage, which can be fed to animals later. The major challenge in feeding guar forage to animals is that it contains hydrocyanic acid (HCN) in its fresh state, which is detrimental to animals (Oke 1964). Mainly mature beans are fed to animals as HCN is mostly found in young seeds.

In one study by Rao and Shahid in which they evaluated ten accessions of guar, they have reported average dry matter yield of 9.5 t ha^{-1}, fresh matter yield of 27.6 t ha^{-1}, and mean seed yield of 2.2 t ha^{-1}. In Jhansi, Uttar Pradesh, India, Singh et al. (1997) investigated 526 guar genotypes, reporting averages of 17% crude protein (CP) and 70% digestible dry matter. Guar provides good source of iron when fed, and other micronutrients like zinc and copper are also present in reasonable quantities The most abundant fatty acids in guar bean are cis-linolenic acid, cis-oleic acid, and palmitic (Badr et al. 2014). High protein is present in crop residues (stubble and header rubbish), and it can be utilized to make an excellent hay.

Early selections and cultivar development in guar were relied on characterizing the genotypes according to morphological traits for yield components and quality traits (Dwivedi et al. 1999). Most forage varieties are developed by hybridization, selection, and pedigree method up to the date as mentioned in (Table 2). Although 9% outcrossing has been reported but due to the flower morphology majorly it results in self-pollination and other methods like caging are reported for hybridization (Gill 2009). For hybridization, it has been found that the best time to pollinate the flowers is between 8 and 9 a.m., and there is a 7% success rate of pollination that has been reported (Chaudhary et al. 1974). Making crosses in guar is a very cumbersome process with slow success rate although inheritance of partial male sterility has been studied by Stafford (1989). Various genetic markers have been used in the past to assess the genetic diversity in guar, but studies have been conducted on very small germplasm panels (Tribhuvan et al. 2019; Santonoceto et al. 2019). Large germplasm panels should be assembled to capture the genetic variation, and then those genomic resources can be used to improve the yield and quality of the fodder. As we know from phenotypic evaluation studies, guar is tolerant to heat, drought, and mild salinity but the genetic base of this tolerance has not been much exploited. Transcriptomic analysis studies have been reported to get some insight into these abiotic stress tolerance by Tanwar et al. (2017); Rawal et al. (2017); and Al-Qurainy et al. (2019). Alshameri et al. (2020) recently reported candidate genes regulating heat stress using RNA-seq technology. Segregation population development, QTL mapping, and association studies combined with SNP genotyping can reveal the genomic regions controlling these traits, which, in result, can be used in marker-assisted selection for developing efficient cultivars with high yield and quality. There are not many institutions and breeding programs that are focused on guar as it is an alternative forage. The potential breeding objectives for breeding guar as a forage crop are breeding for high yield, biomass, nutritional and quality, breeding for less HCN content, exploiting germplasm resources for male sterility for easier hybridization, ability to fix nitrogen under drought stress, breeding against biotic stresses (resistant against diseases like *Alternaria* leaf spot and bacterial blight is required for better production) and abiotic stresses, specially frost tolerance, so that crop can be extended to colder parts of the world, incorporating heat, drought, salinity tolerance to better yielding cultivars and developing varieties which fit better to cropping systems. Evaluation and genetic and phenotypic characterization of the germplasm are much needed for a successful breeding program. Kapoor (2014) has evaluated 76 cluster bean genotypes for genetic variability for green fodder yield, and quality traits and various correlation and path coefficients have been reported. The major advancement in guar breeding, which recently has been reported by Grigoreva et al. (2021), is that they developed an SNP set for marker-assisted selection of guar based on reference genome assembly, and six markers has been found to be linked with agronomic trait percentage of pods matured to the harvest date under long light day conditions by genome-wide association study.

Inheritance patterns of various traits was studied by Chaudhary and Lodhi (1981), and they reported that clustering pattern, growth habit, leaf size, and hairiness in clusters are controlled by single gene, and branching behavior is governed

Table 2 Database for forage cluster bean/guar obtained from All India Coordinated Research Project on Forage Crops and Utilization

Sr. no.	Cultivar	Nature of cultivar	Pedigree	Method of breeding/ selection	If cross-pollinated, extent of natural out-crossing (specify in %)	Special attributes if any	Potential yield (green forage yield q/ ha)
1	Agalta Guara–111	A forage variety	Derivative of the cross G.325 × F.S.277	Hybridization followed by pedigree method of selection	Self-pollinated	It is non-branching type	230.0
2.	Ageta Guara–112	An early maturing variety	Derivative of the cross between (325 × F.S.277) × 315	Hybridization followed by pedigree method	Self-pollinated	An early maturity variety	298.0
3.	Bundel Guar-1 (IGFRI 212-1)	Dual-purpose type	Selection from an indigenous material of Rajasthan (Acc. No. B.5/54)	Single plant selection from indigenous material of Rajasthan	Self-pollinated	Resistant to *Xanthomonas cyamopsis*	310.0
4.	Bundel Guar-2 (IGFRI 2395-2)	Dual-purpose type	Selection from indigenous material collected from Jodhpur as accession no. 5496-2	Selection from indigenous material of Rajasthan	Self-pollinated	This is a dual-purpose variety. Cultivated for gum and fodder	280.0
5.	Bundel Guar-3 (IGFRI 1019-1)	Dual-purpose type	An indigenous material collected from Durgapura (Rajasthan) accession no. RGC-19-1	Single plant selection from the accession no. RGC-19-1	Self-pollinated	Besides the production of medium bold seeds with high gum and protein, this variety maintains good green foliage and high protein content	345.0

(continued)

Table 2 (continued)

Sr. no.	Cultivar	Nature of cultivar	Pedigree	Method of breeding/ selection	If cross-pollinated, extent of natural out-crossing (specify in %)	Special attributes if any	Potential yield (green forage yield q/ ha)
6.	Durgajay	Variety dual-purpose type	A local collection from Nagaur (Rajasthan)	Single-plant selection from Nagaur local material	Self-pollinated	Suitable for late sown areas	265.0
7.	Durgapura Safed	Variety dual-purpose type	An indigenous material adapted from Rajasthan local state	Single-plant selection from Rajasthan local material	Self-pollinated	Suitable for late sown areas	250.0
8.	FS-277	Dual-purpose variety	Indigenous material	Pure line selection from indigenous material	Self-pollinated	Fodder cum seed type	242.0
9.	Guar Kranti (RGC-1031)	Variety dual-purpose type	A derivative of the cross between RGC-936 × RGC-986/P-10	Pedigree method of selection	Self-pollinated	Suitable for late sown areas	340.0
10.	Guara-80	Single-cut, dual-purpose type	A derivative of the cross between FS-277 × strain no. 119	Hybridization followed by pedigree method of selection	Self-pollinated (out-crossing 10–12%)	Resistant to *Xanthomonas cyamopsis*	370.0
11.	HFG 156	Fodder variety	Selection from germplasm	Selection	Self-pollinated	–	110–125 q/ha dry fodder yield
12.	HFG-119	Variety dual-purpose type	Selection	Pure line	Self-pollinated	Dual-purpose type	350.0

(continued)

Table 2 (continued)

Sr. no.	Cultivar	Nature of cultivar	Pedigree	Method of breeding/ selection	If cross-pollinated, extent of natural out-crossing (specify in %)	Special attributes if any	Potential yield (green forage yield q/ ha)
13.	HG 2-20	Grain variety	HG 365 × FS 277	Pedigree selection	Self-pollinated	Moderately tolerant to bacterial blight disease and resistant to *Alternaria* blights and root rot	Grain yield, 21–23 q/ ha
14.	HG-884	Grain variety	HG 75 × FS 296	Pedigree selection	Self-pollinated	Moderately tolerant to bacterial blight disease and resistant to *Alternaria* blights and root rot	Grain yield, 20–22 q/ ha
15.	HG-182	Dual-purpose type	Acc no. HFG-182	Single plant selection from the genetic line acc. No. HFG-182	Self-pollinated	–	295.0
16.	HG-75	Dual-purpose type	Selection from local genetic material	Pure line	Self-pollinated	Suitable for dry conditions	320.0
17.	Maru Guar (2470/12)	Dual-purpose type	Derivative of the germplasm obtained from NBPGR, New Delhi—110,012	Selection from the germplasm	Self-pollinated	This variety is most suitable for the late sown conditions	225.0

by two-gene inheritance. Santhosha et al. (2017) evaluated 43 guar genotypes and reported medium to high broad-sense heritability for traits such as number of branches at 45 days after sowing (DAS), plant height at 90 DAS, number of branches at 90 DAS, stem girth, plant spread (E-W), plant spread (N-S), number of pods per cluster, number of clusters per plant, number of pods per plant, pod length, vegetable pod yield per plant, vegetable pod yield per plot, weight per ten pods, hundred seed weight, seed yield per plant, seed endosperm gum content, and seed protein content, which means that these traits can be manipulated in the breeding programs.

There is very little known about domestication of guar and less wild genetic resources available for this crop, which is a major bottleneck in diversity studies and eventually for the breeding cultivars (Krishnan et al. 2011). Mutation breeding is a potential tool to generate the variation although there are some studies in which they generated mutants for various agronomic traits (Vig 1969; Rao and Rao 1982; Singh 1986; Arora and Pahuja 2008). But more thorough research is required, and mutation breeding can be utilized to generate new efficient cultivars with high yield and fodder quality. Central Arid Zone Research Institute, Jodhpur, and National Bureau of Plant Genetic Resources (NBPGR), New Delhi, have collectively evaluated 4869 germplasm lines for various morphological and yield traits, but genotyping of the germplasm is required to utilize them better in breeding programs. Two wild species, *C. serrata* and *C. senegalensis*, have been introduced to India from the USA (Bhatt et al. 2017). Failed attempt was made by using traditional breeding approaches to transfer the earliness trait from wild species to cultivated one (Sandhu 1988; Kumar and Rodge 2012). This clearly shows the reason for less progress in guar breeding in general, and the deployment of genomic and molecular breeding approaches will hasten the progress of guar forage breeding to efficiently use all the resources.

7 Genome Editing

As discussed earlier, the narrow genetic base in guar crop has compelled scientists to look for different alternative ways to create variability in germplasm. Different approaches have been harnessed and scrutinized to broaden the germplasm such as hybridization, marker-assisted selection, and mutation breeding. Mutation breeding, with the help of physical and chemical mutagens, has been explored long before in guar with mixed responses of success depending on type and number of mutations being used each time (Patil and Rane 2015). The mutation breeding is however, one of the ways to artificially induce mutations in DNA by additions, deletion, translocation of nucleotides. Some treatments have showcased exemplary success in the form of increased number of pods, seeds, branches, reducing plant size, early maturity, and flowering (Arora and Pahuja 2008). However, some mutations have also proved to be detrimental, causing abnormalities, laggards, and fragments, rendering breeding unsuccessful. For instance, Pusa Navbahar variety showed beneficial effect at low doses while inhibitory at high dose of 0–60 kR of X-rays (Lavanya et al. 2020). These physical and chemical mutations might not be an efficient source, due to some adverse effects seen at later generations (Ishii and Araki 2016).

The increasing influence of technology in biological fields has created high-tech throughput discoveries such that it is possible today to isolate, synthesize, and mutate DNA and even introduce foreign DNA to genomes. One such technique that is rapidly evolving in the field of biotechnology is genome editing, which involves cutting the nucleases with the help of nucleases such as transcription activator-like effector nucleases (TALEN), zinc-finger nucleases (ZFN) and clustered regularly interspaced short palindromic repeats (CRISPR) (Petolino 2015; Malzahn et al. 2017). There are two components associated with CRISPR genome editing, i.e., RNA guide and CRISPR-associated Cas9 protein. Single RNA guide directs the Cas9 protein to bind at a particular location, thus causing a break at that particular junction. After binding, nuclease causes double-strand break at targeted sites; these sites are generally active for non-homologous end joining (NHEJ) or homology-directed repair (HDR) methods. Usually, a target-specific homologous sequence of interest is inserted at the site of break, thus allowing modification of DNA sequence by choice (Arnoult et al. 2017; Zhang et al. 2021). Since CRISPRs are more amenable, creating space for flexible editing of the genome, they are touted to be one of the reliable methods of breeding for crop improvement.

Additionally, the burgeoning market of guar due to xanthan gum production urges for the development and strategizing of different breeding techniques to increase the yield (Hu et al. 2019). Although no such genome editing project has been undertaken in guar so far except for mutation breeding, it invests in itself a huge potential yet to be explored. Malik (2013) further proposed the tissue-specific CRISPR for guar as a useful crop improvement method, which will also help to eliminate any pleiotropic effect as well. This will help to study tissue-specific genes without harming neighboring cells, leading to a more precise and localized improvement of particular traits of interest (Singha et al. 2021). Malik (2013) suggested it as to be one of the beneficial strategies for grain trait improvement in favorable as well as non-favorable conditions. The genes of interest can be inserted in *Xanthomonas* for delivery into guar cultivars. This further necessitates the requirement of efficient strains of *Rhizobium* for hybrid gum production. Genome editing is, hence, speculated to improve the cultivars of guar for disease resistance, quality, yield and gum production.

8 Prospectus of Genomic Selection and Data Science

Traditional linkage and association mapping studies have been used in guar to a limited extent to explore the genetic architecture of most of the agronomic, quality, biotic, and abiotic stress-related traits, but these techniques have not achieved much success due to the complex nature of quantitative traits and limited resource availability (Tribhuvan et al. 2019; Grigoreva et al. 2021). However, with the advent of recent genomic techniques, genomic selection (GS) offered a potential advantage for studying complex traits owing to reduction in genotyping cost and maintenance of diversity panels in guar (Saini et al. 2022). GS is a breeding approach that uses the whole genome-wide markers to predict the genomic-estimated breeding values (GEBVs) of the plants. The best candidates can be then selected based on these GEBVs. To make predictions, GS uses the previous year's phenotypic and genotypic datasets for a set of lines with statistical models to train the GS model, which is subsequently used to predict the performance of lines with only genotypic information. The model's performance is assessed by the Pearson correlation coefficient between the predicted GEBVs and actual phenotypic values. The predicted GEBVs aid the breeders either to use them as a parent in the hybrid development or next-generation advancement in the breeding program (Fig. 1) (Sandhu et al. 2022a, b). GS has proven superior to traditional phenotypic selection and marker-assisted selection to improve quantitative traits and accelerate crop improvement in several crop species (Sandhu et al. 2021d). GS uses the linkage disequilibrium between the molecular marker and associated QTL for predicting its effect. Genetic gain can be increased with the help of GS due to off-season selection. Furthermore, the development of better statistical models accounting for additive, dominance, and epistasis variance components opens avenues for predicting performances and better resource allocation (Kaur et al. 2021).

Being a minor crop, guar has not gotten significant benefits in comparison to rice, wheat, and maize from the various policymakers and government officials to develop the upstream science (Samantara et al. 2021). This is particularly seen for GS and genomics, where funding plays an important role in studying many genotypes in a diverse set of environments. Rapid decay of linkage disequilibrium warrants dense genome-wide markers, making the genotyping pipeline more costly. GS prediction accuracies are also affected by high rates of segregation distortion observed in the specific training population. Three-way and top-cross hybrids are more commonly grown in Africa and single-cross hybrids in Asia. However, the GS model development for parental lines demonstrating different heterotic combinations for hybrids is quite challenging. In the coming years, this issue could be resolved with the adoption or development of GS models that can account for genotype by environment interactions and precisely estimate the general combining ability and specific combining ability for different heterotic pools and agro-ecological zones (Merrick et al. 2022).

The ability to analyze and understand data is getting critical for the new inventions and findings. Data science is a multidisciplinary field encompassing computer science, statistics, mathematics, data visualization, domain knowledge, the craft of problem development, artificial intelligence, and machine and deep learning. The experience in all these domains helps data scientists working on genomics for crafting a problem and engineering of the solutions systematically (Poland and Rife 2012; Sandhu et al. 2021a). Our era has witnessed a tremendous development in the plant genomics, resulting from the development of high-throughput genotyping platform with a very less cost, which is even reducing further by new inventions (Poland and Rife 2012). Although genomics data is increasing, it is imperative to develop and integrate some data mining and analysis tools for predicting and

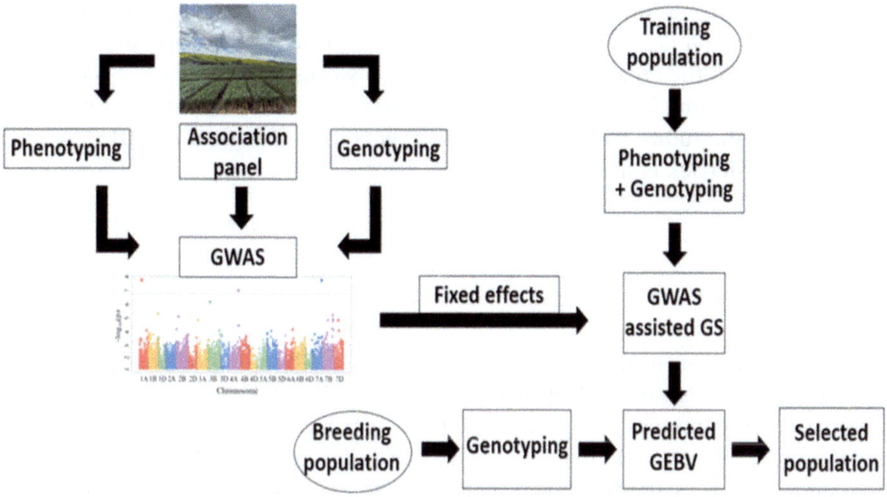

Fig. 1 Outline for the utilization of the results from genomic selection and association mapping for genomic-assisted breeding in the crop plants

explaining the information contained in the sequences. There is a considerable gap between the flow of information between the genome sequence and terminal plant phenotype. There are recent inventions in association analysis and prediction of plant phenotypes, resulting in lowering the bridge between these two domains. The association analysis involves looking at variation in the genome sequence and linking it to the actual plant phenotype using various mixed linear models and machine and deep learning models (Sandhu et al. 2020, 2021b). The prediction of plant phenotypes involves the use of whole-genome sequence information for predicting the real phenotypes by training the model on previous years dataset using the machine and deep learning models.

Machine learning (ML) is a division of artificial intelligence, which is getting attention from plant scientists for the exploitation of massive data in plant genomics. With the increase in genomics dataset, there is a problem with the extraction of useful information without good algorithms (Singh et al. 2021; Sandhu et al. 2021c). In this regard, ML acts a technical basis for digging into the extraction of useful information from the genomics dataset. A major development in the field of ML includes learning information from the data without explicitly trained to do so using the deep neural networks, known as deep learning (DL). The critical difference between ML and DL is that they are more flexible and have a much higher capacity. There are millions of trainable parameters to train the model, and the optimum choice depends upon the dataset used. DL models automatically learn the information from the dataset without any handcrafting. DL models improve the prediction abilities of the models, requiring the collection of large datasets for training the models. The starting point of DL models includes the use of neurons, which learn the information from the input dataset, and weight is associated with each neuron and ultimately performing a nonlinear transformation to provide an output value. The output of each neuron acts as input for the next layer's neurons, and this ultimately results in the creation of dense neural networks. Recently, there have been various studies using DL models in plant genomics and some good opinion papers talking about the future use of DL models in genomics (Singh et al. 2018; Sandhu et al. 2022a; Merrick et al. 2022).

9 Conclusion

This chapter brings forth the elaborative aspects of the comprehensive breeding techniques and programs in the production of guar. The application and the development of novel genomics and genetic tools in guar have been deeply apprehended and discussed, which undoubtedly hold a lot of future potential. Phenotypic evaluation studies have evidently shown guar to be tolerant to heat, drought, and mild salinity. It is also capable of fixing nitrogen in the soil as well as enhancing soil fertility. As a result, techniques involving segregation population development, QTL mapping, and SNP genotyping can be used for unveiling the genomic regions, controlling the useful traits in guar; subsequently, it is followed by the production of efficient cultivars having excellent yield and quality. Apart from this, genome editing has proved to be useful in isolating, synthesizing, mutating, and introducing foreign DNA to genomes, hence making it a beneficial strategy for grain trait improvement in unfavorable conditions too, but surprisingly, it has not been implemented so far. Contemporarily, scientists have mainly emphasized on the traditional breeding methodologies, which have consequently paved the way for the exploration of prospective techniques involving genomic selection, genome editing, and speed breeding. These will, in turn, not only reduce the time needed to produce a cultivar but also aid the smart and efficient gene selection. With the contemplation of scientific data, the gap between the flow of information between genomic sequence and terminal plant phenotype can be traversed considerably. Therefore, the integration of all these strategies and practices with an increased focus on the non-conventional breeding tools will explicitly lead to the development and implementation of the various genetic resources, thus transcending the present-day scenario in the years to come.

References

Alshameri A, Al-Qurainy F, Gaafar AR, Khan S, Nadeem M, Alansi S (2020) Identification of heat-responsive genes in guar (Cyamopsis tetragonoloba (L.) Taub). Int J Genomics 2020:3126592

Al-Qurainy F, Alshameri A, Gaafar AR, Khan S, Nadeem M, Alameri AA, Tarroum M, Ashraf M (2019) Comprehensive stress-based de novo transcriptome assembly and annotation of guar (Cyamopsis tetragonoloba (L.) Taub.): an important industrial and forage crop. Int J Genomics 2019:1

Ansari A, Singh SP, Lade S, Kumar YH, Ranade SA (2016) Genetic diversity clusters in germplasm of cluster bean (Cyamopsis tetragonoloba L. Taub.), an important food and an industrial legume crop. J Agric Sci Technol 18(5):1393–1406
Arnoult N, Correia A, Ma J, Merlo A, Garcia-Gomez S, Maric M et al (2017) Regulation of DNA repair pathway choice in S and G2 phases by the NHEJ inhibitor CYREN. Nature 549(7673):548–552
Arora RN, Pahuja SK (2008) Mutagenesis in guar (Cyamopsis tetragonoloba (L.) Taub.). Plant Mutation Reports 2(1):7–9
Baath GS, Northup BK, Gowda PH, Rocateli AC, Singh H (2020) Summer forage capabilities of tepary bean and guar in the southern Great Plains. Agron J 112(4):2879–2890
Badr SE, Abdelfattah MS, El-Sayed SH, Abd El-Aziz AS, Sakr DM (2014) Evaluation of anticancer, antimycoplasmal activities and chemical composition of guar (Cyamopsis tetragonoloba) seeds extract. Res J Pharm Biol Chem Sci 5(3):413–423
Bhatt RK, Jukanti AK, Roy MM (2017) Cluster bean [Cyamopsis tetragonoloba (L.) Taub.], an important industrial arid legume: a review. Legume Res Int J 40(2):207–214
Brahmi P, Bhat KV, Bhatnagar AK (2004) Study of allozyme diversity in guar [Cyamopsis tetragonoloba (L.) Taub.] germplasm. Genet Resour Crop Evol 51(7):735–746
Chaudhary BS, Lodhi GP (1981) Studies on the inheritance of five qualitative characters in clusterbean [Cyamopsis tetragonoloba (L.) Taub.]. Euphytica 30:161–165
Chaudhary SR, Sindhu SS (2016) Growth stimulation of clusterbean (Cyamopsis tetragonoloba) by coinoculation with rhizosphere bacteria and rhizobium. Legume Res Int J 39(OF):1003–1012. https://doi.org/10.18805/lr.v0iOF.8605
Chaudhary BS, Paroda RS, Solanki KR (1974) A new crossing technique in cluster bean (Cyamopsis tetragonoloba [L.] Taub.). Curr Sci 43
Chudzikowski RJ (1971) Guar gum and its applications. J Soc Cosmet Chem 22(1):43
Dennis RE, Johnson DL (1981) Growing guar in Arizona
Deore PB, Sawant DM (2000) Management of guar powdery mildew by trichoderma spp. culture filtrates. J Maharashtra Agric Univers (India)
Dwivedi NK, Bhandari DC, Bhatnagar N, Dabas BS (1999) Characterization of clusterbean [Cyamopsis tetragonoloba (L) Taub.] germplasm for yield and quality traits. Ann Arid Zone 38:151–156
Gill SL (2009) Evaluation of reciprocal hybrid crosses in guar. Texas Tech University, Texas
Gopala Krishnan S, Dwivedi NK, Singh JP (2011) Primitive weedy forms of guar, Adak guar: possible missing link in the domestication of guar [Cyamopsis tetragonoloba (L.) Taub.]. Genet Resour Crop Evol 58(7):961–966. https://doi.org/10.1007/s10722-011-9728-z
Grigoreva E, Ulianich P, Ben C, Gentzbittel L, Potokina E (2019) First insights into the guar (Cyamopsis tetragonoloba (L.) Taub) genome of the 'Vavilovskij 130' accession, using second and third-generation sequencing technologies. Russian. J Genet 55(11):1406–1416
Grigoreva E, Barbitoff Y, Changalidi A et al (2021) Development of SNP set for the marker-assisted selection of guar (Cyamopsis tetragonoloba (L.) Taub.) based on a custom reference genome assembly. Plants 10. https://doi.org/10.3390/plants10102063
Gurmani ZA, Shafeeque S, Zahid MS, Imran M (2007) Guar (Cyamopsis tetragonoloba L.); a low-input and multipurpose crop of Pakistan. Sc Technol Dev 26(2):45–48
Hu H, Wang H, Zhang Y, Kan B, Ding Y, Huang J (2019) Characterization of genes in guar gum biosynthesis based on quantitative RNA-sequencing in guar bean (Cyamopsis tetragonoloba). Sci Rep 9(1):1–10
Hymowitz T (1972) The trans-domestication concept as applied to guar. Econ Bot 26(1):49–60
Ishii T, Araki M (2016) Consumer acceptance of food crops developed by genome editing. Plant Cell Rep 35(7):1507–1518
Jukanti AK, Pathak R, Mushyam C (2019) Cluster bean [Cyamopsis tetragonoloba (L.) Taub] breeding. In: Advances in plant breeding strategies: legumes. Springer, pp 113–149
Kaila T, Chaduvla PK, Rawal HC, Saxena S, Tyagi A, Amitha MSV, Solanke AU, Kalia P, Sharma TR, Singh NK, Gaikwad K (2017) Chloroplast genome sequence of clusterbean (Cyamopsis tetragonoloba L.) genome structure and comparative analysis. Genes 8(9):212

Kalaskar SR, Acharya S, Patel JB, Sheikh WA, Rathod AH, Shinde AS (2014) Genetic diversity assessment in clusterbean (Cyamopsis tetragonoloba (L.) Taub.) by RAPD markers. J Food Legumes 27(2):92–94

Kaur B, Sandhu KS, Kamal R et al (2021) Omics for the improvement of abiotic, biotic, and agronomic traits in major cereal crops: applications, challenges, and prospects. Plants 10. https://doi.org/10.3390/plants10101989

Krishnan SG, Dwivedi NK, Singh JP (2011) Primitive weedy forms of guar, Adak guar: possible missing link in the domestication of guar [Cyamopsis tetragonoloba (L.) Taub.]. Genet Resour Crop Evol 58(7):961–966

Kumar D, Rodge AB (2012) Status, scope and strategies of arid legumes research in India-a review. J Food Legumes 25(4):255–272

Kumar S, Joshi UN, Sangwan S, Yadav R, Singh JV, Saini ML (2015) Biomolecular characterization of guar (Cyamopsis tetragonoloba) genotypes along with wild species, C. serrata and C. senegalensis. Plant Syst Evol 301(4):1249–1262

Kumar S, Modi AR, Parekh MJ, Mahla HR, Sharma R, Fougat RS, Yadav D, Yadav NR, Patil GB (2017) Role of conventional and biotechnological approaches for genetic improvement of cluster bean. Ind Crop Prod 97:639–648

Kumhar DR, Meena AK, Meena PN (2018) Efficacy of different management modules against bacterial blight of clusterbean under epiphytotic conditions. J Pharm Phytochem 7(3):1505–1509

Kuravadi NA, Verma S, Pareek S, Gahlot P, Kumari S, Tanwar UK, Bhatele P, Choudhary M, Gill KS, Pruthi V, Tripathi SK (2013) Guar: an industrial crop from marginal farms. In: Agricultural sustainability: progress and prospects in crop research

Kuravadi NA, Tiwari PB, Tanwar UK, Tripathi SK, Dhugga KS, Gill KS, Randhawa GS (2014) Identification and characterization of EST-SSR markers in cluster bean (Cyamopsis spp.). Crop Sci 54(3):1097–1102

Lavanya HN, Smaranika Mishra M, Anjanappa TS, Aghora MS, Keshava Rao V, Anjaneya Reddy B (2020) Biological response, efficiency and effectiveness of gamma irradiation as mutagen on Pusa Navbahar cluster bean. Indian J Hortic 77(3):478–483

Lodha SA, Burman U (2000) Efficacy of composts on nitrogen fixation, dry root rot (Macrophomina phaseolina) intensity and yield of legumes. Indian J Agric Sci 70(12)

Lodha S, Kumar R, Mawar R (2009) Diseases and their management in arid legumes: present status. In: Legumes in dry areas. Scientific Publisher, Jodhpur, p 550

Malani S, Ravelombola W, Manley A, Pham H, Brown M, Rahman MM (2025) Genome-wide association study for nodule traits in guar. Curr Issues Mol Biol 47(3):151

Malik VS (2013) Use of genome editing technologies for improving castor bean and guar. J Plant Biochem Biotechnol 22(4):357–358

Malzahn A, Lowder L, Qi Y (2017) Plant genome editing with TALEN and CRISPR. Cell Biosci 7(1):1–18

Merrick LF, Herr AW, Sandhu KS et al (2022) Utilizing genomic selection for wheat population development and improvement. Agronomy 12:522. https://doi.org/10.3390/agronomy12020522

Monier-Williams M (1964) Sanskrit-english dictionary. Рипол Классик

Morris JB (2010) Morphological and reproductive characterization of guar (Cyamopsis tetragonoloba) genetic resources regenerated in Georgia, USA. Genet Resour Crop Evol 57(7):985–993. https://doi.org/10.1007/s10722-010-9538-8

Mudgil D, Barak S, Khatkar BS (2014) Guar gum: processing, properties and food applications-a review. J Food Sci Technol 51(3):409–418. https://doi.org/10.1007/s13197-011-0522-x

Oke OL (1964) Hydrocyanic acid content and nitrogen-fixing capacity of guar. Nature 204(4956):405–406

Pathak R (2015) Plant protection. In: Clusterbean: physiology, genetics and cultivation. Springer, pp 87–106

Pathak R, Singh SK, Singh M, Henry A (2010) Molecular assessment of genetic diversity in cluster bean (Cyamopsis tetragonoloba) genotypes. J Genet 89(2):243–246

Patil BM, Rane GM (2015) Gamma radiation induced chlorophyll mutations in clusterbean (Cyamopsis tetragonoloba (L.) Taub) Var. NCB-12. Int J Allied Pract Res Rev 2:75–85

Petolino JF (2015) Genome editing in plants via designed zinc finger nucleases. In Vitro Cell Dev Biol Plant 51(1):1–8

Poland JA, Rife TW (2012) Genotyping-by-sequencing for plant breeding and genetics. Plant Genome J 5:92. https://doi.org/10.3835/plantgenome2012.05.0005

Rao S, Rao D (1982) Studies on the effect of X-irradiation on Cyamopsis tetragonoloba (L.) Taub. Proc Indian Nat Sci Acad (Biol Sci) 48:410–415

Ravelombola W, Manley A, Adams C, Trostle C, Ale S, Shi A, Cason J (2021) Genetic and genomic resources in guar: a review. Euphytica 217(11):1–19

Rawal H, Kumar S et al (2017) High quality unigenes and microsatellite markers from tissue specific transcriptome and development of a database in clusterbean (Cyamopsis tetragonoloba, L. Taub). Genes 8(11):313

Saharan MS, Saharan GS (2001) Influence of environmental factors on Alternaria blight development on clusterbean. Indian Phytopathol 54(2):188–192

Saharan MS, Saharan GS (2004) Influence of weather factors on the incidence of Alternaria blight of cluster bean (Cyamopsis tetragonoloba (L.) Taub.) on varieties with different susceptibilities. Crop Prot 23(12):1223–1227

Sahu S, Sahu TK, Ghosal S, Gaikwad K, Rao AR (2020) Computational analysis of SNPs and INDELs in cluster bean cultivars involved in multiple trait expression. Indian J Genet 80(2):179–185

Saini DK, Chopra Y, Singh J et al (2022) Comprehensive evaluation of mapping complex traits in wheat using genome-wide association studies. Mol Breeding 42:1. https://doi.org/10.1007/s11032-021-01272-7

Samantara K, Shiv A, de Sousa LL et al (2021) A comprehensive review on epigenetic mechanisms and application of epigenetic modifications for crop improvement. Environ Exp Bot 104479. https://doi.org/10.1016/j.envexpbot.2021.104479

Sandhu HS (1988) Interspecific hybridization studies in genus Cyamopsis. PhD Thesis: CCSHAU, Hissar, India

Sandhu KS, Lozada DN, Zhang Z et al (2020) Deep learning for predicting complex traits in spring wheat breeding program. Front Plant Sci 11:613325. https://doi.org/10.3389/fpls.2020.613325

Sandhu K, Patil SS, Pumphrey M, Carter A (2021a) Multitrait machine- and deep-learning models for genomic selection using spectral information in a wheat breeding program. Plant Genome 14:e20119. https://doi.org/10.1002/tpg2.20119

Sandhu KS, Aoun M, Morris CF, Carter AH (2021b) Genomic selection for end-use quality and processing traits in soft white winter wheat breeding program with machine and deep learning models. Biology (Basel) 10. https://doi.org/10.3390/biology10070689

Sandhu KS, Mihalyov PD, Lewien MJ et al (2021c) Combining genomic and phenomic information for predicting grain protein content and grain yield in spring wheat. Front Plant Sci 12:613300. https://doi.org/10.3389/fpls.2021.613300

Sandhu KS, Mihalyov PD, Lewien MJ et al (2021d) Genomic selection and genome-wide association studies for grain protein content stability in a nested association mapping population of wheat. Agronomy 11:2528. https://doi.org/10.3390/agronomy11122528

Sandhu KS, Merrick LF, Sankaran S et al (2022a) Prospectus of genomic selection and phenomics in cereal, legume and oilseed breeding programs. Front Genet 12. https://doi.org/10.3389/fgene.2021.829131

Sandhu KS, Patil SS, Aoun M, Carter AH (2022b) Multi-trait multi-environment genomic prediction for end-use quality traits in winter wheat. Front Genet 13:831020. https://doi.org/10.3389/fgene.2022.831020

Santhosha GR, Shashikanth E, Gasti VD, Prabhuling G, Rathod VD, Mulge R (2017) Genetic variability studies in cluster bean [Cyamopsis tetragonaloba (L.) Taub.] for growth, yield and quality parameters. Legume research-an. Int J 40(2):232–236

Santonoceto C, Mauceri A, Lupini A, Gresta F, Chiera E, Sunseri F, Mercati F, Anastasi U (2019) Morpho-agronomic characterization and genetic variability assessment of a guar germplasm collection by a novel SSR panel. Ind Crop Prod 138:111568

Sharma P, Sharma V, Kumar V (2014) Genetic diversity analysis of cluster bean [Cyamopsis tetragonoloba (L..) Taub] genotypes using RAPD and ISSR markers. J Agric Sci Technol 16(2):433–443

Sharma S, Singh R, Kumar A (2019) Screening of cluster bean genotypes against Alternaria blight (Alternaria cucumerina var. Cyamopsidis) in field condition. J Pharm Phytochem 8(3):596–599

Singh VP (1986) Induced high yielding mutants in clusterbean. Indian J Agric Sci 56:695–700

Singh UP, Tripathi SN, Natrajan S (1997) Basic information about crop guar or cluster bean (Cyamopsis tetragonoloba L. Taub.). Crop Profi leGUAR. http://www.igfri.res.in/crop%20profile.htm. Accessed 20 Aug 2012

Singh AK, Ganapathysubramanian B, Sarkar S, Singh A (2018) Deep learning for plant stress phenotyping: trends and future perspectives. Trends Plant Sci 23:883–898. https://doi.org/10.1016/j.tplants.2018.07.004

Singh A, Jones S, Ganapathysubramanian B et al (2021) Challenges and opportunities in machine-augmented plant stress phenotyping. Trends Plant Sci 26:53–69. https://doi.org/10.1016/j.tplants.2020.07.010

Singha DL, Das D, Sarki YN, Chowdhury N, Sharma M, Maharana J, Chikkaputtaiah C (2021) Harnessing tissue-specific genome editing in plants through CRISPR/Cas system: current state and future prospects. Planta 255(1):28. https://doi.org/10.1007/s00425-021-03811-0

Tanwar UK, Pruthi V, Randhawa GS (2017) RNA-Seq of guar (Cyamopsis tetragonoloba, L. Taub.) leaves: De novo transcriptome assembly, functional annotation and development of genomic resources. Front Plant Sci 8:91

Thakur O, Prasad R (2020) Engineering resistance to Alternaria cyamopsidis by RNAi mediated gene silencing of chitin synthase export chaperone CHS7 in guar. Physiol Mol Plant Pathol 112:101541

Thakur O, Randhawa GS (2018) Identification and characterization of SSR, SNP and InDel molecular markers from RNA-Seq data of guar (Cyamopsis tetragonoloba, L. Taub.) roots. BMC Genomics 19(1):1–14

Tribhuvan KU et al (2019) Identification of genomic SSRs in cluster bean (Cyamopsis tetragonoloba) and demonstration of their utility in genetic diversity analysis. Ind Crop Prod 133:221–231. https://doi.org/10.1016/j.indcrop.2019.03.028

Tyagi A, Sandhya SP, Saxena S, Sharma R, Amitha MSV, Solanke AU, Singh NK, Sharma TR, Gaikwad K (2019) The genome size of clusterbean (Cyamopsis tetragonoloba) is significantly smaller compared to its wild relatives as estimated by flow cytometry. Gene 707:205–211

Verma S, Dubey N, Dhugga KS, Gill KS, Randhawa GS (2025) Cluster bean: from garnering industrial importance to molecular research interventions for the improvement of commercially viable traits. S Afr J Bot 178:307–317

Vig BK (1969) Studies with Co60 radiated guar [Cyamopsis tetragonoloba (L.) Taub.]. Ohio J Sci 69:18

Whistler RL, Hymowitz T (1979) Guar: agronomy, production, industrial use and nutrition. Purdue University Press, West Lafayette

Zanoni TA, Whistler RL, Hymowitz T (1980) Guar: agronomy, production, industrial use, and nutrition. Brittonia 32(2):143. https://doi.org/10.2307/2806779

Zhang S, Zhang R, Gao J, Song G, Li J, Li W et al (2021) CRISPR/Cas9-mediated genome editing for wheat grain quality improvement. Plant Biotechnol J 19(9):1684

Guinea grass (*Megathyrsus maximus*): Crop Diversity and Genetic Improvement

Subhash Chand, Vijay Kamal Meena, Tejveer Singh, Ashlesha Singla, Meenakshi Goyal, Manoj Kumar Patel, Gayatri Kumawat, Prashant Bisen, and Ajoy Kumar Roy

1 Introduction

Guinea grass (*Megathyrsus maximus* (Jacq.) B.K. Simon & W.S.L. Jacobs), formerly known as *Panicum maximum* (Jacq), is an important perennial range grass in tropical countries, popularly known for its high forage yield, herbivory tolerance, and palatability (Sukhchain 2010; Maciel et al. 2018). It pertains to the botanical family Poaceae, specifically within the sub-family Panicoideae and the tribe Paniceae (Muir and Jank 2004). It is a tetraploid species with a chromosome number of 2n = 4x = 32. Tetraploid guinea grass species are dominant across the globe and mostly apomictic in nature; however, in East Africa, diploid guinea grass (2n = x = 16) is found, which reproduces sexually. It has a basic chromosome

S. Chand (✉) · T. Singh
ICAR-Indian Grassland and Fodder Research Institute, Jhansi, Uttar Pradesh, India

V. K. Meena
Agriculture University, Jodhpur, Rajasthan, India

A. Singla · M. Goyal
Punjab Agricultural University, Ludhiana, Punjab, India

M. K. Patel
ICAR-Indian Agricultural Research Institute, New Delhi, Delhi, India

G. Kumawat
Livestock Feed Resource Management and Technology Center, Rajasthan University of Veterinary and Animal Sciences, Bikaner, Rajasthan, India

P. Bisen
Acharya Narendra Deva University of Agriculture & Technology, Kumarganj, Uttar Pradesh, India

A. K. Roy
ICAR-Indian Grassland and Fodder Research Institute, Jhansi, Uttar Pradesh, India

S. H. Wani, S. S. Gosal (eds.), *Accelerated Plant Breeding, Volume 5*,
https://doi.org/10.1007/978-3-032-04527-0_7

number of 8 (x). In addition, sexual diploid, pentaploid, and hexaploid species are also reported (Combes and Pernes 1970). Despite being an underexplored fodder crop in India, guinea grass has the potential to significantly increase green fodder yield and crop coverage. The average green fodder yield of guinea grass varies from 40 t/ha (Variety PGG-616) to 360 t/ha (Variety TNGG 062), making it the most productive and profitable fodder crop under the silvopastoral system in India. In 2021, it demonstrated a productivity of 16.1 t/ha from three cuttings and provided a benefit-cost ratio of 3.73 in association with *Acacia mangium*, reaffirming its economic benefits.

Guinea grass is a versatile crop with a wide range of uses, which is sure to pique the interest of the readers. It is primarily used as a cultivated forage and pasture and is also effective for silage production in the dairy industry. Beyond its forage uses, it has non-forage applications such as roofing material, crop mulching, soil stabilization, and soil cover. In Kenya and Tanzania, it is even used to make brooms and brushes. Furthermore, it contributes to the ecosystem by providing seeds as a food source for birds, such as *Munias*. Fresh guinea grass has a variable crude protein (CP) content ranging from 5.0 to 5.6%, while guinea grass silage maintains a CP content within the range of 5.0 to 5.5%. Notably, the in vitro Dry Matter Digestibility (IVDMD) varies from 56.9 to 87.7%.

Guinea grass, a clump-forming perennial, thrives in warm, frost-free regions with an annual rainfall exceeding 900 mm (Aganga and Tshwenyane 2004). It exhibits extensive adaptability to various soil and climatic conditions, commonly found along roadsides and riverbanks, spanning altitudes from 0 to 2400 meters. Typically, it is grown in light-textured soils, such as loams and sandy loams, which exhibit excellent drainage properties (Rhodes et al. 2022). Guinea grass generally exhibits tolerance to dry conditions, but extended drought periods exceeding 4 months have been observed to result in higher mortality rates. Moreover, it can survive mild frost and maintain some productivity in cooler climates (Islam and Thakuria 2002). Naturally, Guinea grass grows in grasslands and savannas, mostly beneath woody vegetation or in riparian zones. It has the ability to populate open areas under adequate soil moisture and limited herbivory (Rhodes et al. 2022). It is also known for shade and salinity tolerance.

Guinea grass can be reproduced through both seed and vegetative means such as stem and root cuttings. It has a high seed multiplication rate; a single plant is capable of producing up to 9000 seeds. However, due to seed shattering and small seed size, actual seed yield remains low. Despite a high seed multiplication ratio (SMR), propagation through seed is challenging due to low seed germination and prolonged seed dormancy, which persists for over 3 years (Mishra et al. 2008; Soti and Thomas 2022). However, seed viability may exceed 80% if seeds are gradually dried to 10% moisture (Muir and Jank 2004).

2 Origin and Distribution

M. maximus originated in Africa and was later introduced globally for hay and silage production, resulting in significant ecological impacts. It is native to tropical Africa and is now widely naturalized in the tropics. It has been introduced in the tropics around the world. It became an invasive species in tropical and warm temperate areas, including Sri Lanka, the United States, India, Australia, and Brazil (Daehler and Carino 1998; Sarkar et al. 2018; Roy et al. 2020). By the year 1915, *M. maximus* had established itself in the United States, Mexico, Guatemala, Honduras, El Salvador, Nicaragua, Costa Rica, Panama, Trinidad, Bermuda, Bahamas, Cuba, Jamaica, and Haiti (Hitchcock and Chase 1917; Soti and Thomas 2022).

It was previously known as *Urochloa maxima* and *Panicum maximum*. It was moved to the genus *Megathyrsus* in 2003. Its natural habitat is in open grasslands, woodland, and shady places, which are located between 16.3 °N and 28.7 °S. It thrives well under annual rainfall above 1000 mm, but the dry period should not be more than 4 to 5 months. The average annual daily temperature should range from 19.1 to 22.9 °C. Small types are more tolerant of cooler temperatures than tall types. It prefers well-drained, moist, fertile soils. It is tolerant of light frost, low soil pH if drainage is good, and high Al^{3+} saturation. It is well adapted to sloping, cleared land in rainforest areas.

Guinea grass constitutes an agamic complex alongside two botanical species, *Panicum infestum* (Anders) and *P. trichocladum* (K. Schum). The three aforementioned species share identical chromosome numbers, facilitating compatible crossing among them. Consequently, phenotypic intermediate types were observed within the natural populations of East Africa (Muir and Jank 2004). Guinea grass exhibits variability in terms of size and indumentum on culms, leaves, and panicles. Species featuring pubescent spikelets are often categorized as var. *publiglume* or var. *trichoglume* (Muir and Jank 2004).

3 Crop Diversity

Guinea grass showed wide diversity in morphological, cytological, biochemical, and molecular features (Jain et al. 2003, 2006; de Sousa et al. 2011). It is predominantly tetraploid with an apomictic mode of reproduction, though varied cytotypes and sexually reproducing plants have been reported (Savidan 1983; Nakagawa and Hanna 1992). For example, Nakajima et al. (1979) isolated a diploid (2x) sexual accession GR297 from "Africa collection" collected during 1971 to 1973 from East Africa, specifically in Kenya, Tanzania, Uganda, and Ethiopia. Nakajima et al. (1979) were also able to identify 4x sexual plants from other introductions. Nakagawa and Hanna (1992) successfully developed an auto-tetraploid ($2n = 4x = 32$) sexual plant from the GR297 by colchicine treatment. Hybridization among the newly generated auto-tetraploid plants eventually resulted in the development of the 4x sexual line, "Nekken No. 1" (Nakagawa and Hanna 1992; Nakagawa 1993). A self-compatible obligate sexual guinea grass (*Panicum maximum* Jacq.) line SRP 75 (IC0616375; INGR15055) was also developed in India (Kaushal et al. 2015). These findings set in motion an approach to utilize apomixis gene(s) in the guinea grass breeding. During progeny testing of the open-pollinated GR297 accession, a vigorous hybrid plant that was an apomictic 4x plant which was eventually registered as cv. "Natsukaze (Norin No. 1)". It was the first apomictic hybrid cultivar, which was generated by a cross between a diploid sexual plant from GR297 and an unidentified apomictic line through open pollination (Nakajima 1985; Sato et al. 1990). Ploidy variations, viz., 3x, 4x, 5x, 6x, 7x, 8x, 9x, and 11x in guinea grass have been developed in India (Kaushal et al. 2009, 2015).

Worldwide, there are nearly 1500 gene banks registered in the WIEWS (World Information and Early Warning System on PGR) database (http://apps3.fao.org/wiews/) holding 7.1 million accessions belonging to 53,109 species, including major crop and minor or neglected crop species, as well as trees and wild plants. Of the total germplasm stored, 651,024 accessions belong to forage crops (Anonymous 2010). Conservation of forage germplasm under the Consultative Group of International Agricultural Research (CGIAR) is a part of germplasm conservation activity coordinated on plant genetic resources. Among the international organizations, major forage germplasm repositories are CIAT, Columbia, safeguarding one of the largest and diverse tropical forage collections, with more than 700 different species from 75 countries. The collection primarily focuses on legumes, with over 21,000 accessions, though it also includes nearly 1700 conserved grasses. The ILRI gene bank conserves ~19,000 accessions of forages from over 1000 species, out of which 1820 accessions of grasses belong to 139 species. Similarly, CSIRO-Australia has conserved 2670 accessions of different forage grasses belonging to 252 species (Sandhu et al. 2015).

The ICAR-National Bureau of Plant Genetic Resources (NBPGR) is the nodal agency for characterization, evaluation, maintenance, conservation, documentation, and distribution of germplasm resources in India. Currently, a total of 8107 accessions of cultivated and range grasses and legumes are being maintained at the long-term storage (LTS) module of the National Gene Bank at NBPGR, New Delhi. NBPGR introduced more than 900 accessions of guinea grass in India. Likewise, ICAR-Indian Grassland and Fodder Research Institute (IGFRI) is a unique research and development (R&D) organization in South Asia for sustainable agriculture through quality forage production for improved animal productivity. ICAR-IGFRI, being the National Active Germplasm Sites (NAGS) for forages, also conducts germplasm explorations, acquisitions from various national and international agencies, followed by its conservation under IGFRI's Medium Term Storage (MTS) facility. IGFRI now holds >9940 accessions representing >67 forage genera including guinea grass. The conserved germplasm of guinea grass at IGFRI has been evaluated, and a representative core subset of 23 accessions was developed from 152 diverse *M. maximus* germplasm representing collections from different agro-ecological zones of India as well as those procured from Africa and Brazil (Roy

Table 1 Details of different gene banks involved in the guinea grass germplasm conservation

Gene bank code	Institute/gene bank	Country	Number of accessions conserved
KEN212	GeRRI	Kenya	1950
COL003	CIAT	Columbia	519
BRA010	CNPGC	Brazil	429
AUS167	APG	Australia	288
BRA003	CENARGEN	Brazil	287
USA016	S9	United States	26
ETH013	ILRI-	Ethiopia	158
GBR004	RBGK	United Kingdom	21
USA995	NCGRP	United States	12
LBN002	ICARDA		8
Other			1
Total			3699

et al. 2020). As per the records available on Genesys, a total of 3699 accessions are conserved in different gene banks worldwide (Table 1).

4 Genetic Improvement Constraints

Guinea grass is a valuable forage crop for livestock due to its high yield, better nutritional quality, and tolerance to abiotic stresses (González Marcillo et al. 2021). However, genetic manipulation for desirable traits through traditional breeding approaches faces challenges, and modern tools can be used to tailor the trait of interest. The main constraints of genetic improvement are listed below.

4.1 Apomixis and Polyploidy

Most guinea grass varieties are apomictic, primarily reproducing asexually through seed. This bypasses sexual recombination, limiting genetic diversity and hindering the creation of new varieties through conventional breeding methods. The apomixis type in guinea grass is apospory, where a maternal diploid egg cell develops into an embryo without fertilization (Kaushal et al. 2018). It is a tetraploid species ($2n = 4X = 32$) of autopolyploid origin (Bogdan 1977) that undergoes gametophytic aposporous apomictic reproduction (Savidan 2000). The apomictic seeds reduce the genetic variation from parent to offspring plants due to asexual reproduction. Consequently, offspring are genetically identical to the parent plant (Jank et al. 2011). Nonetheless, a small number of genotypes in natural populations have a sexual reproduction mode and are diploid ($2n = 2x = 16$) (Savidan 1983). Sexual cytotypes are valuable in breeding programs because they are an effective tool for diversification and breeding. In order to enable hybridization with apomictic tetraploid genotypes, tetraploidy can be artificially generated in diploid plants. In guinea grass, apomixis is determined by a gene or set of genes through simple inheritance; hence, the offspring of asexual and apomictic plant crosses will display a 1:1 ratio of apomictic to sexual plants (Savidan 1983). The fixing of desired F_1 hybrids by cloning to produce heterosis and beneficial heterogeneous gene combinations, with subsequent seed production and commercialization, is made possible by the inheritance of apomixis (Savidan 2000).

4.2 *Sterility*

Most guinea grass varieties are sterile triploids, meaning they have three sets of chromosomes instead of the usual two. This sterility makes it challenging to develop new varieties through sexual reproduction. Breeders must rely on techniques like vegetative propagation or inducing chromosome doubling to create new breeding lines (Kaushal et al. 2009).

4.3 *Long Breeding Cycle*

Guinea grass is a perennial grass with a long breeding cycle (Rhodes et al. 2022). It can take several years to evaluate new varieties for important traits like yield, disease resistance, and drought tolerance. This long cycle slows down the development of improved cultivars.

4.4 *Limited Genetic Diversity*

The gene pool for cultivated guinea grass is relatively narrow. This lack of diversity makes it difficult to find desirable traits for breeding programs. Breeders are exploring ways to introduce new genes from wild relatives of guinea grass to increase diversity (Roy et al. 2020).

4.5 *Seed Shattering*

Some guinea grass varieties shatter their seeds readily, making it difficult to collect and store seeds for planting. This can be a challenge for breeders who need to produce large quantities of seeds for new varieties (Sukhchain 2010).

4.6 Cross-Pollination Nature

Guinea grass spikelets are an agamic complex which have two florets. According to Assienan and Noirot (1999), the lower floret is male, while the upper is hermaphrodite. The plant may be hermaphrodite (the lower floret is vestigial) or andromonoecious (one hermaphrodite floret, the other male floret). Because more pollen is generated, the andromonoecious plant may produce more seed. Andromonoecy is inherited in a Mendelian, monogenic, and recessive manner. Plants, which are diploid and sexual, are all andromonoecious. In a 1:1 ratio, all apomicts are andromonoecious/hermaphrodites. In andromonoecious plants, the male bloom appears inside the same spikelet later than the hermaphrodite flower (Assienan and Noirot 1999). This led to cross-pollination in guinea grass. Since guinea grass is cross-pollinated, it is harder to propagate and maintain the identity of lines due to heterozygosity (Noirot 2001).

Apart from cross-pollination in guinea, small floret size (2.5 to 5 mm) creates difficulty in the hybridization program. Seed production is also poor, with low viability of seed, which also hinders commercial seed production. Seed shattering and the range in maturation rate in panicles create difficulty for seed harvesting in guinea grass (Joaquin et al. 2001). Harvesting seed is often difficult due to the range in maturation rates in panicles (Joaquin et al. 2001). Guinea grass is a perennial crop that requires many years to evaluate the germplasm lines. Week seeding is another problem in guinea grass for crop establishment.

Despite these challenges, researchers are making progress in developing improved guinea grass varieties. Techniques like genetic markers and assisted reproductive technologies are accelerating breeding programs. These advancements can help to overcome the constraints and develop guinea grass varieties that are even more productive and resilient.

5 Breeding Methods

Guinea grass is a vital forage crop valued for its high biomass yield, nutritional content, and adaptability to various climates. However, breeding this crop for desirable traits poses unique challenges due to its predominantly apomictic reproduction. Breeding methods in guinea grass aim to improve agronomic traits such as yield, nutritional content, resistance to pests and diseases, and adaptability to different environmental conditions. In India, 17 varieties were developed by various institutes for various agroecological conditions (Table 2). Both conventional and modern breeding methods are employed to achieve these goals.

Table 2 List of released/notified varieties for different agro-climatic zones of India

Variety	Year of release/ notification	Developed by	Area of adaptation
PGG.1	1982	PAU, Ludhiana	Hill, North, North-west, and Central India
PGG 9	1986	PAU, Ludhiana	Hill, North, North-west, and Central India
PGG 13	1987	PAU, Ludhiana	Central India and hills
PGG 14	1988	PAU, Ludhiana	Entire country
PGG 19	1989	PAU, Ludhiana	Punjab
Haritha	1990	KAU, Vellayani	Kerala
Marathakam	1993	KAU, Vellayani	South Zone
CO 1	1993	TNAU, Coimbatore	Tamil Nadu
PGG 101	1995	PAU, Ludhiana	Punjab
PGG 518	1997	PAU, Ludhiana	Punjab
CO 2	2000	TNAU, Coimbatore	Tamil Nadu
PGG 616	2001	PAU, Ludhiana	Punjab
Bundel Guinea 1 (JHGG 96-5)	2004	IGFRI, Jhansi	Northwest and hill zone
Harithasree	2006	KAU, Vellayani	Kerala
Bundel Guinea 2 (JHGG-04-1)	2008	IGFRI, Jhansi	All India
CO-3 (TNGG 062)	2009	TNAU, Coimbatore	Tamil Nadu
Dharwad Guinea Grass-1 (DGG-1)	2016	IGFRI, RRS Dharwad	All of India

5.1 *Conventional Breeding Methods*

Despite the inherent challenges, conventional breeding methods have played a significant role in developing improved guinea grass varieties. Below are a few approaches that cover their concepts.

5.1.1 Clonal Selection

It is the most widely used method for guinea grass due to its apomictic nature. It involves identifying and propagating superior individual plants exhibiting desirable traits like high yield, pest resistance, and palatability (Sukhchain 2010). The selected plants are vegetatively propagated through stem cuttings, tiller splitting, tissue culture, etc. Clonal selection is simple and cost-effective and preserves the desirable traits of the selected plant. However, it has some limitations, such as a limited scope for introducing new genetic variation and susceptibility to pests and diseases existing in the original plant. Variety CO 3 (GG) was developed through this method (Vijayakumar et al. 2010).

5.1.2 Mass Selection

This method involves selecting individuals with desired traits from a large, genetically diverse population over several generations. This allows for the gradual accumulation of favorable alleles in the population. This method is relatively simple to implement and can lead to modest improvements in desired traits. However, it has slow progress due to limited selection pressure and limited control over specific genes being selected (Sandhu et al. 2015).

5.1.3 Polycross Technique

A large number of selected plants are cross-pollinated simultaneously. The goal is to maximize genetic diversity in the offspring. After the polycross, seeds are collected and planted, and the resulting plants are evaluated for desired traits (Kobabe 1983).

5.1.4 Hybridization

This method involves the controlled crossing of parents with complementary desirable traits. However, the apomictic nature and sterility of guinea grass pose difficulties. Breeders may utilize the following methods. (1) Facultative apomictic lines: These lines exhibit both apomictic and sexual reproduction, allowing for some degree of controlled crosses. However, identifying these lines is challenging. (2) Induction of chromosome doubling: Colchicine treatment can double chromosome number in sexually-reproducing diploid plants, creating fertile tetraploids for crossing with apomictic lines. (3) Intergeneric hybridization: Crossing guinea grass with related species can introduce new genetic variation, but compatibility issues and sterility barriers exist. The major advantage of this method, it has the potential for significant genetic gain by combining desirable traits from different parents. However, it requires the identification of suitable parents with compatible flowering times and overcoming sterility barriers in crosses.

5.1.5 Recurrent Selection

This method involves repeated cycles of selection, propagation, and intercrossing of superior individuals within a population. Different variations exist, including (1) half-sib recurrent selection: selected individuals are intercrossed and seeds are used for the next generation, and (2) full-sib recurrent selection: selected individuals are self-pollinated or crossed with siblings, followed by selection in subsequent generations. Recurrent selection can lead to steady improvement in desired traits over time. The limitation of recurrent selection is that it is a time-consuming process, with limited effectiveness due to apomixis (Simeão et al. 2021).

5.2 *Modern Breeding Techniques*

Compared to other crops, tropical grasses have a higher genetic diversity because of their apomixis and polyploidy (Jank et al. 2011), which provide for greater opportunities for hybridization between related species and prevent genetic bottlenecks brought on by domestication. On the other hand, breeding may be more challenging due to a lack of understanding of the biology and genetics of the species, including its autotetraploidy and facultative apomictic mode of reproduction (Warmke 1954). For these reasons, genetic research is necessary.

Biotechnology-based new methods are becoming more widely available. Forage crop enhancement has greatly benefited from genome editing, especially with regard to CRISPR/Cas9 technology. These developments have shown promise in improving qualities including stress tolerance, disease resistance, biomass yield, and feed quality—all essential for the long-term viability of the livestock industry and the generation of biofuel. The development of targeted mutagenesis in a range of fodder crops was also assessed, along with the possible advantages and difficulties of using CRISPR/Cas9 in fodder breeding (Haq and Zheng 2022). For perennial ryegrass, reliable transformation techniques have been created, enabling breakthroughs in functional genomics and breeding for this significant grass species. Still, this technology is not used for improvement in guinea grass.

Functional genomics enables marker-assisted selection and introgression. Nevertheless, a molecular assessment of the germplasm's genetic diversity has not been conducted. The development of breeding techniques, accession selection, and the preservation of these genetic resources all depend on our capacity to understand the genetic variability of the germplasm. Scientific means are now available to effectively characterize the organization of genetic diversity found in germplasm or within cultivars, as well as the diversity within populations, without phenotyping, thanks to the advent of molecular marker technologies (Bolaric et al. 2005). The genetic relationship between molecular genetic markers and a quantitative trait locus (QTL) of interest is the basis for the use of these markers for genetic improvement and selection (Resende et al. 2013).

SNP-based molecular genetic markers, which identify polymorphisms resulting from a single-nucleotide alteration in the genome, have become extensively employed in numerous species in recent times (de Sousa et al. 2011; Elshire et al. 2011; Resende 2015). In general, a polymorphism must be present in at least 1% of the population for an SNP to be deemed genetically derived (Resende 2015). SNPs are favored over alternative genetic markers since they are the most prevalent form of genomic variation. de Sousa et al. (2011) sequenced 576 clones from microsatellite-enriched libraries and designed 116 flanking primers for 116 microsatellite loci and screened using a sample of 25 guinea grass accessions. de Sousa et al. (2011) selected 25 *P. maxima* genotypes from a Brazilian germplasm collection and used 20 new polymorphic microsatellite markers to estimate genetic relatedness. They were also examined in four more kinds of forage grass for cross-species amplification. It was reported that by cross-amplification these microsatellites could be transferred to four tropical species of forage grass.

QTL mapping has shed light on the genetic underpinnings of apomixis, an asexual method of seed reproduction. Accurate identification of the diplospory locus in *Eragrostis curvula* marks a significant step forward in the integration of apomixis into plant-breeding initiatives. Meanwhile, guinea grass's genomic areas linked to the control of complex characteristics and apospory have been thoroughly mapped, adding to our knowledge of the genetic processes underlying apomictic reproduction (Deo et al. 2020). They built a linkage map with allele dosage and generated novel information on the genetic architecture of traits that are important for the breeding of *M. maximus*. They obtained a linkage map containing 858 SNP markers with allele dosage information expected for an autotetraploid from a full-sib progeny. The high genetic variability of the progeny allowed them to map 10 QTLs related to agronomic traits, such as regrowth capacity and total dry matter, and 36 QTLs related to nutritional quality, which were distributed among all homology groups (HGs). Various overlapping regions associated with the quantitative traits suggested QTL hotspots.

However, there are no mapping studies of loci linked to complex variables in guinea grass, like those pertaining to nutritional quality and forage output. Therefore, the mapping of these and other significant quantitative trait loci (QTLs) may help develop new breeding program techniques for *M. maximus* as well as provide information about the genetic architecture of attributes. For non-model species, like *M. maximus*, whose genomes have not been sequenced, linkage maps have served as the main source of genetic information. Building dense linkage maps facilitates the mapping of features with polygenic and monogenic inheritance, allowing one to identify the structure and evolution of the genome. It may even aid in the assembly of a species' genome (Doerge 2002; Flint-Garcia et al. 2003; Luo et al. 2004).

With the use of genomic selection (GS), agricultural productivity may be raised and the breeding interval cycle shortened to at least half of its traditional duration. The primary benefit of GS in pasture breeding is a shorter breeding cycle. In order to quickly select new cultivars to meet the demand for food and fodder in the future, genome-wide association studies (GWAS) and GS have great potential for use in the selection of complex traits like yield, disease resistance, and insect resistance. Applying GS in forages presents a number of challenges, such as creating training populations of sufficient quantity and quality; coping with polyploidy; creating a high-quality, affordable, and reliable genotyping platform; and learning about the genetic makeup of desired characteristics. With so many focused qualities, GS will permit a higher number of evaluations of individuals in the early generation. There is still more work to be done, as we find out while dealing with GS in the enhancement of tropical forages. Plant breeders have a big job when it comes to creating forage breeding programs, mostly for apomictic and polyploid grasses, and suggesting breeding methods that maximize GS utilization. While this is a barrier, it also presents a fantastic opportunity to expedite genetic gain in forage breeding. Some advanced biotechnological approaches are as follows, which offer promising avenues for overcoming limitations and accelerating guinea grass breeding:

5.2.1 Molecular Marker-Assisted Selection (MAS)

This technique utilizes DNA markers linked to desirable traits to identify plants with those traits without requiring field evaluation. This allows for more efficient selection in breeding programs (Simeão et al. 2021).

5.2.2 Doubled Haploid (DH) Technology

This technique involves inducing the formation of haploid plantlets directly from pollen grains (Kaushal et al. 2009; Sukhchain 2010).

5.2.3 Genomic Selection

Genomic selection relies on advanced genomic technologies to analyze the entire genome of a plant. This enables breeders to predict the performance of a plant based on its genetic profile. By understanding the genetic architecture of desirable traits, breeders can make more accurate selections, speeding up the breeding process (Simeão et al. 2021; Chen 2024; Younas et al. 2025).

Genome Editing
Techniques like CRISPR/Cas9 have revolutionized plant breeding by enabling precise modifications to the genome. This allows for the targeted introduction or removal of specific genes associated with desired traits. Genome editing can be used to enhance disease resistance, nutritional content, and other important characteristics (Chen 2024; Ghane et al. 2024; Farooq et al. 2023; Bilal et al. 2025).

5.2.4 Challenges and Considerations of Modern Biotechnological Approaches

Ethical Concerns
The use of modern breeding methods, especially genome editing, raises ethical considerations regarding the release of genetically modified organisms (GMOs) into the environment.

Regulatory Approval
Breeders must adhere to strict regulatory frameworks governing the release of genetically modified crops. Obtaining approval for new varieties can be a time-consuming process.

6 Balancing Traits

Breeding programs need to strike a balance between different traits. For example, increasing yield should not compromise nutritional content or other essential characteristics. Breeding guinea grass involves combining conventional and modern methods to improve various agronomic traits. Conventional methods continue to play a crucial role in accumulating and selecting desirable traits over generations. At the same time, modern techniques leverage cutting-edge technologies to accelerate the breeding process and achieve more precise results. The integration of these methods allows for the development of guinea grass varieties that are not only high-yielding but also resilient, nutritious, and well-adapted to diverse environmental conditions. As technology continues to advance, the future of guinea grass breeding holds exciting possibilities for further enhancing its performance and contribution to sustainable agriculture.

7 Fodder Quality Traits

Animal husbandry practices are inherently correlated to animal health, production, and welfare. To maintain optimal animal health and to produce high-quality animal products safely and efficiently, management of diet composition arguably represents the easiest strategy that can be implemented at the farm level. In the ruminant industry, feed costs can contribute up to 70% of total production costs; therefore, improving the efficiency of feed conversion into milk or meat can have a significant impact on the profitability of a ruminant enterprise. Forages provide important nutritional components for the healthy development of a wide range of ruminants and pseudo-ruminants. The quality and digestibility of any crop reside in the nutrients it holds. The ratio of protein, lipid, and carbohydrate determines the nutritional status of the crop.

7.1 Protein

Among all the nutritional components in forages, protein is the most important for the health and growth of the ruminants. High crude protein (CP) in forages makes rumen microflora more active and increases the efficiency of livestock to digest and assimilate the nutrients. This helps in improving the growth, development, and reproduction of cattle. Crude protein represents the quantity of nitrogen (N) present in the plant. Crude protein is an additive of true protein and non-protein nitrogen. True protein represents the nitrogen present in the protein portion of the plant. Non-protein nitrogen refers collectively to the compounds such as amides, amines, ammonia, urea, and biuret. These compounds are not proteins but could be a source of nitrogen to ruminants. The feeding value of forages mainly depends on CP and total digestible nutrients. The CP percentage in guinea grass ranged from 8.2% to 10.8%, which was above the minimum CP level required for rumen function (7.5%). Since guinea grass is multi-cut in nature, many agronomic and environmental variations may influence the nutritional quality of its forage.

7.2 Carbohydrates

The major source of digestible energy in forages comes from carbohydrates. Forage carbohydrates can be divided into structural carbohydrates and non-structural carbohydrates. The cell wall of forages represents large reserves of potential energy. Cellulose is the most abundant polysaccharide in forage cell walls. The primary types of carbohydrate are the insoluble structural saccharides, cellulose and hemicellulose, or the storage forms such as starch and water-soluble polymers (e.g., fructans). These are degraded into simple sugars through cleavage of glycosidic bonds, either by the animal itself (non-ruminant and ruminants) or via microbial digestion and subsequent animal absorption (ruminants only). Different ratios of carbohydrates within the forage crop alter downstream digestibility of the animals, especially if the cell-wall structure constrains digestion by the microbial population or limits plant cell wall penetration (Weimer 1996). In general, acid detergent fiber (ADF) and neutral detergent fiber (NDF) are the chief variables used for the estimation of fiber fraction of forages. ADF and NDF fractions are the insoluble portions remaining after the treatment with acid detergent solution (ADS) and neutral detergent solution (NDS). ADS dissolves the cell contents, and the leftover is cellulose, lignin, and silica which represents ADF (Van Soest and Robertson 1985). Similarly, NDS dissolves the cell contents and the remaining portion of cellulose, hemicellulose, lignin, and silica represents NDF. The cell content fraction dissolved in ADS and NDS includes the proteins, lipids, nucleic acids, nutrient ions, plus non-structural carbohydrates and some structural carbohydrates. The ADF and NDF content in *P. maximum* ranged from 32.5% to 41.4% and 58.5% to 74.3%, respectively.

7.3 Lignin

Lignin is a class of complex organic polymers that form key structural materials in the support tissues of most plants. It is particularly important in the formation of cell walls and lends rigidity. Chemically, lignin is a polymer made by cross-linking phenolic precursors. It has a dramatic impact on the digestibility of cellulose and hemicellulose. Lignin binds with structural carbohydrates and cell wall proteins and reduces nutrient availability. The lignin content in guinea grass ranged from 4.0% to 8.0%.

7.4 Ash

Ash content in guinea grass ranged from 8.0% to 11.0% (Hindoriya et al. 2019). In general, ash is the inorganic mineral matter required for the bone formation of cattle. These nutrients or minerals play a vital role in the physio-biochemical activities of the animals. For instance, calcium is necessary for blood clotting, muscle contraction, and various biochemical reactions in cattle. Phosphorous is the principal component of energy currency (ATP) of the cell. Many biochemical reactions in the cattle system need energy in the form of ATP. Potassium and zinc maintain the acidic conditions in body fluids and are also required for several enzyme reactions of protein and carbohydrate synthesis. Magnesium helps in bone growth, and iron is the prime component of hemoglobin, the oxygen carrier in the blood. Manganese is a necessary nutrient helpful for the utilization of carbohydrates. Copper and selenium maintain the fertility of cattle. Iodine is necessary for the thyroid gland. The *P. maximum* comprises 0.16–0.71% calcium, 0.04% magnesium, 0.08% potassium, 43.4–44.5 mg kg^{-1} manganese, 104–112 mg kg^{-1} iron, and 41.5–42.3 mg kg^{-1} zinc (Sodeinde et al. 2007).

7.5 Crude Fat

Ether extract or crude fat content is another significant metric important for dairy animals. It acts as the reserve energy and forms the structural tissues of the body. Lipids in forage crops are mostly found as polyunsaturated fatty acids (PUFAs) in the range of 10–30 g kg^{-1} (Hatfield et al. 2007) These dietary lipids are important in the final animal product quality. Forage diets with lower PUFA levels than cereal diets can produce leaner meat (Wood et al. 2004; Van Elswyk and McNeill 2014). Studies have been used to profile PUFAs across forage species, with grasses tending to have more α-linolenic acid when compared to legumes (Boufaïed et al. 2003). Striking differences in PUFA content were found within species through profiling cultivars; moreover, the harvest period and its environmental conditions influence the PUFA level (Elgersma et al. 2003; Clapham et al. 2005).

7.6 Tannin

Tannins are polyphenolic compounds containing free hydroxyl and carboxyl groups to form strong complexes with various macromolecules. Tannins are abundant in forages, trees, shrubs, and legumes often ingested by ruminants. Tannin is further classed into either condensed or hydrolysable forms. The condensed tannins are the flavonoid polymers and thus are products of phenylpropanoid metabolism. Radical reactions are probably involved in their synthesis, but few details of the biosynthesis pathways are known. The hydrolysable tannins are made up of gallic acids and are synthesized from the shikimate pathway in plants. Many of these gallic acids are linked to hexose molecules. Condensed tannin levels in excess of 50 g/kg dry matter reduce digestibility, lower feed intake, reduce palatability, inhibit digestive enzymes, and are toxic to rumen microorganisms (Kumar and Vaithiyanathan 1990).

7.7 Phytic Acid

Phytic acid, also called inositol hexaphosphate (IP6), occurs naturally in plant roots, tubers, and seeds (Brooks 2001), with considerable quantities in many oilseeds and legumes (Akande et al. 2010). Phytic acid is the major storage compound for phosphorus (P) in plants. They accumulate in the seeds throughout the ripening period as phytic acid. Phytic acid is a negatively charged molecule that forms complexes with positively charged metal ions, including calcium, iron, magnesium, and zinc, limiting their availability in the body through decreased absorption rates (Samtiya et al. 2020). The accumulation site of phytic acid in monocotyledonous seeds is the aleurone layer, particularly the aleurone grain. Corn differs from other cereals as more than 80% of phytic acid is concentrated in the germ. Phytic acid content of cereals varies from 0.5% to 2.0%. While accounting for up to 90% in many seeds, usually only 10% of total P is found in phytic acid in green leaves (Alkarawi and Zotz 2014). Phytic acid is considered an anti-nutrient for humans and other monogastric animals such as swine, poultry, and fish (Silva et al. 2021). The monogastric animals lack phytase required for separating phosphorous from phytic acid, and additionally, the phosphorous released from undigested phytic acid causes environmental pollution such as eutrophication.

8 Digestibility

The digestibility mainly refers to the extent to which the feed is digested by an animal. The process of digestion differs with the type of biomolecule (carbohydrates, proteins, and lipids) which along with minerals and vitamins provides the amount of energy which can be derived by the animal. Cattle are a polygastric animal having a well-developed four-chambered stomach. The four chambers are the rumen, the reticulum, the omasum, and the abomasum. The majority of digestion takes place in the rumen which is the home of a vast array of microbes.

In forage crops, carbohydrates comprise 50–80% of DM and the primary types of carbohydrate are the insoluble structural saccharides cellulose and hemicellulose, or the storage forms such as starch and water-soluble polymers. High concentration of cell wall carbohydrates in mature plants constrains digestion by limiting plant cell wall penetration of the microbial population (Weimer 1996). Animal digestion of simple carbohydrates produces monosaccharides which can be readily metabolized. In ruminants, only microbial digestion of structural carbohydrates produces simple sugars which are subsequently metabolized to pyruvate. Pyruvate is absorbed by the animal and is metabolized further into volatile fatty acids (VFAs) which are a major energy source (Bergman 1990). Ruminants absorb VFAs in their rumen, and the rate of this is dependent on the concentration of individual VFAs, rumen pH, and the absorptive area in the ruminal lining. Lignin has a dramatic impact on the digestibility of cell wall components. It binds with structural carbohydrates and cell wall proteins and reduces nutrient availability. Higher lignin concentration in the growing forage crop increases the percentage of indigestible DM. Of the major forage grasses grown globally, *Lolium perenne* has high digestibility due to its high soluble sugar content and low lignin content (Ruckle et al. 2017). In guinea grass, dry matter, acid detergent fiber, and lignin contents increased linearly with cutting age, reducing digestibility and nutritive value. The main reason behind the decline is the progression from a leafy crop to a mature crop with a high ratio of stem portion.

Protein in the ruminant animals is digested (degradation) in either the reticulorumen or the abomasum and small intestine. Two sources of protein are available for the ruminant to use: protein from feed and microbial protein from the microbes that inhabit its rumen. A ruminant is unique in that it has a symbiotic relationship with these microbes. Like other living creatures, these microbes have requirements for protein and energy to facilitate growth and reproduction. Each feedstuff has different proportions of degradable intake protein (DIP) and undegradable intake protein (UIP, also called "rumen bypass protein"). Rumen microbes break down the DIP into ammonia (NH_3), amino acids, and peptides, which are used by the microbes along with energy from carbohydrate digestion for growth and reproduction. During digestive contractions, some of these microorganisms are "washed" out of the rumen into the abomasum, where they are digested like other proteins, thereby creating a source of protein for the animal. The digestibility of protein is affected by the presence of tannin in the forages. Some micronutrients like proanthocyanidins or condensed tannins (CT) also change the digestibility of protein. Tannin molecules are oligomeric, consisting of repeating structural units containing phenolic groups (Hagerman and Butler 1991) that are free to complex with proteins, starch, cellulose, hemicellulose, pectin, and minerals (Hassanpour et al. 2011). The concentration of condensed tannins above 4% is toxic for ruminants as they are harmful to a variety of rumen microbes. Condensed tannins generally occur in leaves, roots, wood, bark, fruit, buds, and flowers (Kraus et al. 2003), and they can play an important role in regulating nutrient dynamics in ecosystems (Schweitzer et al. 2008). CT has the ability to bind protein in feeds, enzymes, and microbial cells, therefore disrupting microbial digestion and slowing ruminal protein and dry matter (DM) digestion. This can be advantageous as rapid protein degradation is causative of bloat; however, too high tannin content indicates that protein passing through the digestive tract is unabsorbed, resulting in a loss of nutritional value (Piluzza et al. 2014). All grasses contain little or no proanthocyanidins, whereas many legumes, especially big trefoil (*Lotus pedunculatus*) and Sericea lespedeza (*Lespedeza cuneata*), can have levels as high as 18% DM (Mueller-Harvey 2006).

In ruminants, P plays key functions than any other mineral nutrient. The enzyme phytase catalyzes the stepwise hydrolysis of phytate. In ruminants, the phytase enzyme could be available from ruminal microbial phytase, endogenous mucosal phytase, large intestinal microbial phytase, plant phytase, and exogenous microbial phytase. In general, ruminal phytase is significant, whereas endogenous mucosal phytase is incapable of hydrolyzing sufficient amounts of phytate-bound P (Sehested and Lund 2007) Thus, although continuous P supply is of crucial importance, dietary P supply should not exceed the requirement of the animal. This makes them unavailable for absorption during digestion causing micronutrient malnutrition. The levels of phytate in the present work are unlikely to have any adverse consequence in ruminant nutrition according to the previous work of Okoli et al. (2003).

9 Effect of Agronomic Interventions on the Nutritive Value

Tropical grasses are normally low in CP content. Cutting interval influences the chemical composition and nutritional quality of guinea grass. Physiological and morphological development strongly influence the nutritional quality of grasses. A progressive advancement from the leafy vegetative to the maturity stage decreases the digestibility of forage plants. The advancing cutting stages from 49 to 91 increased the dry matter, acid detergent fiber, and lignin contents linearly whereas the opposite effect was observed for the crude protein content in guinea grass. A higher fibrous to cellular content ratio with increased cutting interval decreased the digestibility of plants (Neely et al. 1988; Junges et al. 2024). Total ash, CP, hemicellulose, and IVDMD contents decreased with decreasing frequency of defoliation, whereas DM, ADF, ADL, and cellulose contents showed an increasing trend (Tessema et al. 2010) in guinea grass. The CP percentage of all the treatments ranged from 11.1 to 15.3%, which was above the minimum CP level required for rumen function (7.5%) and the NDF percentages recorded in this experiment ranged from 54.4 to 91%, which is below and above the average value for tropical grasses (66.2%) and within the threshold level of NDF beyond which DM intake of cattle is negatively affected (60.0%).

Further, the CP content varied with plant spacing and was higher with increasing plant distances of 75 × 20 and 100 × 20 cm compared to broadcasting in *P. maximum* (Ramírez-Ordoñes et al. 2022). The CP value of the guinea grass *cv*. Mombasa harvested at 30 d of regrowth was recorded as 7% (Ramírez-Ordoñes et al. 2022), a minimum requirement for adequate digestion in a bovine (Van Soest and Robertson 1985). Among the performance of the guinea grass *cv*. Mombasa in different seasons, CP was the highest in the dry season (13.9%), intermediate in the cool season (10.8%), and lowest in the rainy season (9.3%). The higher biomass accumulation with rainfall input might have caused greater dilution of the nitrogen (Muñoz-González et al. 2016). In vitro dry matter digestibility (IVDMD) and in vitro neutral detergent fiber digestibility (IVNDFD) were also higher in the dry season which could be attributed to lower contents of ADF and ADL during that season. Verdecia et al. (2012) reported a negative correlation between the availability of water and the CP content in the fodder.

Soil texture also serves as a controlling factor for the growth of the plant. Guinea grass performed best in well-drained, light-textured sandy loam or sandy clay soil but could not withstand heavy clay soil. In another similar study, two guinea grass cultivars grew best in well-drained sandy loam soil (Sodeinde et al. 2007). Guinea grass could also be preserved by the silage technique; the nutritive value could further be enhanced by incorporating additives. A well-designed study was conducted to investigate the effect of silage additives on the 5-week re-growth of guinea grass in goats. The silages were made without additive (T1), with cassava peels additive (T2), and with fermented epiphytic juice of lactic acid bacteria in *Panicum maximum* (FEJPM) additive (T3). All silages had firm textures with varying degrees of firmness. Cassava peels used as an additive in silage production are known to enhance nutrient digestibility, as is reported by Red Sokoto goats fed Napier grass ensiled with cassava peels. The use of cassava peels as an additive has shown higher values of propionic and acetic acids that imply the highest energy contents in the silage. The concentration of lactic acid and buffering capacity of the crop are two of many factors affecting the final pH of silage. In maize, T3 silage had the lowest pH. This low pH is indicative of the efficiency of the FEJPM as an additive that acts as an inoculant for LAB known to increase the lactic acid content of silage.

10 Diseases and Insect Pests and Their Management

There is less research reported on guinea grass for biotic stresses, particularly diseases and insect pests. In general, guinea grass is hardy and shows tolerance to most diseases or insect pests; however, some disease symptoms can be observed on leaves during the rainy season. Globally, few reports are available systematically covering diseases and insect pests of guinea grass. For instance, in Florida, *Bipolaris yamadae* causes leaf spot disease (Adhikari et al. 2021) (Table 3). Aphid-borne (potyvirus group Y) Guinea Grass Mosaic Virus was reported in the Ivory Coast (Kukla et al. 1984). Blast which is caused by *Pyricularia*-like fungus (*Pyricularia* sp. LS group) was reported in Japan (Tsukiboshi et al. 2009).

As far as insect pests are concerned, Spittlebug (*Aeneolamia reducta*), a pest of guinea grass, was reported in Colombia (Thomas and Laponte 1989), which was

Table 3 List of major diseases in guinea grass

S. no.	Disease	Causal organism	Location	References
1.	Leaf spot	*Bipolaris yamadae*	Florida	Adhikari et al. (2021)
2.	Guinea Grass mosaic virus	Potyvirus group: Aphid-borne	Ivory Coast	Kukla et al. (1984)
3.	Blast	*Pyricularia*-like fungus: Pyricularia sp. LS-group	Japan	Tsukiboshi et al. (2009)

Table 4 List of major insect pests in guinea grass

S. no.	Insect	Family	Species	Location	References
1.	Spittlebug (*Aeneolamia reducta*)	Cercopidae	Homoptera	Columbia	Thomas and Laponte (1989)
2.	*Heteropsylla cubana*	Hemiptera	Psyllida	Cuba	Alonso et al. (2005)
3.	*Epitrix* sp.	Ceoleoptera	Chrysomelida	Cuba	Alonso et al. (2005)
4.	*Wasmannia auropunctata*	Hemenoptera	Formicida	Cuba	Alonso et al. (2005)
5.	*Doru taeniatum*	Dermaptera	Forficulida	Cuba	Alonso et al. (2005)

controlled by bio-control agents (insect predators, nematodes, and the fungus *Metarhizium anisopliae*) and genetic resistance. In Cuba, a number of arthropods were identified in a silvi-pastoral system, Leucaena-guinea grass (Table 4) (Alonso et al. 2005). Being a potential fodder crop, chemical control measures are not preferred as they are not eco-friendly and cost-effective. The most suitable and eco-friendly approach is to develop varieties resistant/tolerant to biotic stresses. Major diseases and their control measures have been mentioned below:

10.1 Blast

Blast is mainly a fungal disease that attacks the leaf sheath, rachis, culms, and glumes. Severe leaf infection leads to the destruction of the whole foliage. The disease first occurred in guinea grass in Yamaguchi Pref. in 1997 and Okinawa Islands in the southern region of Japan in 2002 (Tsukiboshi et al. 2009). This disease occurs in regions with high humidity or high rainfall.

Symptoms

On the leaves, symptoms first appear as small, round or ellipsoid lesions of 2 to 5 × 1 to 2 mm in size. Lesions are almost spindle-shaped with greyish-white centers along with dark brown or black-colored margins. In older leaves, the center becomes almost straw-colored, and the lesions remain circular, but on younger leaves, spots enlarge to cover a larger area to give the leaf a blighted appearance. Similar symptoms appear on the leaf sheath and other parts of the plant.

The Causal Organism

The causal organism of blast is caused by *Pyricularia* sp. Sacc. The LS group belongs to the imperfect fungi. Earlier, all the isolates of *Pyricularia* obtained from grasses except rice were identified as *P. grisea*. According to the new taxonomy proposed by Couch and Kohn, only the isolates obtained from *Digitaria* were placed in *P. grisea* and other grasses were placed in *P. oryzae*. Reclassification of *Pyricularia* isolates was done by Hirata et al. (2007), according to their classification, *Pyricularia* isolates other than *P. oryzae* and *P. grisea* isolates were grouped into LS and CE groups. Therefore, *Pyricularia* causing blast in guinea grass has been placed in *Pyricularia* sp. LS group by Tsukiboshi et al. (2009). Conidiophores emerge singly or in groups, pale brown in color, flexuous or straight, geniculate toward the apex, and 36 to 197 × 2 to 5 μm in size. Conidia are hyaline to light brown, straight, pyriform, smooth, 2–3 celled, and 19 to 30 × 5 to 10 μm in size.

Disease Cycle and Epidemiology

The fungus overwinters in seeds or straw of guinea grass. Both mycelium and conidia are considered primary sources of inoculum. The most likely source of perpetuation is through survival in grasses such as *Leersia* and *Setaria* grasses. The disease cycle of this disease is short, and most of the damage is caused by secondary infections. Conidia are carried by air currents and are the cause of secondary spread and infection. Mostly conidia are released at night in the presence of dew or rain. Weather parameters, especially temperature and relative humidity, play a vital role in the development of blast disease. As guinea grass is grown in tropical and subtropical regions, the temperature of those areas is optimum for the fungus to produce spores, spore germination, and infection process. This disease requires high relative humidity, more than 88%, for the production of conidia. High relative humidity between 93 and 99% and low night temperature between 15 and 20 °C or less than 26 °C favors spore production and spread. Other factors like intermittent drizzles, cloudy weather, longer duration of dew, excessive doses of nitrogenous fertilizers, and availability of collateral hosts also favor disease development.

Management

Use seeds from a disease-free crop. Remove and destroy the weed hosts from field bunds and channels. Apply nitrogen in split doses and use nitrogenous fertilizers judiciously. Treat seeds with captan, thiram, carbendazim, carboxin, or tricyclazole (2 g/kg seed). Alternatively, treat seeds with biocontrol agents such as *Trichoderma viride* (5 g/kg seed) or *Pseudomonas fluorescens* (10 g/kg seed). For foliar application, use fungicides like tricyclazole (0.06%) and thiophanate methyl (0.1%)..

10.2 Smut

Earlier, guinea grass was not the host of the smut pathogen. Recently, guinea grass has been found infected with smut disease, which was first reported from Okinawa Islands in the southern region of Japan in 2010. Later, it was also noticed in multiple locations in Saint Lucie and Okeechobee counties in Florida, USA, infecting spikelets of guinea grass in 2015 (Rosskopf et al. 2019). However, yield losses due to smut have not been studied well in this crop.

Symptoms

Initially, an individual grain or even a whole spikelet is infected with the disease, depending upon the susceptibility of the host variety. The spikelets become swollen and are replaced by masses of grey or black-colored spores. The sori in the infected inflorescences are inflated, cylindrical, or ovoid; no grains are formed, and seed production is significantly reduced.

The Causal Organism

The guinea grass smut is caused by a fungal pathogen *Conidiosporomyces ayresii* (Berk.) Vánky and Bauer (1992) (formerly *Tilletia ayresii*) belongs to the smut family Tilletiaceae. The sori of fruiting structures of *C. ayresii* are brown or pale brown in color. These are globose to sub-globose swollen masses of spores surrounded by a sac-like structure with an apical opening. Their size varies from 14 to 16 × 15 to 18 μm in diameter with 2- to 3-μm-thick walls. Later on, short and dense warts are also formed with pointed tips. Spores germinate to produce basidia which bear basidiospores. These basidiospores are cylindrical, long, aseptate, straight, to slightly curved, and 20 to 37 × 2 to 3 μm in size. The conidia are thin-walled, colorless, and Y-shaped, loosely connected into balls of conidia.

Disease Cycle and Epidemiology

The teliospores of smut overwinter in seeds or sometimes in the soil. These teliospores germinate to produce basidia and basidiospores. Ballistospores infect spikelets of the host plant, and balls of conidia are formed asexually during the main season. These conidia germinate more quickly than thick-walled teliospores. Conidia are the secondary source of smut infection in guinea grass. Smut infection is high during the rainy season due to the presence of flowering host plants. High humidity favors the teliospore or conidia germination, growth of mycelium, and infection of susceptible host plants.

Management

Crop rotation, field sanitation, and seed treatment help reduce disease incidence. Rogueing diseased plants before the rupture of smut balls is an important cultural practice against the disease. Seed treatment with fungicides like carboxin + thiram or benomyl is also useful to reduce systemic infection.

10.3 Brown Leaf Spot

Brown leaf spot was first observed on the leaves of guinea grass in Gainesville, Florida, USA (Wang 2019; Adhikari et al. 2021). Later on, similar leaf spots were reported on guinea grass from Brazil.

Symptoms

Symptoms initially appear as small brown to black necrotic spots on the leaves, 1–2 mm in size. Lesions on the leaves are oblong to irregular in shape, with dark grey to brown centers and brown to black margins. Later, under favorable climatic conditions, necrotic lesions coalesce to form larger lesions and blight, primarily affecting the lower leaves, then other parts of the plant, resulting in canopy collapse. Diseased plants exhibit defoliation and senescence.

The Causal Organism

The brown leaf spot is caused by *Bipolaris yamadae* (Y. Nisikado) Shoemaker (earlier known as *Helminthosporium yamadae*). *Bipolaris* genus has economic importance as it can cause devastating diseases in staple crops such as southern corn leaf blight caused by *B. maydis* resulted in the loss of 710 million bushels of maize in the USA in the 1970s and the Great Bengal famine in India caused by *B. oryzae* (1943–1944) resulted in 40–90% losses which led to the death of 2 million people by starvation (Manamgoda et al. 2011). Conidiophores arise individually or in groups from grey to dark brown mycelium. They are straight or cylindrical, septate, smooth, and geniculate with swollen upper tips. The size varies from 74.8 to 215.5 μm in length and 3.4 to 7.5 μm in width. Conidia are brown to black in color, and the shape varies from straight to curved, cylindrical or elliptical, and obclavate with three to ten septations.

Disease Cycle and Epidemiology

The pathogen survives in soil, seed, or crop residues as mycelium and conidia. Both mycelium and conidia are carried over from one season to another. The initial infection is caused by inocula present in debris or seed, and the secondary spread of the disease is through air-borne asexual spores conidia. High relative humidity of more than 90% favors conidia formation, and the optimum temperature for conidia production is between 21 and 36 °C. Nutritional imbalances also predispose the plants to brown leaf spot infection.

Management

Apply adequate, balanced doses of fertilizers to the crop. Use disease-free seeds for planting. Sanitize fields properly. Treat seeds with thiram or captan (2 g/kg seed). Apply foliar fungicides like mancozeb (0.3%) or copper oxychloride (0.2%). Maintain proper water management and use soil amendments.

10.4 Summer Blight

The disease first occurred on guinea grass in Yamaguchi Prefecture of Japan in 1997 (Inoue and Sumida 1999).

Symptoms
The symptoms first appear on the plant parts near the soil surface and spread towards the upper parts. Initially, lesions are white or ash green, oval with brown borders. Later, these lesions fuse together, giving them a banded appearance. This results in the rotting of leaves or whole plants and death. On infected plant parts, light brown-colored mycelium and brown to dark brown sclerotia are formed.

The Causal Organism
The summer blight is caused by *Rhizoctonia solani* Kühn AG-1 (anastomosis group 1). The mycelium of *Rhizoctonia* is brown in color with septate hyphae. Fungus also produces brown- or black-colored sclerotia, and septa are dolipore type. The lateral branches are constricted at the point of origin from the main hypha.

Disease Cycle and Epidemiology
The fungus survives as mycelium or sclerotia in the diseased plant debris in the field. Sclerotia or mycelium in the soil come in contact with seedlings, germinate, and cause fresh infections. Conditions like high humidity and high temperature favor disease development and spread. The optimal temperature range for fungal growth and formation of sclerotia is 23–35 °C, with relative humidity at 100% and at 85–88%. An abrupt increase or decrease in the temperature accelerates the production of sclerotia.

Management
Follow crop rotation, destroy infected crop plants, and maintain field sanitation. Use biocontrol agents like *Pseudomonas fluorescens* or *Trichoderma harzianum* as foliar applications on crop plants. Use fungicides like carbendazim for foliar spray or soil application.

11 Guinea Grass Mosaic

The Guinea grass mosaic was first observed in 1976 in the Ivory Coast, Ghana, and France (Niblett and Paulsen 1975; Thouvenel et al. 1976).

Symptoms
The plants infected with Guinea grass mosaic disease show dwarfing and a light green mosaic or patterns of green and yellow color on the leaves. Initially, on young plants, symptoms appear as green or yellow eyespots. On dry leaves, eyespots become more elongated and spindle-shaped. Later on, the eyespots combine to form a striped mosaic.

The Causal Organism

Guinea grass mosaic is caused by the Guinea grass mosaic virus, which belongs to the virus group—Potato virus Y. The virus particles are flexuous rods about 15 nm in diameter. The virus is transmitted mechanically.

Disease Cycle and Epidemiology

The primary source of inoculums is the vegetative material used for sowing. As the virus is mechanically transmitted, it is transmitted through sap when plants come in contact with each other or indirectly through some means, like humans or implements, and also by environmental contamination. The epidemiology of the guinea grass virus has not been studied so far in detail.

Management

Obtain virus-free planting material or seeds for sowing. Rogue out the infected plants as soon as they are located in the fields.

12 Abiotic Stresses

12.1 Drought

Climate change has accelerated the frequency of drought and reduced rainfall in arid and semi-arid regions globally (Njarui et al. 2015; Siele et al. 2023) Drought is one of the most important factors that limit the growth of guinea grass. Genetic variability plays a vital role to alleviate the adverse effects of climatic conditions. Populations with high genetic variability perform better in relation to forage yield and drought adaptability. Threshold leaf water potential plays a critical role in selecting the lines for drought tolerance (Klar et al. 1978). Drought stress has an effect on morphological traits including growth and yield, and water use efficiency. It has a significant increase in leaf rolling score and relative water leaf ratio but reduces green forage, plant height, and water use efficiency. However, drought stress has no correlation with drought tolerance, tiller number, and leaf dying score (Purbajanti et al. 2012). Its persistence and productivity over the years are affected by drought frequency in semi-arid tropics. Drought stress significantly reduces guinea grass total biomass and membrane stability but upregulates the glutathione reductase and proline accumulation. Plant growth-promoting bacteria (PGPB) have a potential role to ameliorate drought stress by modulating the antioxidant system (Moreno-Galván et al. 2020). Further, *Bacillus* sp. XT13 and XT14 were used to formulate inoculants as alginate macro-beads. Mendoza-Labrador et al. (2021) reported biodegradability of inoculants in soil with slow release (two prototypes, namely 70:30 XT13 and 80:20XT14) and improved plant growth, nutritional parameters, and morphological traits under drought stress.

12.2 Heat Stress

Heat is an important abiotic stress affecting crop productivity significantly. Usberti and Jain (1978) studied the role of sexuality in guinea grass populations to heat stress provided a seedling stage. Sexual lines with high heterogeneity and polymorphism responded differently, mostly due to interlocus gene frequency associations. By 2050, global atmospheric CO_2 will reach 600 μmol mol^{-1} and the temperature will rise by 2%. The crop growth and productivity would be affected greatly, and food and fodder security would be at top risk. In guinea grass, elevated temperature and CO_2 affect the transcripts and metabolite profiles associated with stomatal function, secondary metabolism, and environmental responses.

12.3 Salinity Stress

Soil salinity is increasing gradually over time and can affect crop productivity significantly if not managed properly. Ion toxicity and imbalance, and osmotic stress are common in saline soil. It affects the crop's nutritive value and physiological, biochemical, and morphological traits (Benabderrahim and Elfalleh 2021; Ganie and Wani 2024). Mostly *Panicum* species are known to be salinity-tolerant. A low or moderate level of salinity has a stimulating effect on germination in guinea grass (Malaviya et al. 2019). Mostly genotypes showed higher germination percentage at EC 12 and 16 than the control under salinity stress. Under salinity, tolerant plants improve the water use efficiency and reduce the transpiration rate. However, the molecular mechanism involved in salinity tolerance must be investigated thoroughly.

12.4 Shade Stress

Variations in light intensity and quality affect the morphological and physiological responses of plants (Valladares et al. 2016; Malaviya et al. 2020). C4 plants have higher photosynthetic rates than C3 and CAM when exposed to higher temperatures and irradiance, as there is no CO_2 limitation (Kluge et al. 2015). Grasses are mostly studied under shade tolerance as they can undergo physiological and morphological changes and might be susceptible to insects and diseases. In guinea grass, electron transport rate and other physiological parameters were higher in 25 and 40% of shade (Lima et al. 2022). Pigment content was reported to be positively associated with shade level. Photosynthetic performance of plants was higher under intermediate and higher levels of shade.

13 Conclusion

Guinea grass stands out as a valuable forage resource for tropical and subtropical regions, owing to its high productivity, adaptability, and nutritional quality. Despite breeding challenges related to apomixis, polyploidy, and limited genetic diversity, both conventional and modern breeding strategies offer significant potential for genetic improvement. Integrating advanced molecular tools with traditional methods can accelerate the development of superior varieties. Addressing biotic and abiotic stresses through genetic resistance, agronomic management, and ongoing research will be key to maximizing the crop's resilience and productivity. Sustained efforts in germplasm conservation, trait enhancement, and breeding innovation will ensure guinea grass continues to support sustainable livestock systems and forage-based agriculture.

References

Adhikari A, Wang X, Lane B, Harmon PF, Goss EM (2021) First report of *Bipolaris yamadae* leaf spot disease on Guinea grass (*Panicum maximum*) in Florida. Plant Dis 105:1208

Aganga AA, Tshwenyane S (2004) Potentials of Guinea grass (Panicum maximum) as forage crop in livestock production. Pak J Nutr 3:1–4

Akande EK, Doma UD, Agu HO, Adamu HM (2010) Major anti-nutrients found in plant protein sources: their effect on nutrition. Pak J Nutr 9:827–932

Alkarawi HH, Zotz G (2014) Phytic acid in green leaves of herbaceous plants – temporal variation in situ and response to different nitrogen/phosphorus fertilizing regimes. AoB Plants 6. https://doi.org/10.1093/aobpla/plu048

Alonso J, Valenciaga N, Achang G, Mora C (2005) Study of the arthropods associated with a silvopastoral system leucaena-Guinea grass in different exploitation times. Cuban J Agric Sci 39:217

Anonymous (2010) The state of ex-situ conservation: the second report on the state of the world's PGRFA. FAO, Rome. http://www.fao.org/docrep/013/i1500e/i1500e03.pdf

Assienan B, Noirot M (1999) Inheritance, distribution and biology of andromonoecy in the agamic complex of the Maximae (Panicoideae). Theor Appl Genet 98:622–627

Benabderrahim MA, Elfalleh W (2021) Forage potential of non-native Guinea grass in north African agroecosystems: genetic, agronomic, and adaptive traits. Agronomy 11:1071. https://doi.org/10.3390/agronomy11061071

Bergman EN (1990) Energy contributions of volatile fatty acids from the gastrointestinal tract in various species. Physiol Rev 70:567–590. https://doi.org/10.1152/physrev.1990.70.2.567

Bilal M, Geng J, Chen L, García-Caparros P, Hu T (2025) Genome editing for grass improvement and future agriculture. Hortic Res 12(2):uhae293

Bogdan AV (1977) Panicum maximum. In: Bogdan AV (ed) Tropical pasture and fodder plants. Longmead, London, pp 181–191

Bolaric S, Barth S, Melchinger AE, Posselt UK (2005) Molecular genetic diversity within and among German ecotypes in comparison to European perennial ryegrass cultivars. Plant Breed 124:257–262

Boufaïed H, Chouinard PY, Tremblay GF, Petit HV, Michaud R, Belanger G (2003) Fatty acids in forages. Factors affecting concentrations. Can J Anim Sci 83:501–511

Brooks SPJ (2001) Proposed phytic acid standard including a method for its analysis. J AOAC Int 84:1125–1129

Chen S (2024) Advances in molecular breeding of forage crops: technologies, applications and prospects. Agriculture 14:279. https://doi.org/10.3390/agriculture14020279

Clapham WM, Foster JG, Neel JPS, Fedders JM (2005) Fatty acid composition of traditional and novel forages. J Agric Food Chem 53:10068–10073

Combes D, Pernes J (1970) Variations dans les nombres chromosomiques du Panicum maximum en relation avec le mode de reproduction. CR Acad Sci Paris 270:782

Daehler CC, Carino DA (1998) Recent replacement of native pili grass (Heteropogon contortus) by invasive African grasses in the Hawaiian islands. Pac Sci 52:220–227

de Sousa ACB, Jank L, de Campos T, Sforça DA, Zucchi MI, de Souza AP (2011) Molecular diversity and genetic structure of Guinea grass (Panicum maximum Jacq.), a tropical pasture grass. Trop Plant Biol 4:185–202. https://doi.org/10.1007/s12042-011-9081-6

Deo TG, Ferreira RC, Lara LA (2020) High-resolution linkage map with allele dosage allows the identification of regions governing complex traits and apospory in Guinea grass (Megathyrsus maximus). Front Plant Sci 11:15

Doerge RW (2002) Mapping and analysis of quantitative trait loci in experimental populations. Nat Rev Genet 3:43–52

Elgersma A, van der Horst EG, van Dijk H, van Soest LJM (2003) Influence of cultivar and cutting date on the fatty acid composition of perennial ryegrass (Lolium perenne L.). Grass Forage Sci 58:323–331. https://doi.org/10.1046/j.1365-2494.2003.00384.x

Elshire RJ, Glaubitz JC, Sun Q, Poland JA, Kawamoto K, Buckler ES, Mitchell SE (2011) A robust, simple genotyping-by-sequencing (GBS) approach for high diversity species. PLoS One 6:e19379. https://doi.org/10.1371/journal.pone.0019379

Farooq S, Shahzadi A, Razzaq A, Saleem F, Wani SH, Sandhu K (2023) Advances in genome editing for maize improvement. In: Maize improvement: current advances in yield, quality, and stress tolerance under changing climatic scenarios. Springer International Publishing, Cham, pp 181–194

Flint-Garcia SA, Thornsberry JM, Buckler ES (2003) Structure of linkage disequilibrium in plants. Annu Rev Plant Biol 54:357–374

Ganie SA, Wani SH (2024) Genetics of salt tolerance in plants: a central dogma perspective and strategies for enhancement. CABI

Ghane A, Malhotra PK, Sanghera GS, Verma SK, Jamwal NS, Kashyap L, Wani SH (2024) CRISPR/Cas technology: fueling the future of biofuel production with sugarcane. Funct Integr Genomics 24(6):205

González Marcillo RL, Castro Guamán WE, Guerrero Pincay AE, Vera Zambrano PA, Ortiz Naveda NR, Guamán Rivera SA (2021) Assessment of Guinea grass Panicum maximum under silvopastoral systems in combination with two management systems in Orellana Province. Ecuador Agric 11:117. https://doi.org/10.3390/agriculture11020117

Hagerman AE, Butler LG (1991) Tannins and lignins. In: Rosenthal GA, Berenbaum MR (eds) Herbivores: their interactions with secondary plant metabolites, pp 355–383

Haq SI, Zheng D (2022) Progresses of CRISPR/Cas9 genome editing in forage crops. J Plant Physiol 279:153860

Hassanpour S, Maherisis N, Eshratkhah B (2011) Plants and secondary metabolites (tannins): a review. Int J Soil Eros 1:47–53

Hatfield RD, Jung HJG, Broderick G, Jenkins TC (2007) Nutritional chemistry of forages. In: Barnes RF, Nelson CJ, Moore KJ, Collins M (eds) Forages: the science of grassland agriculture, 6th edn. Blackwell Publishing, pp 467–485

Hindoriya PS, Meena R, Singh M, Kumar R, Ram H, Meena VK, Kushwaha M (2019) Evaluation of kharif forage crops for biomass production and nutritional parameters in indo-Gangetic plains of India. Indian J Anim Nutr 36:25–29

Hirata K, Kusaba M, Chuma I, Osue J, Nakayashiki H, Mayama S, Tosa Y (2007) Speciation in Pyricularia inferred from multilocus phylogenetic analysis. Mycol Res 111:799–808. https://doi.org/10.1016/j.mycres.2007.05.014

Hitchcock AS, Chase A (1917) Grasses of the West Indies. Smithsonian Institution Press, Washington

Inoue T, Sumida Y (1999) The new leaf deca rot disease of Panicum maximum Jacq. (Guinea grass) caused by Rhizoctonia solani Kühn AG-1 and pathogenicity to the other plants. Bull Yamaguchi Agric Expt Stn 50:54–61

Islam M, Thakuria K (2002) Seasonal variation in green fodder production of important perennial grasses and legumes intercropping systems. J Agric Sci Soc North-East India 15:192–195

Jain A, Zadoo SN, Roy AK, Kaushal P, Malaviya DR (2003) Meiotic system and probable basic chromosome number of Panicum maximum Jacq. Accessions. Cytologia (Tokyo) 68:7–13. https://doi.org/10.1508/cytologia.68.7

Jain A, Roy AK, Kaushal P et al (2006) Isozyme banding pattern and estimation of genetic diversity among Guinea grass germplasm. Genet Resour Crop Evol 53:339–347. https://doi.org/10.1007/s10722-004-0574-0

Jank L, Valle CB, Resende RMS (2011) Breeding tropical forages. Crop Breed Appl Biotechnol 11:27–34

Joaquin BM, Hernandez A, Perez J et al (2001) Fertilizacion nitrogenada y momento de consecha en la produccion de semilla de pasto Guinea (Panicum maximum Jacq.): parametros y componentes de rendimiento. Pasturas Trop 23:10–15

Junges L, Oliveira Gomes EN, Ramires Candido A et al (2024) Longer cutting intervals on the characteristics of Guinea grass: morphogenetic, productive, and nutritional traits. Acta Sci 46:e60358

Kaushal P, Agrawal A, Malaviya DR, Siddiqui SA, Roy AK (2009) Ploidy manipulation in Guinea grass (Panicum maximum Jacq., Poaceae) utilizing a hybridization-supplemented apomixis-components partitioning approach (HAPA). Plant Breed 128:295–303. https://doi.org/10.1111/j.1439-0523.2008.01567.x

Kaushal P, Paul S, Saxena S, Dwivedi KK, Chakraborti M, Radhakrishna A, Roy AK, Malaviya DR (2015) Generating higher ploidies (7x and 11x) in Guinea grass (Panicum maximum Jacq.) utilizing reproductive diversity and uncoupled apomixis components. Curr Sci 109:1392–1395

Kaushal P, Dwivedi KK, Radhakrishna A, Saxena S, Paul S, Srivastava MK, Baig MJ, Roy AK, Malaviya DR (2018) Ploidy dependent expression of apomixis and its components in Guinea grass (Panicum maximum Jacq.). Euphytica 214:152. https://doi.org/10.1007/s10681-018-2232-1

Klar AE, Usberti JA Jr, Henderson DW (1978) Differential responses of Guinea grass populations to drought stress. Crop Sci 18:853–857. https://doi.org/10.2135/cropsci1978.0011183X001800050044x

Kluge RA, Tezotto-Uliana JV, da Silva PP (2015) Physiological and environmental aspects of photosynthesis. Rev Virtual Quím 7:56–73. https://doi.org/10.5935/1984-6835.20150004

Kobabe G (1983) Heterosis and hybrid seed production in fodder grass. In: [place unknown], pp 124–137. https://doi.org/10.1007/978-3-642-81977-3_5

Kraus TEC, Dahlgren RA, Zasoski RJ (2003) Tannins in nutrient dynamics of forest ecosystems – a review. Plant Soil 256:41–66

Kukla B, Thouvenel JC, Fauquet C (1984) A strain of Guinea grass mosaic virus from pearl millet in the Ivory Coast. J Phytopathol 109:65–73

Kumar R, Vaithiyanathan S (1990) Occurrence, nutritional significance and effect on animal productivity of tannins in tree leaves. Anim Feed Sci Technol 30:21–38

Lima DDP, Torre FD, Cordeiro ÁR, Scotti MR, França MGC (2022) Shade induces contrasting light photosynthetic performance between signal and Guinea grasses. Acta Sci Agron 44:e53561

Luo ZW, Zhang RM, Kearsey MJ (2004) Theoretical basis for genetic linkage analysis in autotetraploid species. Proc Natl Acad Sci USA 101:7040–7045

Maciel GA, Braga GJ, Guimarães R Jr, Ramos AKB, Carvalho MA, Fernandes FD, Jank L (2018) Seasonal liveweight gain of beef cattle on Guineagrass pastures in the Brazilian Cerrados. Agron J 110:480–487. https://doi.org/10.2134/agronj2017.05.0262
Malaviya DR, Roy AK, Anand A, Choubey RN, Baig MJ, Dwivedi K, Kushwaha N, Kaushal P (2019) Salinity tolerance of Panicum maximum genotypes for germination and seedling growth. Range Manag Agrofor 40:227–235
Malaviya DR, Baig MJ, Kumar B, Kaushal P (2020) Effects of shade on Guinea grass genotypes Megathyrsus maximus (Poales: Poaceae). Rev Biol Trop 68:563–572. https://doi.org/10.15517/rbt.v68i2.38362
Manamgoda DS, Cai L, Bahkali AH, Chukeatirote E, Hyde KD (2011) Cochliobolus: an overview and current status of species. Fungal Divers 51:3–42
Mendoza-Labrador J, Romero-Perdomo F, Abril J, Hernández JP, Uribe-Vélez D, Buitrago RB (2021) Bacillus strains immobilized in alginate macrobeads enhance drought stress adaptation of Guinea grass. Rhizosphere 19:100385. https://doi.org/10.1016/j.rhisph.2021.100385
Mishra S, Sharma S, Vasudevan P (2008) Comparative effect of biofertilizers on fodder production and quality in Guinea grass (Panicum maximum Jacq.). J Sci Food Agric 88:1667–1673. https://doi.org/10.1002/jsfa.3267
Moreno-Galván AE, Cortés-Patiño S, Romero-Perdomo F, Uribe-Vélez D, Bashan Y, Bonilla RR (2020) Proline accumulation and glutathione reductase activity induced by drought-tolerant rhizobacteria as potential mechanisms to alleviate drought stress in Guinea grass. Appl Soil Ecol 147:103367. https://doi.org/10.1016/j.apsoil.2019.103367
Mueller-Harvey I (2006) Unravelling the conundrum of tannins in animal nutrition and health. J Sci Food Agric 86:2010–2037
Muir JP, Jank L (2004) Guineagrass. In: Moser LE, Burson BL, Sollenberger LE (eds) Warm-season (C4) grasses, vol 45, pp 589–621. https://doi.org/10.2134/agronmonogr45.c17
Muñoz-González JC, Huerta-Bravo M, Lara BA, Rangel SR, la Rosa AJL (2016) Producción y calidad nutrimental de forrajes en condiciones del trópico húmedo de México. Rev Mex Cienc Agríc Pub Esp 16:3315–3327
Nakagawa H (1993) Cytological study and breeding of some tropical grasses. Bull Hiroshima Agric Res Cent
Nakagawa H, Hanna WW (1992) Induced sexual tetraploids for breeding guineagrass. J Jpn Grassl Sci 38:152–159
Nakajima K (1985) Isolation of sexual plants in guineagrass and the potential of its hybridization breeding in Japan. In: Proc XV Int Grassl Congr, [place unknown], pp 258–259
Nakajima K, Komatsu T, Mochizuki N, Suzuki S (1979) Isolation of diploid and tetraploid sexual plants in guineagrass. Jpn J Breed 28:228–238
Neely WC, Martin JM, Barker SA (1988) Products and relative reaction rates of the oxidation of tocopherols with singlet molecular oxygen. Photochem Photobiol 48:423–428. https://doi.org/10.1111/j.1751-1097.1988.tb02840.x
Niblett CL, Paulsen AQ (1975) Purification and further characterization of Panicum mosaic virus. Phytopathology 65:1157–1160
Njarui DM, Gatheru M, Mwangi DM, Keya GA (2015) Persistence and productivity of selected Guinea grass ecotypes in semiarid tropical Kenya. Grassl Sci 61:142–152. https://doi.org/10.1111/grs.12092
Noirot M (2001) Guinea grass. In: Tropical plant breeding. CRC Press, Boca Raton, pp 267–282
Okoli LC, Maureen OA, Obuam BE, Enemuo DC (2003) Studies on selected browses of southern Nigeria with particular reference to their proximate and some endogenous anti-nutritional constituent. Livest Res Rural Dev 15(9):123
Piluzza G, Sulas L, Bullitta S (2014) Tannins in forage plants and their role in animal husbandry and environmental sustainability: a review. Grass Forage Sci 69:32–48
Purbajanti ED, Anwar S, Wydiati KF (2012) Drought stress effect on morphology characters, water use efficiency, growth and yield of Guinea and Napier grasses. Int Res J Plant Sci 3(4):47–53

Ramírez-Ordoñes S, Rueda JA, Antonio-Cisneros CM, Sánchez-Hernández MA, Hernández-Bautista J (2022) Yield and quality of Guinea grass cv. Mombasa according to plant spacing and season of the year. Agro Productividad 59:2286

Resende MD (2015) Genética quantitativa e de populações. Suprema, Visconde do Rio Branco

Resende RMS, Casler MD, de Resende MDV (2013) Selection methods in forage breeding: a quantitative appraisal. Crop Sci 53:1925–1936

Rhodes AC, Plowes RM, Martins DJ et al (2022) The invasiveness of Guinea grass (Megathyrsus maximus) is characterized by habitat and differing herbivore assemblages in its native and invaded range. NeoBiota 78:25–44. https://doi.org/10.3897/neobiota.78.87069

Rosskopf EN, Abbasi M, Aime MC (2019) First report of Guinea grass smut caused by Conidiosporomyces ayresii in North America. Plant Dis 103:760. https://doi.org/10.1094/PDIS-08-18-1410-PDN

Roy AK, Malaviya DR, Kaushal P, Mahanta SK, Tewari R, Chauhan R, Chandra A (2020) Diversity study among Guinea grass (Megathyrsus maximus Jacq.) (Poales: Poaceae) genotypes and development of a core germplasm set. Plant Genet Resour 18:470–482. https://doi.org/10.1017/S1479262120000507

Ruckle M, Meier M, Frey L, Eicke S, Kölliker R, Zeeman S et al (2017) Diurnal leaf starch content: an orphan trait in forage legumes. Agronomy 7:16. https://doi.org/10.3390/agronomy7010

Samtiya M, Aluko RE, Dhewa T (2020) Plant food antinutritional factors and their reduction strategies: an overview. Food Prod Process Nutr 2:1–14

Sandhu JS, Kumar D, Yadav VK, Singh T, Sah RP, Radhakrishna A (2015) Recent trends in breeding of tropical grass and forage species. In: Roy MM, Malaviya DR, Yadav VK, Singh T, Sah RP, Vijay D, Radhakrishna A (eds) Proceedings of the XXIII International Grassland Congress. University of Kentucky, New Delhi

Sarkar D, Meitei CB, Das A, Ghosh PK, Mandal B (2018) Changes in soil organic carbon pools in a long-term trial with perennial fodder crops in acid soils of north-East India. Grass Forage Sci 73:473–481. https://doi.org/10.1111/gfs.12311

Sato H, Shimizu N, Nakagawa H, Nakajima K (1990) A new registered cultivar "Natsukaze" of guineagrass. JARQ 23:196–201

Savidan Y (1983) Genetics and utilization of apomixis for the improvement of guineagrass (Panicum maximum Jacq.). In: Smith JA, Hayes VW (eds) Proceedings of the 14th International Grassland Congress. Westview Press, Lexington, pp 182–184

Savidan Y (2000) In: Janick J (ed) Apomixis: genetics and breeding. Wiley, Oxford

Schweitzer JA, Madritch MD, Bailey JK, LeRoy CJ, Fischer DG, Rehill BJ, Whitham TG (2008) From genes to ecosystems: the genetic basis of condensed tannins and their role in nutrient regulation in a Populus model system. Ecosystems 11:1005–1020

Sehested J, Lund P (2007) Ruminal degradation of inositol bound P is affected by feed composition and external phytase. In: 58th annual meeting of the European association for animal production, vol, 13, pp 328

Siele SC, Nzuve FM, Cheboi JJ, Kirwa EC, Mwangi DM, Wanjala B (2023) Effects of drought on morpho-physiological and biomass yield of Guinea grass accessions. East Afr Agric For J 87(3–4):15

Silva VM, Putti FF, White PJ, Reis AR (2021) Phytic acid accumulation in plants: biosynthesis pathway regulation and role in human diet. Plant Physiol Biochem 164:132–146

Simeão RM, Resende MDV, Alves RS, Pessoa-Filho M, Azevedo ALS, Jones CS, Pereira JF, Machado JC (2021) Genomic selection in tropical forage grasses: current status and future applications. Front Plant Sci 12:665195

Sodeinde FG, Oladiop MA, Odunsi AA, Asaolu VO, Amao SR (2007) Mineral and feeding values of two cultivars of Guinea grass as influenced by soil type in the derived savanna zone of Nigeria. Agric J 2(2):226–230

Soti P, Thomas V (2022) Review of the invasive forage grass, Guinea grass (Megathyrsus maximus): ecology and potential impacts in arid and semi-arid regions. Weed Res 62(1):68–74

Sukhchain S (2010) Breeding Guinea grass – a review. Forage Res 31(2):109–115

Tessema ZK, Mihret J, Solomon M (2010) Effect of defoliation frequency and cutting height on growth, dry matter yield and nutritive value of Napier grass (Pennisetum purpureum (L.) Schumach). Grass Forage Sci 65:421–430

Thomas D, Laponte S (1989) Testing new accessions of Guinea grass (Panicum maximum). Trop Grassl 23(4):233

Thouvenel JC, Givord L, Pfeiffer P (1976) A new member of the potato virus Y group. Phytopathology 66:954–957

Tsukiboshi T, Okabe I, Sugawara K (2009) First report of blast of Guinea grass caused by Pyricularia sp LS-group in Japan. Plant Dis 93(12):135

Usberti JR, Jain SK (1978) The role of sexuality in the response of Guinea grass populations to heat stress. J Hered 69(3):188–190

Valladares F, Laanisto L, Niinemets Ü, Zavala MA (2016) Shedding light on shade: ecological perspectives of understorey plant life. Plant Ecol Divers 9(3):237–251

Van Elswyk ME, McNeill SH (2014) Impact of grass/forage feeding versus grain finishing on beef nutrients and sensory quality: the U.S. experience. Meat Sci 96:535–540

Van Soest PJ, Robertson JB (1985) Analysis of forages and fibrous foods. In: AS 613 Manual, Department of Anim Science. Cornell University, Ithaca

Vánky K, Bauer R (1992) Conidiosporomyces, a new genus of Ustilaginales. Mycotaxon 43:426–435

Verdecia DM, Herrera RS, Ramírez JL, Leonard I, Bodas R, Andrés S, Giráldez FJ, Álvarez Y, López S (2012) Valoración nutritiva del Panicum maximum vc. Mombasa en las condiciones climáticas del Valle del Cauto, Cuba. Rev Cubana Cienc Agric 46(1):97–101

Vijayakumar G, Babu C, Velayudham K (2010) A high yielding new Guinea grass variety CO (GG) 3. Madras Agric J 97:97–98

Wang X (2019) First report of Bipolaris yamadae leaf spot disease on Guinea grass in Florida. University Florida, Gainesville

Warmke HE (1954) Apomixis in Panicum maximum. Am J Bot 41(1):5–11

Weimer PJ (1996) Why don't ruminal bacteria digest cellulose faster? J Dairy Sci 79:1496–1502

Wood JD, Richardson RI, Nute GR, Fisher AV, Campo MM, Kasapidou E, Enser M (2004) Effects of fatty acids on meat quality: a review. Meat Sci 66:21–32

Younas A, Riaz N, Rashid M, Fiaz S, Tufail A, Noreen Z, Aslam M, Khan MU, Tabassum M (2025) Modification in conventional methods and modern plant breeding techniques to enhance genetic gain for future food security. In: Crop biofortification: biotechnological approaches for achieving nutritional security under changing climate, pp 377–394

Genomic Approaches for Alfalfa Breeding: Advances and Future Prospects

Urmila Devi, Jatinder Paul Kaur, Monika Bansal, Dolly Sharma, Shabir Hussain Wani, and Arafat Abdel Hamed Abdel Latef

1 Introduction

Medicago sativa L. is known as the queen of forages (Michaud et al. 1988). This perennial, outcrossing, and autotetraploid (2n = 4× = 32) feed crop known as cultivated alfalfa (*Medicago sativa* ssp. *sativa*) was chosen from the *M. sativa* complex that includes both diploid and tetraploid interfertile subspecies complex, originated in and around Transcaucasia and probably in central Asia, and is cultivated throughout globe in a total area of around 30 million hectares (Annicchiarico 2015). The cultivated alfalfa has a chromosome count of eight and a genome size of 800–1000 Mb. However, alfalfa's high heterozygosity, polysomic inheritance, and outcrossing nature make breeding difficult and prevent the development of inbred lines. As a result, Alfalfa breeding now depends on recurrent phenotypic selection; however, alternatives that use molecular marker-assisted breeding could speed up

U. Devi · J. P. Kaur
Department of Plant Breeding and Genetics, Punjab Agricultural University, Ludhiana, Punjab, India

M. Bansal
Monika Bansal, ChemBioBridge Research and Scientific Innovations Private Limited, Greater Noida, Uttar Pradesh, India

D. Sharma
Department of Biotechnology, Noida Institute of Engineering and Technology, Greater Noida, Uttar Pradesh, India

S. H. Wani
National Seed Project (NSP), Sher-e-Kashmir University of Agricultural Sciences and Technology – Kashmir (SKUAST-K), Srinagar, Jammu and Kashmir, India

A. A. H. A. Latef (✉)
Faculty of Science, Department of Botany and Microbiology, South Valley University, Qena, Egypt

S. H. Wani, S. S. Gosal (eds.), *Accelerated Plant Breeding, Volume 5*,
https://doi.org/10.1007/978-3-032-04527-0_8

alfalfa traits improvement and increase genetic gain per unit time and expense. Alfalfa also boosts the productivity of subsequent crops through rotation and increases legume nitrogen credits. It improves the soil environment by reducing erosion and remediating soil. Additionally, compared to other fodder crops, alfalfa has comparatively higher protein and less fiber, which contributes to better forage intake. It also serves alternative uses as a natural biofuel, essential proteins, and a source of industrial enzymes, including lignin peroxidase, cellulase, and phytase, and new medicinal chemicals. Current issues with alfalfa fodder cultivation and use include biotic and abiotic stressors, decreased seed yield, anti-quality characteristics, inefficient nutrient utilization, and low hay digestibility. In the present decade, genomics-based techniques like molecular markers, genomic selection, genome-wide association strategies, and genome editing facilitate more accurate, effective, and focused methods of creating novel plant varieties and have completely changed the field of plant breeding. The most pertinent developments in the creation of genomic tools are presented and discussed in this chapter, along with instances of their use in plant breeding. Significant advances in next-generation sequencing technology over the past 20 years have also made genomics-assisted breeding a viable substitute for traditional breeding methods for complicated traits (Razzaq et al. 2021). In order to improve crop production, disease resistance, and stress tolerance, important approaches such as Genome-Wide Association study (GWAS), Marker-Assisted Selection (MAS), and Genomic Selection (GS) help in precise trait selection and breeding outcome prediction (Wani et al. 2020; Choudhary et al. 2022; Singh et al. 2023; Kumar et al. 2024; Wasudeo et al. 2025) embodying a paradigm shift by replacing trial-and-error methods with data-driven, predictive approaches.

2 Alfalfa: Genome Sequence and Assembly

Alfalfa has a highly complex genome that makes it difficult to understand its genome and improvements of its traits. Previous studies concerning alfalfa genetic and genomic resources have substantially utilized the sequenced diploid *Medicago truncatula* with 2n = 2× = 16 = 860 Mb; it is considered a closely related species with cultivated alfalfa. Chen et al. (2018) prepared an allele-aware chromosome-level genome assembly for the cultivated alfalfa using PacBio CCS (circular consensus sequencing) and Hi-C (high-throughput chromosome conformation capture) technologies. Recent advances in algorithms and technologies for sequencing and scaffolding have made it possible to sequence and assemble five distinct alfalfa accessions (Mudge and Farmer 2021). These assemblies consist of three tetraploid assemblies (NECS-141, Zhongmu No.1, and Xinjiang DaTe) and two diploid assemblies (CADL and PI464715). In the roughly 5 years that these assemblies were created, technological advancements have made it possible for assemblies to become more contiguous and for scaffolding to get better, leading to chromosome-level assemblies that enable the detection of extensive structural rearrangements.

Long et al. (2022) revealed a thorough database of genomic variants based on resequencing of 220 germplasms and the de novo assembly of the allele-aware chromosome-level genome of Zhongmu-4, a cultivar that is commonly grown in China. 2.56-Gb contigs were anchored to 32 pseudo-chromosomes, and an estimated 2.74-Gb contigs were assembled, representing 88.39% of the genome size.

3 Transcriptomics and Gene Expression Studies

Transcriptome sequencing is a high-resolution, sensitive, and high-throughput next-generation sequencing (NGS) technique for researching non-model plants and other species. It is a collection of RNA transcripts from single or whole samples of developmental and functional phases. RNA-Seq is an important method for finding gene predictions and mining functional analysis that improves understanding mechanisms of gene ontogeny, biological processes, molecular functions, and other cellular components (Tyagi et al. 2022). The differentially expressed genes (DEGs) linked to salt stress in two alfalfa cultivars, the salt-tolerant "Halo" and the salt-intolerant "Vernal," were found using an RNA-Seq approach. Samples of leaf and root tissues were taken for RNA extraction at 0–27 hours under a NaCl-induced salt stress at a level of 12 dSm^{-1}. The differential expression pattern of the two cultivars under salt stress at different times is particularly emphasized by the unequal expression pattern during this research (Bhattarai et al. 2021). Transcriptome profiling of the leaves from two genotypes of alfalfa was performed for freezing-tolerant "Gannong NO.3" and freezing-sensitive "WL326GZ" exposed to −10 °C (Wang et al. 2022). Identification and examination of drought-responsive genes in the roots of several alfalfa root types helped us to learn more about the mechanism by which alfalfa (*Medicago sativa*) responds to drought stress; using PEG-6000, the transcriptome data and physiological traits of the root systems of the rhizomatous-rooted *M. sativa* cv. "Qingshui" (QS), tap-rooted *M. sativa* cv. "Longdong" (LD), and creeping-rooted *M. varia* cv. "Gongnong No. 4" (GN) were examined under simulated drought stress (Wang et al. 2024). Transcriptome sequencing investigation of alfalfa in response to freezing and cold stress was conucted by Shu et al. (2017), identified many transcripts that are sensitive to cold and/or freezing, which were found to be primarily involved in signal transduction, transcriptional control, hormone signaling pathways, stress response, etc. Liu et al. (2013) studied transcriptomic data from 15 tissue types of the "Golden Queen" alfalfa cultivar, including germinated seeds, cotyledons, leaves, stems, inflorescences, and pods. A total of 40,433 unigenes and 1694 SSR markers (2–6 nucleotides, ≥5 repeats) were identified. The transcriptomes of tissue from elite and unimproved populations of alfalfa were sequenced using RNA-Seq by Li et al. (2012), and because the study was centered on using alfalfa as a feedstock for biofuel, stems were employed. The study employed principal components analysis (PCA) to examine the correlations between the various varieties and found 31,760 indels and 871,384 SNPs. Luo et al. (2019) investigated salt stress as a significant environmental element influencing alfalfa development and productivity using PacBio RSII. Zhang et al. (2015) demonstrated that the response to salt stress includes both osmotic and ionic components. They created the first full-length transcriptome database of alfalfa root tips treated with both mannitol and NaCl. They discovered 8861 DEGs that were regulated by NaCl and 8016 that were regulated by mannitol.

4 Marker-Assisted Selection (MAS) in Alfalfa Breeding

Quantitative trait loci (QTL) mapping and genome-wide association studies (GWAS) are good techniques that have helped discover QTLs regulating traits such as yield, forage quality, and stress tolerance in plants such as alfalfa (Jiang et al. 2022). In QTL mapping, a biparental mapping population is applied that results in longer linkage regions. Multiple QTLs linked to alfalfa forage and crown root (CR) biomass during drought were assessed for their ability to increase elite germplasm's forage yield in water-limited environments. DNA marker alleles linked to ten biomass QTL were introduced into three elite backgrounds using marker-assisted selection (Singh et al. 2022). Zhang et al. (2014a, b) studied 352 individuals from two biparental crossings that were phenotyped for vulnerability to Verticillium wilt (VW) and genotyped using high-resolution melting (HRM) for 172 previously reported SNP markers. There was a strong correlation between 21 markers and VW resistance. Ray et al. (2015) obtained genotypes of the progeny from an original set of backcrosses involving drought-resistant and drought-susceptible plants. Sledge et al. (2005) phenotyped a total of 253 plants for yield under drought. Using both new SNPs found through a candidate-gene approach and pre-existing SSR markers, genotyping was carried out. A linkage map was created using a total of 600 markers, including 557 SSR markers and 43 candidate-gene SNPs. McCord et al. (2014) conducted a study to map quantitative trait loci (QTL) related to yield, lodging resistance, and spring vigor. They created a mapping population of 128 plants by backcrossing a hybrid of lodging-resistant and susceptible plants to the susceptible parent line. To construct the linkage map, they utilized 58 SRAP, 142 AFLP, and 36 SSR markers. The research identified three significant QTLs associated with lodging resistance, seven related to forage yield, and four linked to spring vigor. Khu et al. (2013) identified QTL associated with aluminum tolerance using a biparental cross between a susceptible and a resistant parent. To phenotype petiole-derived callus growth and the total root length ratio (TRLR) of stem cuttings cultured in medium, the study employed 1024 SSR markers that were genotyped by PCR. Twenty of the 257 markers that were discovered to be polymorphic between the parent plants were significantly correlated with the formation of calluses under aluminum exposure, while 41 were correlated with TRLR. Jia et al. (2017) used a panel of 336 individuals from 75 accessions to perform association mapping for protein and mineral content. The plants were phenotyped for protein content and various micronutrients and genotyped for 85 SSR markers. There were 85 correlations between markers and traits. Mackie et al. (2007) determined the locations of QTLs for reaction to *C. trifolii* races 1, 2, and 4 in the pest-resistant autotetraploid alfalfa clone W126. For three *C. trifolii* characteristics, interactions between specific QTLs and phenotypic variants have been reported. The markers that were obtained could be used to introduce numerous sources of resistance in alfalfa breeding.

5 Genomic Selection (GS)

A breeding technique known as genomic selection selects individuals (Heffner et al. 2009; Meuwissen et al. 2001) for breeding by genotyping a large number of polymorphic sites using a statistical model which connects these markers for one or more required traits. To get the maximum coverage of the genome, the model must first be trained by using a training population that is phenotyped and genotyped at several loci. The increased ability to choose many traits at once is another possible benefit of GS. This is particularly crucial for perennial forages due to the high expense of phenotyping and the requirement to choose at least 8–10 parent genotypes for a synthetic variety (Annicchiarico et al. 2021), and finding correlations between alleles at the genotyped loci and the phenotype or phenotypes being evaluated is done using the statistical model. Using GBS, Li et al. (2015) genotyped a population of alfalfa derived from a blend of commercial cultivars, finding 175,000 SNPs, then phenotyped it for dry yield. A mixed linear model was used to test for associations. Annicchiarico et al. 2015 discovered 68,972 and 77,610 SNP markers after performing GBS on 278 individuals from two populations (Po Valley and Mediterranean, 124 and 154 plants, respectively). Using biomass production data from the populations, the study employed a cross-validation process to test for the precision of several prediction models using different marker imputation techniques under varied degrees of simulated missing genotypes and filtering settings. Jia et al. (2018) used three Bayesian statistical approaches (BayesA, BayesB, and BayesCπ) to predict 25 agronomic and quality traits using the genome of *Medicago sativa*. They showed that the prediction accuracy varied depending on the trait under study. Another strategy is genomic selection (GS), which uses genome-wide markers to calculate each individual's genomic estimated breeding value (GEBV) within a population. In three field trials spanning 7 years, the current study assessed two full-sib populations of 149 and 392 progeny lines that were segregated for yield-related features. SNP markers covering the complete genome were subsequently obtained by genotyping the two F1 populations using GBS and RAD-seq. Eight GS prediction models were developed to examine the impact of the size of populations and SNP markers on genome prediction accuracy after comparing the number of SNP markers from the two library generation techniques.

6 Genetic Engineering in Alfalfa for Trait Improvement

Alfalfa was among the first few plants for which the methods of regeneration in tissue culture were established. There are currently two transgenic features which are available for commercial use in alfalfa: the first one is with decreased lignin (HarvXtra®), and the second one is with resistance to the herbicide glyphosate (Roundup Ready®) (L.-X. Yu and C. Kole). Studies on edible vaccines that can be given to protect livestock were performed in alfalfa which expressed an antigen for the avian reovirus (Huang et al. 2006). According to earlier research, alfalfa's tolerance to salinity stress was increased when stress-related genes were overexpressed (Luo et al. 2019). Li et al. (2014) used the strong stress-inducible SWEETPOTATO PEROXIDASE ANIONIC 2 (SWPA2) promoter to successfully transform CHOLINE OXIDASE A (CODA) cDNA, which was derived from *Agrobacterium globiformis*, transformed to alfalfa chloroplasts. Zhang et al. (2014a, b) used agrobacterium-mediated transformation for transferring the exogenous gene *SeNHX1* (*Salicornia europaea* NHX1) into alfalfa, and transformed plants demonstrated increased tolerance to salt stress. Enhancing resistance to abiotic stresses was the major goal of early alfalfa genetic engineering initiatives. The protein-coding gene *QQS* (Qua-Quine Starch; At3g30720) is an orphan gene that is unique to *Arabidopsis thaliana* and has been shown to control carbon and nitrogen partitioning by binding to the conserved transcription factor nuclear factor Y, subunit C4 (NF-YC4) in a variety of species. Without changing the morphology of the plants, ectopic expression of QQS in corn, rice, and soybeans also increased the protein content in the grains and leaves (Li et al. 2009, 2015). Wang et al. (2023) carried out research and proved that QQS overexpression in alfalfa enhances tolerance to low-nitrogen stress and improves a number of critical agronomic characteristics, such as biomass production, soluble sugar content, crude protein content, and fodder digestibility. Alfalfa affected by root lesion nematodes and leaf beetles was transformed by Rice ORYZACYSTATIN-I (OC-I) and ORYZACYSTATIN-II (OC-II) genes, driven by a potato wound-inducible promoter. These transgenic plants showed increased mortality of *Phytodecta fornicata* larvae and decreased *Pratylenchus penetrans* population (Ninković et al. 1995; Samac and Smigocki 2003). A fungal endochitinase (ECH42) was released by transgenic alfalfa plants generated by Tesfaye et al. (2005). The chitinase activity in the root exudates and vegetative organs of these transgenic plants was up to 25.7 times higher. These released endochitinases demonstrated antifungal efficacy by preventing the spore germination of two fungal diseases, *Phoma medicaginis* and *C. trifolii*, in addition to maintaining their lytic action against glycol chitin. According to García et al. (2014), SNAKIN proteins are antibacterial substances that are part of plant innate immunity. Alfalfa transgenic plants with SNAKIN-1 (MsSN1) under constitutive promoter control demonstrated enhanced resistance to harmful fungi. When treated with *C. trifolii* and the oomycete *P. medicaginis*, three separate transgenic lines containing the CaMV35S: MsSN1 construct displayed noticeably fewer infected leaves than wild-type plants (García et al. 2014).

7 CRISPR and Genome Editing Technologies

New improvements in gene-editing technologies have made it possible to genetically modify a wide range of crops that can be accessed through genetic transformation, which has the potential to create new traits of interest (Mushtaq et al. 2021; Wani and Hensel 2022; Wani et al. 2022; Fan et al. 2025). Breeding plans for crop plants have been altered by this method, which is better suited for manipulating plants with numerous allele copies, such as alfalfa. Alfalfa is a long-day (LD) plant that is sensitive to photoperiod and can complete both its vegetative and reproductive stages quickly. Rapid flowering, however, may reduce the quality and production of feed biomass. Using multiplex CRISPR/Cas9-mediated mutagenesis of FLOWERING LOCUS Ta1 (MsFTa1), a crucial floral integrator and activator gene, we tried to postpone alfalfa flowering. With increases of up to 78% in fresh weight and 76% in dry weight as compared to controls, six distinct MsFTa1 lines with mutations in all four copies of MsFTa1 accumulated noticeably higher forage biomass output (Wolabu et al. 2023). A key component of GA-regulated plant architecture is MtGA3ox1. Two *MtGA3ox* genes, *MtGA3ox1* and *MtGA3ox2*, were studied. Crop breeding could be accelerated through genome editing. The advantages of the MsGA3ox1 gene-edited semidwarf and prostrate alfalfa include a high leaf/stem ratio, numerous branches, and a high crude protein content. These traits of alfalfa may provide the benefits of improved moisture retention, weed control, and ground coverage (Zheng et al. 2022). The gene PALM1, linked to a distinctive multifoliate phenotype, was targeted, which leads to editing at two locations in the wild-type Gongnong 1 using the CRISPR_2.0 toolset; as a result, tetra-allelic homozygous *palm1* mutations exhibited a clear multifoliate phenotype (Zhao et al. 2024). It was discovered that alfalfa SQUAMOSA PROMOTER-BINDING PROTEIN-LIKE8 (MsSPL8) improved drought tolerance and promoted branching, regrowth, and biomass. Singer et al. (2022) used CRISPR/Cas9-induced mutations in the gene to investigate the role of MsSPL8 in alfalfa, assessing the technology's efficiency in inducing phenotypic changes by targeting the MsSPL8 gene across a range of mutational doses. Plants with the greatest number of mutant MsSPL8 alleles also showed notable reductions in the length of internodes, plant height, biomass of root and shoot, and root length, suggesting that phenotypic impacts were influenced by mutational dosage. A ubiquitin-conjugating E2 enzyme (UBC24), encoded by PHO2, has been previously identified in *Medicago truncatula*. CRISPR/Cas9 technology was used in tetraploid alfalfa to cause mutations in PHO2 that causes Pi homeostasis to be disrupted, leading to Pi hyper-accumulation. In addition to hyper-accumulating Pi, the resultant alfalfa mutants show promise as a bioremediation technique for phosphate-contaminated soils (Miller et al. 2022).

8 Current Challenges

Alfalfa plants are very heterozygous and display significant inbreeding depression, which prevents the generation of inbred lines (Julier et al. 2003). Identification of genetic diversity and linkages within and between distinct alfalfa populations is critical for alfalfa genetic improvement (Wang et al. 2023). Information on local adaptation under long-term grazing is especially important, as alfalfa stands are frequently influenced by grazing (Smith et al. 1989), environmental conditions, and their interactions, all of which can result in genetic alterations over time (Jefferson and Cutforth 1997). Long-term continuous grazing can deplete alfalfa's total non-structural carbohydrates (TNC) and nitrogen reserves in the root, as well as eliminate photosynthetically active leaf areas, making the plant more susceptible to biotic and abiotic stresses such as winter injuries and disease infections (Berdahl et al. 1986).

References

Annicchiarico P (2015) Alfalfa forage yield and leaf/stem ratio: narrow-sense heritability, genetic correlation, and parent selection procedures. Euphytica 205:409–420

Annicchiarico PN, Nazzicari X, Li Y, Wei L, Pecetti EC (2015) Brummer, accuracy of genomic selection for alfalfa biomass yield in different reference populations. BMC Genomics 16:1020

Annicchiarico P, Nazzicari N, Pecetti L (2021) Genomic selection for higher yield and quality in alfalfa. In: Yu LX, Kole C (eds) The alfalfa genome. Compendium of plant genomes. Springer, Cham

Berdahl J, Wilton A, Lorenz R, Frank A (1986) Alfalfa survival and vigor in rangeland grazed by sheep. Rangeland Ecol Manag/J Range Manag Arch 39:59–62

Bhattarai S, Fu YB, Coulman B et al (2021) Transcriptomic analysis of differentially expressed genes in leaves and roots of two alfalfa (*Medicago sativa* L.) cultivars with different salt tolerance. BMC Plant Biol 21:446

Choudhary JR, Get S, Tripathi A, Kaldate R, Rana M, Mehta S, Ahlawat J, Bansal M, Zaid A, Wani SH (2022) Breeding efforts for crop productivity in abiotic stress environment. In: Augmenting crop productivity in stress environment. Springer Nature Singapore, Singapore, pp 63–103

Fan S, Jia L, Wu J, Zhao Y (2025) Harnessing the potential of CRISPR/Cas in targeted alfalfa improvement for stress resilience. Int J Mol Sci 26(7):3311

García AN, Ayub ND, Fox AR, Gómez MC, Diéguez MJ, Pagano EM et al (2014) Alfalfa snakin-1 prevents fungal colonization and probably coevolved with rhizobia. BMC Plant Biol 14:248. https://doi.org/10.1186/s12870-014-0248-9

Huang L-K, Liao S-C, Chang C-C, Liu H-J (2006) Expression of avian reovirus rC protein in transgenic plants. J Virol Methods 134:217–222. https://doi.org/10.1016/j.jviromet.2006.01.013

Jefferson P, Cutforth H (1997) Sward age and weather effects on alfalfa yield at a semi-arid location in southwestern Saskatchewan. Can J Plant Sci 77:595–599

Jia CXW, Chen M, Wang Y, Liu X, Gong P, Xu Q, Wang X, Gao H, Wang Z (2017) Identification of genetic loci associated with crude protein and mineral concentrations in alfalfa (Medicago sativa) using association mapping. BMC Plant Biol 17:97

Jia C, Zhao F, Wang X, Han J, Zhao H, Liu G, Wang Z (2018) Genomic prediction for 25 agronomic and quality traits in alfalfa (*Medicago sativa*). Front Plant Sci 9:1220

Julier B, Flajoulot S, Barre P, Cardinet G, Santoni S, Huguet T, Huyghe C (2003) Construction of two genetic linkage maps in cultivated tetraploid alfalfa (*Medicago sativa*) using microsatellite and AFLP markers. BMC Plant Biol 3:9

Khu DM, Reyno R, Han Y, Zhao PX, Bouton JH, Brummer EC, Monteros MJ (2013) Identification of aluminum tolerance quantitative trait loci in tetraploid alfalfa. Crop Sci 53:148–163

Kumar R, Das SP, Choudhury BU et al (2024) Advances in genomic tools for plant breeding: harnessing DNA molecular markers, genomic selection, and genome editing. Biol Res 57:80

Li LC, Foster Q, Gan D, Nettleton M, James A, Myers E (2009) Wurtele, identification of the novel protein QQS as a component of the starch metabolic network in Arabidopsis leaves. Plant J 58(3):485–498

Li XA, Acharya AD, Farmer JA, Crow AK, Bharti RS, Kramer Y, Wei Y, Han J, Gou GD, May MJ, Monteros EC (2012) Brummer, prevalence of single nucleotide polymorphism among 27 diverse alfalfa genotypes as assessed by transcriptome sequencing. BMC Genomics 13:568

Li H, Wang Z, Ke Q, Ji CY, Jeong JC, Lee HS et al (2014) Overexpression of codA gene confers enhanced tolerance to abiotic stresses in alfalfa. Plant Physiol Biochem 85:31–40

Li XY, Wei A, Acharya JL, Hansen JL, Crawford DR, Viands R, Michaud A, Claessens EC (2015) Brummer, genomic prediction of biomass yield in two selection cycles of a tetraploid alfalfa breeding population. Plant Genome 8:eplantgenome2014.12.0090

Liu Z, Chen T, Ma L, Zhao Z, Zhao PX, Nan Z et al (2013) Global transcriptome sequencing using the Illumina platform and the development of EST-SSR markers in autotetraploid alfalfa. PLoS One 8(12):e83549

Long R, Zhang F, Zhang Z, Li M, Chen L, Wang X, Liu W, Zhang T, Yu LX, He F, Jiang X, Yang X, Yang C, Wang Z, Kang J, Yang Q (2022) Genome assembly of alfalfa cultivar Zhongmu-4 and identification of SNPs associated with agronomic traits. Genomics Proteomics Bioinformatics 20(1):14–28

Luo D, Zhou Q, Wu Y, Chai X, Liu W, Wang Y et al (2019) Full-length transcript sequencing and comparative transcriptomic analysis to evaluate the contribution of osmotic and ionic stress components towards salinity tolerance in the roots of cultivated alfalfa (*Medicago sativa* L.). BMC Plant Biol 19:32

Mackie JM, Musial JM, Armour DJ, Phan HTT, Ellwood SE (2007) Identification of QTL for reaction to three races of *Colletotrichum trifolii* and further analysis of inheritance of resistance in autotetraploid lucerne. Theor Appl Genet 114:1417–1426. https://doi.org/10.1007/s00122-007-0527-z

McCord P, Gordon V, Saha G, Hellinga J, Vandemark G, Larsen R, Smith M, Miller D (2014) Detection of QTL for forage yield, lodging resistance and spring vigor traits in alfalfa (Medicago sativa L.). Euphytica 200:269–279

Meuwissen THE, Hayes BJ, Goddard ME (2001) Prediction of total genetic value using genome-wide dense marker maps. Genetics 157:1819–1829

Michaud R, Lehman WF, Rumbaugh MD (1988) World distribution and historical development. Alfalfa Alfalfa Improv 29:25–91

Miller SS, Dornbusch MR, Farmer AD, Huertas R, Gutierrez-Gonzalez JJ, Young ND, Samac DA, Curtin SJ (2022) Alfalfa (Medicago sativa L.) pho2 mutant plants hyperaccumulate phosphate. G3 12(6):jkac096

Mudge J, Farmer AD (2021) Sequencing, assembly, and annotation of the alfalfa genome. In: Yu LX, Kole C (eds) The alfalfa genome. Compendium of plant genomes. Springer, Cham

Mushtaq M, Pir HA, Zaid A, Wani SH (2021) CRISPR/Cas system: a powerful approach for enhanced resistance against rice blast. In: Biocontrol agents and secondary metabolites. Woodhead Publishing, pp 649–658

Ninković S, Miljuš-Đukić J, Nešković M (1995) Genetic transformation of alfalfa somatic embryos and their clonal propagation through repetitive somatic embryogenesis. Plant Cell T Organ Cult 42:255–260. https://doi.org/10.1007/BF00029996

Ray IM, Han Y, Meenach LECD, Santantonio N, Sledge MK, Pierce CA, Sterling TM, Kersey RK, Bhandari HS, Monteros MJ (2015) Identification of quantitative trait loci for alfalfa forage biomass productivity during drought stress. Crop Sci 55:2012–2033

Razzaq A, Kaur P, Akhter N, Wani SH, Saleem F (2021) Next-generation breeding strategies for climate-ready crops. Front Plant Sci 12:620420

Samac D, Smigocki A (2003) Expression of oryzacystatin I and II in alfalfa increases resistance to the root-lesion nematode. Phytopathology 93:799–804. https://doi.org/10.1094/PHYTO.2003.93.7.799

Shu Y, Li W, Zhao J, Zhang S, Xu H, Liu Y, Guo C (2017) Transcriptome sequencing analysis of alfalfa reveals CBF genes potentially playing important roles in response to freezing stress. Genet Mol Biol 40(4):824–833

Singer SD, Burton Hughes K, Subedi U, Dhariwal GK, Kader K, Acharya S, Chen G, Hannoufa A (2022) The CRISPR/Cas9-mediated modulation of SQUAMOSA PROMOTER-BINDING PROTEIN-LIKE 8 in alfalfa leads to distinct phenotypic outcomes. Front Plant Sci 12:774146

Singh L, Pierce C, Santantonio N, Steiner R, Miller D, Reich J, Ray I (2022 Mar) Validation of DNA marker-assisted selection for forage biomass productivity under deficit irrigation in alfalfa. Plant Genome 15(1):e20195

Singh B, Wani SH, Kukreja S, Kumar V, Goutam U (2023) Genome-wide association studies (GWAS) for agronomic traits in maize. In: Maize improvement: current advances in yield, quality, and stress tolerance under changing climatic scenarios. Springer International Publishing, Cham, pp 83–98

Sledge MK, Ray IM, Jiang G (2005) An expressed sequence tag SSR map of tetraploid alfalfa (*Medicago sativa* L.). Theor Appl Genet 111:980–992

Smith S, Bouton J, Hoveland C (1989) Alfalfa persistence and regrowth potential under continuous grazing. Agron J 81:960–965

Tesfaye M, Denton MD, Samac DA, Vance CP (2005) Transgenic alfalfa secretes a fungal endochitinase protein to the rhizosphere. Plant Soil 269:233–243. https://doi.org/10.1007/s11104-004-0520-0

Tyagi P, Singh D, Mathur S, Singh A, Ranjan R (2022) Upcoming progress of transcriptomics studies on plants: an overview. Front Plant Sci 13

Wang X, Kang W, Wu F, Miao J, Shi S (2022) Comparative transcriptome analysis reveals new insight of alfalfa (Medicago sativa L.) cultivars in response to abrupt freezing stress. Front Plant Sci 13

Wang H, Coulman B, Bai Y et al (2023) Genetic diversity and local adaption of alfalfa populations (*Medicago sativa* L.) under long-term grazing. Sci Rep 13:1632

Wang K, Nan LL, Xia J et al (2024) Transcriptome analysis reveals candidate genes for different root types of alfalfa (*Medicago sativa*) after water stress induced by PEG-6000. Chem Biol Technol Agric 11:107

Wani SH, Hensel G (eds) (2022) Genome editing: current technology advances and applications for crop improvement. Springer International Publishing, Cham

Wani SH, Choudhary JR, Choudhary M, Rana M, Gosal SS (2020) Recent advances in genomics assisted breeding for drought stress tolerance in major cereals. J Cereal Res 12(1):1–2

Wani SH, Choudhary M, Barmukh R, Bagaria PK, Samantara K, Razzaq A, Jaba J, Ba MN, Varshney RK (2022) Molecular mechanisms, genetic mapping, and genome editing for insect pest resistance in field crops. Theor Appl Genet 135(11):3875–3895

Wasudeo NG, Kumar MA, Gautam BK, Subhasri K, Wani SH (2025) Metabolomics and breeding for crop improvement. In: Plant specialized metabolites: phytochemistry, ecology and biotechnology. Springer Nature Switzerland, Cham, pp 859–892

Wolabu TW, Mahmood K, Jerez IT, Cong L, Yun J, Udvardi M, Tadege M, Wang Z, Wen J (2023) Multiplex CRISPR/Cas9-mediated mutagenesis of alfalfa FLOWERING LOCUS Ta1 (MsFTa1) leads to delayed flowering time with improved forage biomass yield and quality. Plant Biotechnol J 21(7):1383–1392. https://doi.org/10.1111/pbi.14042. Epub 2023 Mar 25

Zhang LQ, Niu YD, Huridu H, Hao JF, Qi Z, Hasi A (2014a) Salicornia europaea L. Na+/H+ antiporter gene improves salt tolerance in transgenic alfalfa (*Medicago sativa* L.). Genet Mol Res 13:5350–5360

Zhang T et al (2014b) Identification of molecular markers associated with Verticillium wilt resistance in alfalfa (*Medicago sativa* L.) using high-resolution melting. PLoS One 9:e115953

Zhang S, Shi Y, Cheng N, Du H, Fan W, Wang C (2015) *De novo* characterization of fall dormant and nondormant alfalfa (*Medicago sativa L.*) leaf transcriptome and identification of candidate genes related to fall dormancy. PLoS One 10:e0122170. https://doi.org/10.1371/journal.pone.0122170

Zheng L, Wen J, Liu J, Meng X, Liu P, Cao N, Dong J, Wang T (2022) From model to alfalfa: gene editing to obtain semidwarf and prostrate growth habits. Crop J 10(4):932–941

Utilization of Genetic and Genomic Resources for Accelerated Breeding for Millet Improvement

Rahul Kapoor, Ankita K. Kawadiwale, Rutuparna Pati, Harpreet Kaur Oberoi, Pawan Kumar Malhotra, and Shabir Hussain Wani

1 Introduction

Small-seeded, nutrient-rich, edible cereals with usage as livestock feed belonging to the Graminae (Poaceae) Family grown under semi-arid tropical regions of developing nations are referred to as millets. Since millets are utilized as important food from semi-arid tropical regions of emerging nations, they are also referred to as food for people in the low-income group. They can be categorized into Sorghum (*Sorghum bicolor* L.) and pearl millet (*Pennisetum glaucum*) (major millets) and Finger millet (*Eleusine coracana*), Foxtail millet (*Setaria italica*), Proso millet (*Panicum miliaceum*), Barnyard millet (*Echinochloa* spp.), Kodo millet (*Paspalum scrobiculatum*) (minor millets) (Ajeesh Krishna et al. 2022). Millets are found to be rich in minerals, particularly iron, calcium, phosphorus, zinc, magnesium, and potassium, along with high contents of dietary fiber, polyphenols, fatty acids, proteins, and carbohydrates. Richness of nutrients and longer storage life make millets a famine reserve crop, truly referred to as nutria millets/nutria-cereals/nutraceuticals (Swaminaidu et al. 2015). Globally, millet area constitutes 74.00 million hectares with production of 89.17 million metric tons. The area, production, and productivity for different millets are given in Fig. 1. Among countries, India stands

R. Kapoor · A. K. Kawadiwale · R. Pati · H. K. Oberoi
Department of Plant Breeding and Genetics, Punjab Agricultural University, Ludhiana, Punjab, India

P. K. Malhotra (✉)
School of Agricultural Biotechnology, Punjab Agricultural University, Ludhiana, Punjab, India
e-mail: pkmalhotra-soab@pau.edu

S. H. Wani
National Seed Project (NSP), Sher-e-Kashmir University of Agricultural Sciences and Technology – Kashmir (SKUAST-K), Srinagar, Jammu and Kashmir, India

S. H. Wani, S. S. Gosal (eds.), *Accelerated Plant Breeding, Volume 5*,
https://doi.org/10.1007/978-3-032-04527-0_9

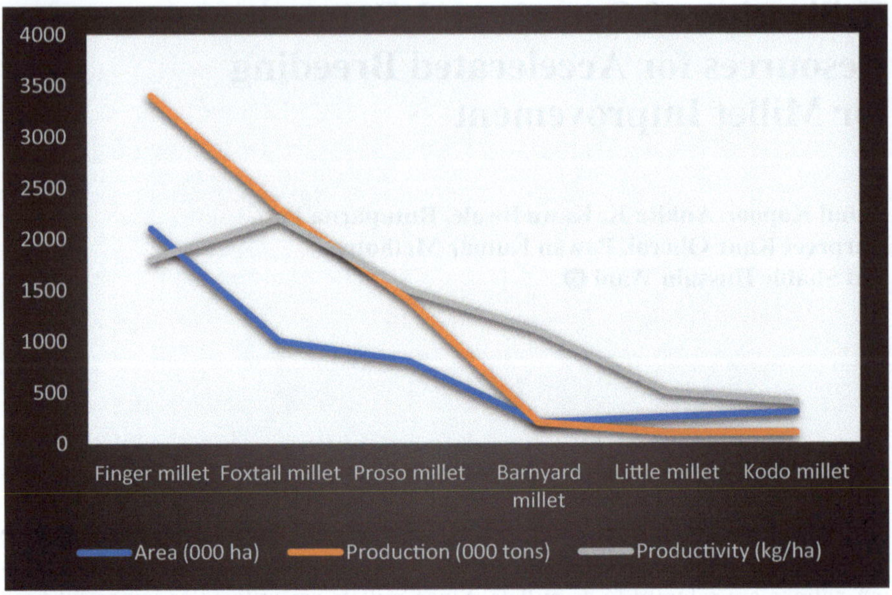

Fig. 1 World area, production, and productivity scenario of small millets

first in millet production (15.53 million tonnes) with a large area of about 12.45 million hectares (FAOSTAT).

Pearl millet (*Pennisetum glaucum* L.R. Br.) is a vital C_4 cereal crop found in tropical regions. Belonging to the Poaceae family and the subfamily Panicoideae (2n = 2x-14), this highly tillering crop is cross-pollinated with a substantial genome size of approximately 2352 Mbp (Bennetzen et al. 2012). Typically cultivated in marginal environments characterized by arid or semi-arid conditions, such as Sub-Saharan Africa and the Indian subcontinent, pearl millet exhibits resilience to sparse and unpredictable rainfall, subpar soil quality, and high temperatures. Due to its ability to endure extreme conditions and its inherent adaptation to infertile soil, coupled with high photosynthetic efficiency, it stands out as an excellent choice for shorter growing seasons (Yadav and Rai 2013; Serba et al. 2020). Recently, Pearl millet breeding efforts have witnessed significant advancements, often considered a notable success story in agriculture (Yadav et al. 2019). In the coming years, it is anticipated that pearl millet will assume a more significant role in ensuring food and nutritional security. Additionally, pearl millet is poised to serve as a valuable genomic reservoir for identifying candidate genes associated with tolerance to environmental stresses, thereby expediting the genetic enhancement of this crop.

Sorghum (*Sorghum bicolor* (L.) Moench) is the fifth-ranked cereal after maize, rice, wheat, and barley (area-wise) and is mostly cultivated under semi-arid tropics with limited resources. It is also known as the "camel of grains" or "great millet," due to its ability to thrive in challenging environments. As a fodder crop, it can be replaced with maize for silage making. Sorghum green stalks serve a purpose in the production of products like ethanol for biofuel, jaggery as a food, industrial alcohol, animal fodder, and in the paper and cardboard industry. It is a priority crop for many industries involved in the production of alcoholic beverages like beer, malt, alcohol, along with industries producing economic products like starch, adhesives, and packaging materials.

Finger millet (*Eleusine coracana* (L.) Gaertn.), belonging to the Poaceae family and Eleusine genus, is an allotetraploid (2n = 4x = 36, AABB) and self-pollinating crop having a genome size of 1593 Mb (Goron and Raizada 2015). The herb nature of this annual crop makes it cultivable widely in parts of Asia and Africa. It has high levels of nutrients such as calcium (Gupta et al. 2017; Sharma et al. 2017). Therefore, the genetic and genomic resources are crucial for enhancing crop plants for specific traits, as they serve as a reservoir of many genes contributing to the adaptation and yield of the crop. Recent advancements, such as Illumina sequencing, have enabled the sequencing of the whole genome along with transcriptome analysis in various plants (Bolger et al. 2014). Consequently, whole-genome sequences (WGS) are now available for crops, even those with complexity in genomes like polyploidy. However, the sequencing of the finger millet genome has lagged behind other major cereals, making it one of the last cereals to have its WGS completed. The first draft of the whole genome sequence in finger millet was recently made available (Hittalmani et al. 2017). This milestone is expected to significantly accelerate studies aimed at improving finger millet in the near future, serving as a crucial resource for researchers.

Barnyard millet (*Echinochloa species*) is a dual-purpose, temperate and warm season crop widely grown in Asia, Majorly Hindustan, China, Korea, and Japan. As per the reports, India is the largest cultivator of barnyard millet, both in terms of area (0.146 m/ha) as well as production (0.147mt), with an average productivity of 1034 kg/ha.

Proso millet (*Panicum miliaceum* L.) is one of the oldest domesticated cereal crops in the semiarid regions of northern China. But today it is widely cultivated across the high plains of the USA. Proso millet is a valued crop for its short growing season of 6090 days and high water-use efficiency because of a shallow root system (Khound et al. 2023). It also shows high resilience towards drought and temperature stress (Boukail et al. 2021). Nutritionally, the prolamin fraction present in proso millet protein makes it a suitable diet for celiac disease patients. (McSweeney et al. 2017). While certain seed phenolic compounds protect humans against devastating diseases like cancer and heart disease. Not only this, but proso millet is a good food option for people to overcome type-2 diabetes mellitus as well (Santra et al. 2019).

Foxtail millet (*Setaria italica*) is a grass which is emerged as a C_4 model crop plant for genetic studies as it performs NADP malic enzyme type C4 photosynthesis (Sachdev et al. 2021). The popular name "foxtail" arises from the distinct shape of its panicle, which appears as a fox's tail. Foxtail millet gives high yields, higher productivity, and sustainable production under the conditions of nutrition as well as water scarcity since it has high photosynthetic, water use efficiency, and nitrogen use efficiency as well (Brutnell et al. 2010).

Kodo millet (*Paspalum scrobiculatum*), an ancient grain native to Africa and cultivated in India a millennium ago, is known for its resilience to drought. Thriving in arid and semi-arid regions across Africa and Asia, this crop can endure on marginal soils and demands minimal water for growth, showcasing excellent drought resistance (Deshpande et al. 2015).

2 Origin and Taxonomy

The genus *Echinochloa* (Barnyard millet) contains 250 annual and perennial species (Bajwa et al. 2015). The popular species of *Echinochloa* that are available worldwide constitute Indian barnyard millet (*E. frumentacea*) and Japanese barnyard millet *(E. esculenta)* (Renganathan et al. 2020). The Japanese barnyard is predicted to have originated from its wild counterpart *E. crus-galli*, "barnyard grass," while Indian Barnyard Millet (*E. frumentacea*) is originated from its wild counterpart *E. colona*, "jungle rice" (Hilu 1994). The traits like large spikelets having awns with glumes and giving a papery appearance differentiate the Japanese Barnyard Millet (*Echinochloa esculenta*) from the Indian species (De Wet et al. 1983). The studies by de Wet categorized proso millet into five races on the basis of panicle morphology as follows: *miliaceum*, *patentissimum*, *contractum*, *compactum*, and *ovatum*. The primitive races of proso millet are predicted to be *miliaceum* and *patentissimum*. Race *miliaceum* possesses open inflorescences and sub-erect branches, while *patentissimum* possesses narrow and diffused panicle branches. *Contractum*, *compactum*, and *ovatum* have more compact inflorescences, which are drooped, cylindrical, and curved, respectively.

The chloroplast and nuclear phylogenetic studies revealed a close relative association between foxtail and green millet (*Setaria viridis*) (Doust et al. 2017). This supports the conclusion of foxtail millet being evolved through the domestication of green millet (Le Thierry d'Ennequin et al. 2000). Among more than 100 different *Setaria* species, the *Setaria italica* was popular as a major principal cereal crop plant of ancient Chinese civilization.

3 Genetic Resources and Germplasm Banks

ICRISAT has a collection of 37,949 sorghum germplasm accessions, including cultivated and wild relatives, from 92 countries. Core collection in sorghum consists of 3475 accessions, 2247 accessions, or 3011 accessions, while the mini core collection 242 accessions (Upadhyaya et al. 2014a, b).

Within the genus Pennisetum, the bifurcation of 55 taxa as 53 original species (wild) and 2 intraspecific taxa has been done categorically based upon genetic affinity with cultivated species of *Pennisetum. glaucum*, employing the classification model proposed by Maxted and Kell (2009). They are further delineated into primary gene pool, secondary gene pool, and tertiary gene pool depending upon the crossing among wild relatives. The primary wild relative consists of *P. glaucum* cultivars, landraces such as *P. glaucum subsp. stenostachyum* and *P. glaucum subsp. monodii*. Secondary wild relatives include species like *P. purpureum* and *P. squamulatum*, while all other taxa within the genus fall under tertiary wild relatives.

A collection of about 26,000 accessions of pearl millet germplasm has been done by the National Plant Germplasm System of the United States Department of Agriculture, along with the International Crops Research Institute for Semi-Arid Tropics (ICRISAT), making these institutions primary repositories for pearl millet germplasm. The accessions include landraces, wild relatives with counterparts, historically important specimens, and breeding lines from ongoing programs. Along with the above institutions, the NBPGR gene bank of India possesses 8291 collections, IRD in France holds 3986 collections or accessions, and 3821 collections in the Canadian Genetic Resources Program at Saskatoon supporting the availability of germplasm of pearl millet (Yadav et al. 2017).

Utilization of the core collection having germplasms from diverse germplasm sets is an effective strategy for enhancing the scope of crop improvement and breeding programs through the use of germplasm. The core collection development in pearl millet was undertaken by ICRISAT Gene Bank in 1998 by characterizing a pool of 16,000 accessions to prepare a core collection with 1600 accessions. This was increased by the incorporation of 501 accessions, resulting in a total of 2094 revised collections (Upadhyaya et al. 2009). A representative mini core collection was organized, consisting of 238 accessions after precise evaluation of the original 2094 collections for 18 important agro-morphological traits of pearl millet. The mini group of core collection so developed consisted of principal variation present in the larger group of core collection set, which could prove for direct utilization of genetic and genomic resources of pearl millet for crop breeding program (Upadhyaya et al. 2011).

Chinese National Genebank in Beijing holds ~28,000 accessions of foxtail millet since 1920, which makes 90% of landraces reported to be the largest germplasm collection. Apart from this, The institutes like ICRISAT, India (1534 collections from 26 nations), the NIAS (National Institute of Agrobiological Sciences), Japan (1279 collections), and the Plant Genetic Resources Conservation Unit of the USDA-ARS (766 accessions) holds smaller germplasm collections supporting the germplasm exploitation in foxtail millet research (Reddy et al. 2006). The representation of variation in foxtail millet accessions in the way of forming core and mini-core collections proves to be a potential resource for biotic (disease) and abiotic (climate resilience and salinity tolerance) stresses, and they have been developed by Krishnamurthy et al. (2014).

The largest genetic resource collection of proso millet is held by the N.I. Vavilov All-Russian Scientific Research Institute of Plant Industry in St. Petersburg, with roughly 8778 accessions as of 2012 followed by Institute of Crop Germplasm Resources, Chinese Academy of Agricultural Sciences (ICGR-CAAS) Beijing, China (6517 accessions), Ustymivka Experimental Station of Plant Production S. Ustymivka, Ukraine (3976 accessions), Yuryev Plant Production Institute UAAS

Kharkiv, Ukraine (1046 accessions), Botanical Garden of the Plant Breeding and Acclimatization Institute Bydgoszcz, Poland (721 accessions), USDA Agricultural Research Service (USDA-ARS) Griffin, USA (719 accessions), North Central Reg. Plant Introd. Station, USDA-ARS Ames, USA (713 accessions), Estación de Iguala, Instituto Nacional de Investigaciones Agrícolas Iguala, Mexico (400 accessions), NIAS, Kannondai, Japan (302 accessions) (Dwivedi et al. 2012). Apart from this, in India, ICRISAT has 842 collections possessing great diversity in all 5 races of proso millet (Reddy et al. 2007).

The collection of both species of barnyard millet is held by the International Crops Research Institute for the Semi-Arid Tropics (ICRISAT), Patancheru, India (743 accessions) from nine nations, along with a basic core collection consisting of 89 accessions established recently (Upadhyaya et al. 2014a, b). Other important accessions are present at NIAS (159 accessions) and the USDA (2321 accessions) (Hilu 1994).

Kodo millet, which was domesticated approximately 3000 years ago in India, remains primarily cultivated in this country, notably at the Deccan plateau, India (De Wet et al. 1983). This millet possesses a vast array of proteins with high quality (Kulkarni and Naik 2000) and exhibits significant antioxidant properties, particularly in comparison to other millets, making it potentially beneficial in anti-cancer strategies (Chandrasekara and Shahidi 2011). Similar to finger millet, kodo is abundant in fiber, suggesting potential benefits for individuals with diabetes (Geervani and Eggum 1989). Being a climate-resilient crop, it has the ability to survive in various types of soil conditions (De Wet et al. 1983; M'Ribu and Hilu 1996). Cultivation of kodo millet as a weed along with other cultivated cereals was the way of domestication events which took place across its present cultivable area like parts of Africa (De Wet et al. 1983). The wide diversity has been observed in kodo millet genotypes originating from African and Indian region cultivars through the application of molecular markers (M'Ribu and Hilu 1996). Classification of kodo millet into three races of regularis, irregularis, and variabilis has been done based upon panicle morphology (De Wet et al. 1983), along with two varieties that exist in the southern part of India: small (karu varagu) and bold (peru varagu) seeded. All the races have considerable morphological variability in various phenotypic traits, including yield (Subramanian et al. 2010).

Kodo millet is often characterized as being incompletely domesticated, with being labeled as "pseudo-cultivated" (De Wet 1992; Blench 1997). Consequently, systematic breeding efforts for kodo millet have been neglected, although there have been some promising initiatives. The potential breeding objectives for kodo millet consists of traits like disease resistance against smut and rust, morphological characters like green fodder yield, tallness, panicle lodging before harvest, and seed-grain yield (Upadhyaya et al. 2014a, b) along with resistance to microorganisms like fungus *Aspergillus*, which produces acid (cyclopiazonic), causing symptoms such as sleepiness, tremors, and giddiness upon consuming diseased grain, called as "kodua poisoning" (Rao and Husain 1985). Interestingly, the ability of some commercial landraces to retain the perennial nature, just like their ancestors, continuing to produce new shoots after older shoots mature (De Wet et al. 1983) can be promoted through breeding and hybridization.

There is a limited in information on genetic or molecular maps that are available for the genome of kodo millet (Dwivedi et al. 2012) due to the consistent challenge of persistent introgression of DNA from its ancestors. However, rice genomic sequence was utilized for binding site prediction for the generation of a molecular map since it is closely related to kodo millet (Ceasar and Ignacimuthu 2010). Few molecular markers have been employed in assessing diversity and phylogeny (Kushwaha et al. 2015). The prediction of target sites for miRNA using ESTs from online sources of kodo millet has also been attempted to be developed (Babu et al. 2013). Apart from this, investigations for callus regeneration media protocols have also been conducted for tissue culture and successful plantlet regeneration, and soil transfer has also been reported in kodo millet (Ceasar and Ignacimuthu 2010).

With over 656 collections of kodo millet, ICRISAT ranks one in germplasm availability for further research. They have further established a core collection to represent the morphological diversity of all the wider collections (Upadhyaya et al. 2014a, b). An appreciable number of 336 accessions are present in the National Plant Germplasm System of the USDA. Additionally, some universities possess extensive kodo millet seed banks, with the University of Agricultural Sciences in Bangalore being a notable example (Ceasar and Ignacimuthu 2010).

As per the data available, more than 28,041 finger millet germplasms have been contributed by various organizations worldwide. From these 28,041 germplasms, 10,507 germplasms have been maintained in the NBPGR gene bank of India, while ICRISAT, India, holds the second largest collection with 5957 germplasms. Comparatively smaller collections have also been maintained at KARI, Kenya (2875), IBC, Ethiopia (2156), USDA-ARS, U.S. (1452), and the SAARI in Uganda which contribute significantly to the germplasm pool (Gupta et al. 2017). Despite the richness of germplasm resources, crop improvement advancements, and genomic information for finger millet is comparatively restricted in comparison to other important cereals, hindering further crop improvement (Saha et al. 2016). However, the role of different strategies for the improvement of finger millet is given in Fig. 2.

4 Improvement of Millet in Post-genomic Era Using NGS for Developing Millet Genetic Resources

From past times, millet improvement has been centrally focused by conventional breeding strategies based upon exploiting phenotypic selection having high environmental effect. Through this way, the development of variety required almost 5–12 years, along with less effectivity for traits with high complexity and less heritability. The breeding programs for crop improvement in millets have been hampered majorly due to the narrow genetic base of the pool and poor selection tactics

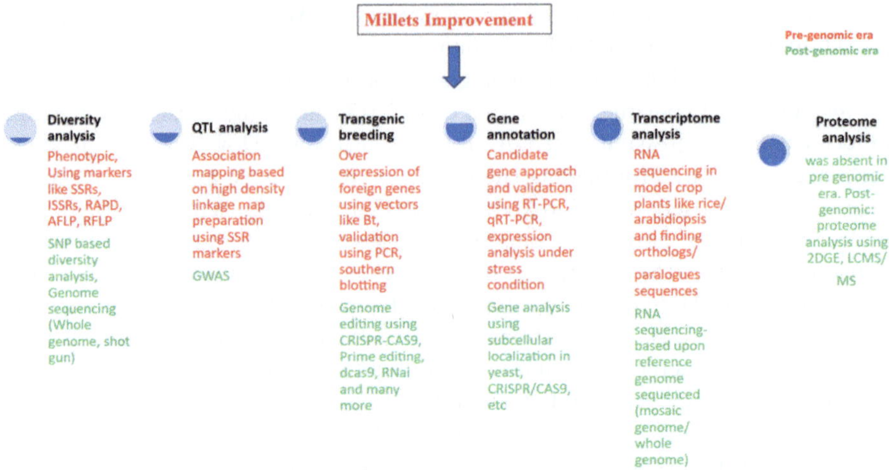

Fig. 2 Strategies for improvement of millets in the pre- and post-genomic era

applied for genetic enhancements (Tuberosa 2012). Crop improvement of neglected and underutilized millets needs to be prioritized using the omics-based modern tools. Whole Genome Sequencing (WGS) provides deeper insights for studying and better understanding the genes controlling traits, organization of genes to constitute a genome, and thereby leading to genomics-assisted crop improvements in millets through using this genomic data. Identification of SNP"s, discovery of new alleles, development of linkage and association maps, candidate gene identification, genomic selection (GS), genome-wide association studies (GWAS) and marker assisted breeding (MAB) programs are the further studies that can be undertaken using Next-generation sequencing (NGS) genome sequence data (Ajeesh Krishna et al. 2022).

Precision in crop improvement has been achieved by different NGS applied technologies which helped in correlating the phenotypes with genotypes at the nucleotide level or genomic level (Chhapekar et al. 2016). NGS technologies have cracked the web of markers and linkage, genetic, consensus maps through high-throughput genome sequencing (Serba and Yadav 2016). Genome sequencing via resequencing has enabled the identification of SNPs throughout the genome that can be utilized for millet improvement. The allelic diversity exploitation through fine mapping of QTLs using multi- and bi-parental populations, development of high- and ultra-high-density maps through linkage and association mapping has led to the exploitation of NGS technologies to the utmost possible level (Poland and Rife 2012).

Understanding practical application can show that the abovementioned NGS strategies are found to be used in proso millet by different scientists for the characterization of allelic pools, through genome-wide diversity (Jayakodi et al. 2019), characterization of gene expression (Yue et al. 2016), and forward genetic approaches to identify quantitative trait loci (QTL) and marker-trait associations (MTAs) (Rajput et al. 2016). The published proso millet genome sequence and chromosome assembly (Zou et al. 2019; Shi et al. 2019) projected this species into modern genomics having great potential for gene mining (Shan et al. 2020) and large-scale genotyping applications like GWAS.

5 Genome Sequence

The information on the number of genes along with their annotated functions within the genome constitutes the genome sequence. Unfortunately, gene annotation and sequencing studies of millet genomes are still lagging which could have otherwise hastened the millet improvement (Wu et al. 2019). As per the data available, genome sequences (either complete and fully annotated or partial) are available for barnyard millet, finger millet, pearl millet, proso millet, and foxtail millet so far in millets (Fig. 3). Genome sequencing has resulted in better genotyping through genetic diversity analysis, the development of high-density linkage, genetic and physical maps along with their annotation, and the detection of the major and minor QTLs involved in controlling traits linked with micronutrients, and important economic characters. For instance, single-nucleotide polymorphism (SNP) marker-based genome development and in barnyard core collection study of diversity for target traits (Wallace et al. 2015). Although the genome sequences of some of the millets (barnyard millet, pearl millet, and proso millet) are available, they are not yet annotated which could have further enabled the comparison of genomic studies in other millets. Further, comparative genomics can be utilized to gain genomic insights into other millets where little information is available. The smallest genome size is reported for foxtail millet to be 510 Mb, and the largest for pearl millet ~1.79 Gb among the millets.

The genome assembly of pearl millet revealed that the genome was about 1.76 Gb having 77.2% of repetitive sequences (Varshney et al. 2017). The proportion of repetitive DNA sequence in pearl millet was similar to the proportion from other cereal crop genome sequences, as found in approx. 42% of the rice genome (Yu et al. 2020), approx. 46% of foxtail millet (Bennetzen et al. 2012), approx. 61% sorghum (Paterson et al. 2009), and more than 85% of maize (Schnable et al. 2009). Rajaram et al. 2013 assembled 69,398 contigs in Pennisatum using RNA sequencing. According to CEGMA analysis, 437 complete genes were found from 458

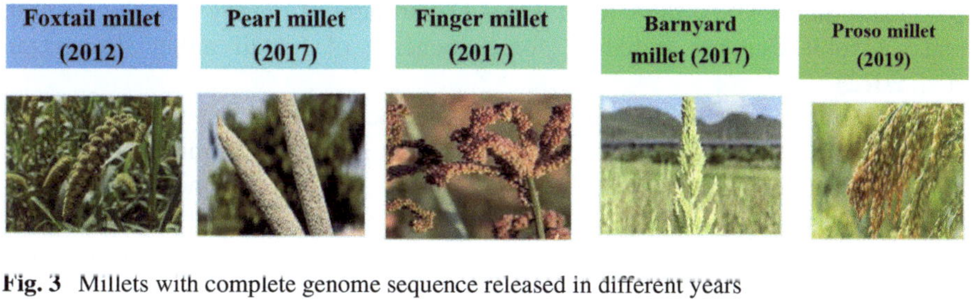

Fig. 3 Millets with complete genome sequence released in different years

conserved genes, which constituted 96.7% and were annotated using BUSCO analysis.

Sorghum genome sequencing using Sanger sequencing in inbred line BTx623 revealed the presence of ~10.5 million reads with the coverage of ~8× along with complete annotation. The genome assembly and alignment revealed the presence of more than 97% protein-coding genes with the overall assembly length of 732 Mbs (Roche GS FLX n.d.). Out of this, 24% of genes were reported to be grass family specific and 7%were reported to be sorghum specific (Paterson et al. 2009). Further fine-mapping of the genome using Sanger, 454, and Illumina sequencing released as v3.0 shows incorporation of 351 Mb of the fully sequenced sorghum genome. The above refining closed 4426 gaps with the addition of a total of 4.96 Mb of sequence to the assembly making it more precise to study (Boatwright et al. 2023).

No comprehensive gene or unigene sequences have been identified for finger millet thus far. As we have observed several genome assemblies for many cereals, the same is not the case with finger millet as it has only one genome assembly (ASM218045v1) reported till date. Not only this, but the variety of proteins recorded for finger millet is limited in comparison to other important cereals. Presently, no single-nucleotide polymorphisms (SNPs) have been generated within the genome of finger millet. However, the recent release of the finger millet whole-genome sequence (WGS) is expected to catalyze the development of these resources in the forthcoming years, thereby expediting research across all aspects of finger millet. Furthermore, finger millet possesses a minimal number of RNA sequences, primarily acquired under specific stress situations and for seed-grain calcium content. Some studies have been conducted for transcriptome analysis of particular landraces subjected to various stresses such as drought, saline conditions, and blast. Nonetheless, most of these studies have not yet validated sequence read information or further characterized key genes; instead, they have merely been submitted as raw reads. The presently released whole genome sequence of finger millet is anticipated to play a pivotal role in the development of various genomic studies for facilitating further investigations. For instance, the RNA sequencing reads can be corroborated using the finger millet WGS to identify the role of important genes participating in different processes (Fig. 2).

Foxtail is the initial millet whose genome has been sequenced along with annotation for two commercial genotypes which served as a model crop for conducting comparative and functional genomic studies in other related millets (Ramakrishnan et al. 2016; David et al. 2021). In China, the Beijing Genome Initiative (Zhang et al. 2012) and, in the USA, the United States Department of Energy were involved to make a Joint Genome Initiative, for sequencing of foxtail millet (Bennetzen et al. 2012).

The genome sequencing and repeated sequence annotations of foxtail millet reported 46 percent of different retroelements along with DNA transposons which are transposable elements (Zhang et al. 2012). The genome sequence of green foxtail (*Setaria viridis*), the wild relative of foxtail millet, is found to be small (~ 510 Mb) and a completely annotated genome (Brutnell et al. 2010). Thus, the completely annotated genomes of domesticated foxtail and green foxtail millet are currently a model system in millets and all related grass families to understand and correlate evolutionary, physiological, molecular, and architectural aspects (Ceasar 2019). The contrasting WGS (whole-genome sequence) data of millets based on various parameters like genotype used, genome size, and so on is presented in Table 1.

The genome sequencing in cultivar Longmi 4 of broomcorn millet reveled generation of ~103.4 Gb reads using Illumina with paired-end reads of 150 bp. Along with the reads, 15 responsive genes for ABA or water deficit stress were also identified which indicated their utilization through integrating them in various millet crops through techniques like MAS or transgenic approach (Shi et al. 2019).

Japanese barnyard millet (*Echinochloa crus-galli*) cultivar STB08 was genome sequenced by the Illumina HiSeq 2000 system which predicted the size of the genome to be 1.27 Gb, with 108,771 annotated genes (Guo et al. 2017). The encoding genes data can show its utility in other related crop millets and grasses to study the metabolic pathways. In millets like kodo millet which have genome complexity and lesser explored genomics, the genomic structure revelation and the analysis of downstream processes associated with it have been minimally attempted by scientists (Renganathan et al. 2020).

The long-awaited whole-genome sequencing of the ML-365 genotype of finger millet was accomplished using Illumina and SOLiD sequencing technologies through the characterization of its drought tolerance, resistance to blast disease, and desirable cooking qualities which are key traits in the adaptation of finger millet for regular consumption (Hittalmani et al. 2017). This endeavor yielded approximately 45 Gb of paired-end data and 21 Gb of mate-pair data. The resultant genome assembly comprised 525,759 scaffolds exceeding 200 base pairs, with an average length of 2275 bp having an N50 length of 23.73 kilobases (Hittalmani et al. 2017). Moreover, scientists also assembled the sequences of ML-365 plants under well-watered and low-moisture stressed conditions revealing 53,300 unigenes and 100,046 unigenes, respectively, which can be proven to be a valuable asset in breeding stress-tolerant millet crops. Among the assembled unigenes, approximately 64% received functional annotations using *Viridi* plantae protein sequences (UniProt

Table 1 Sequencing (WGS) data of various millet crops

Crop	Pearl millet	Finger millet		Foxtail millet		Barnyard millet	Proso millet
Genotype used for sequencing	Tift 23D2B1-P1-P5	ML-365	PR 202	cv. Zhang gu × A10	cv. Zhang gu × A2	STB08	Landrace (Ac. No: 00000390)
Genome size (number of genes)	~ 1.79 Gb (69,398)	1196 Mb (85,243)	1189 Mb (62,348)	~400 Mb (24,000-29,000)	~423 Mb (38,801)	1.27 Gb (108,771)	~ 923 Mb (55,930)
References	Varshney et al. (2017)	Hittalmani et al. (2017)	Hatakeyama et al. (2018)	Bennetzen et al. (2012)	Zhang et al. (2012)	Guo et al. (2017)	Zou et al. (2019)

database). A large number of unigenes were unveiled using differential gene expression analysis; out of those, 2267 unigenes were exclusively controlled in well-watered conditions, 12,893 unigenes were involved in less-moisture stressed conditions, and 111,096 unigenes were found to be present in both well-watered and less moisture stressed conditions for their regulation. The homology modeling based upon protein-protein interactions identified plant transcription factors (TFs), revealing 11,125 genes showing similarity with 56 TF families, out of which 2866 were related to drought response linked with 19Pfam domains, 1766 genes were designated as R-genes providing resistance to various diseases, and the remaining 330 genes reported to be involved in the accumulation and transport of calcium (Hittalmani et al. 2017).

Strong co-linearity has been demonstrated from the WGS data of finger millet with foxtail millet and rice as compared to other grass family cereals. Furthermore, similar reports for equal genome sizes have been reported in other subspecies of *Elusine coracana* like *africana*, etc. (Hittalmani et al. 2017), depicting and supporting the potential fact of cultivated finger millet's domestication from *E. coracana subspecies africana* (Dida et al. 2008). Hatakeyama et al. (2018) reported the genome size of finger millet to be 1.5 Gb, with the assembled genome spanning 1189 Mb, covering an estimated 78.2% of the genome through the combined study of WGS and assembly analysis of finger millet genotype PR-202 using an innovative polyploidy genome assembly workflow. The study revealed that genotype PR-202 comprised 2387 scaffolds, with an N50 length of 905.318 Kb and 5 Mb being the highest sequence length. The final scaffolds and annotations in FASTA file format are publicly accessible at NCBI under the BioSample number SAMD00076255. In total, this study identified 62,348 genes, with nearly 91% of them being functionally annotated and 96.5% identified as single-copy genes. In PR-202, NCBI BLAST analysis revealed that all the 57,913 genes were duplicated, having more than two copies in the whole genome (Hatakeyama et al. 2018). The availability of WGSs for both ML-365 and PR-202 presents significant opportunities for further research endeavors. These include SNP identification, next-generation sequencing (NGS)-based allele discovery, construction of linkage and association maps, identification of candidate genes associated with agronomically important traits, functional characterization of these genes using reverse genetic approaches, and the implementation of marker-assisted breeding programs (Fig. 2).

The primary challenge in the molecular characterization of kodo millet is the lack of genomic resources such as DNA markers, saturated linkage maps, and sequences of genomes. However, genomic resources from related species like foxtail millet, which has two reference genome sequences available (Bennetzen et al. 2012; Zhang et al. 2012), can be utilized to enhance the genomic resources for kodo millet. DNA markers, developed from foxtail millet genome sequences, such as SSR (simple sequence repeat), EST-SSR (expressed sequence tag-simple sequence repeat), ILP (intron length polymorphic), and microRNA-based molecular markers, have shown over 85% transferability to other millets like proso, barnyard, little, and kodo millets, as well as to non-millet species (Yadav et al. 2014). Additionally, agrobacterium-mediated transformation methods for kodo millet (Bhatt et al. 2021) are now available, which are expected to enhance our understanding of the genomics of small millets. Nevertheless, there are still research gaps in small millets, particularly regarding important nutritional and medicinal traits that could be leveraged to improve our diets.

Omics-based technologies which lead to the study of genes, mRNA involved in controlling those genes, proteins produced in pathways, metabolic pathways controlling protein and gene interaction, and epigenetic changes involved have enabled us to understand the complex interactions between the metabolic pathways from genes to protein and their interaction resulting in any phenotype of millets. The correlated omics-derived data can be interpreted as comparative genomics.

5.1 Pangenome Reconstruction

The pan genome is a comprehensive characterization of a species' gene content and provides better insights into its genetic diversity. The pangenome reconstruction by Ruperao et al. (2021) in sorghum, using 176 genetically diverse and race-specific accessions along with the reference genome, revealed that integrating 13 sorghum genomes generated a pangenome of 44,079 gene families with 222.60 Mb of new sequence (Tao et al. 2021).

5.2 Comparative Genomics

The initial linkage map of pearl millet, reported by Liu et al. (1994), covered a genetic length of 303 cM. Comparative genetic mapping by Devos et al. (2000) aimed to elucidate the homoeology between the genomes of pearl millet, foxtail millet, and rice. Despite several structural rearrangements distinguishing the pearl millet genome from that of rice, a strong genomic relatedness was revealed between pearl and foxtail millets. Within the genome of pearl millet, one or possibly two duplications were identified in linkage groups 1 and 5. Additionally, internal gibberellic acid levels indicated that the genes d1, d2, and d4 behave as recessive dwarfing genes, analogous to the ct1 and ct2 dwarfing genes of rye, located in the long arm of centromeric regions of rye chromosomes 7R and 5R, respectively, as reported by Devi et al. (1994). These regions correspond to Linkage Group 4 and Linkage Group 2 of pearl millet, respectively, as detailed by Devos et al. (2000). The application of comparative genomics for crop improvement is promising, providing valuable insights for enhancing these climate-resilient crops, as emphasized by Hu et al. (2015). Furthermore, genomic analysis of *Cenchrus purpureus* (elephant grass) has revealed enzyme-coding gene families responsible for the biosynthesis of anthocyanidins and flavonoids. Evolutionary analysis suggests that the sub-genome A of elephant grass and pearl millet may share a common ancestor, as described by Yan et al. (2021).

First, comparative mapping to study the synteny between foxtail and related crops by using functional markers like SSRs was attempted by Pandey et al. (2013), followed by EST-SSRs reported by Kumari et al. (2013) and then by ILP (intron length polymorphic markers) as reported by Muthamilarasan et al. (2014). Such comparative genomic studies revealed a higher degree of synteny between sorghum and foxtail millet, succeeded by maize and foxtail millet as depicted through DNA markers. The synteny study between foxtail and other crop genomes revealed 71.8% similarity with sorghum, 72.1% with rice, 61.5% with brachypodium, and 86.7% with maize (Muthamilarasan et al. 2017). The generated comparative genomics data can be used to study orthologous and paralogous genes present across the different millets and verify in the remaining millets to accelerate gene identification and gene annotation.

6 GWAS (Genome-Wide Association Studies) and Genomics-Assisted Breeding for Millet Improvement

Genotyping by sequencing (GBS) can be utilized to generate repositories of SNPs which can further be utilized in studying genetic diversity, developing linkage maps, and establishing associations between DNA markers and traits of plants (Sehgal et al. 2012). Puranik et al. (2020) used GWAS in sorghum to identify 37 putative genes associated with epicuticular wax content which shows adaptation to biotic and abiotic stresses. The sorghum genome sequence information was utilized in mining a total of 459,304 SNPs related to the genetic variation of complex traits. These SNP data were used further for GWAS analysis in determining candidate SNPs linked with different useful traits (Hu et al. 2019). In sorghum, GWAS has been applied by different researchers for varying traits such as plant height and inflorescence (Morris et al. 2013), grain size (Tao et al. 2021), and grain quality (Kimani et al. 2020) in sorghum. Two SNPs in sorghum were found to be significantly associated with *Macrophomina phaseolina* and *Fusarium thapsinum* through research by Adeyanju et al. (2015) by using 79,132 SNP markers in a panel of 300 genotypes.

Crucial studies on establishing diversity groups in cross-pollinated crops, through both genotyping and phenotyping for clustering restorer lines, are vital in pearl millet breeding programs. These studies streamline the breeding process by reducing the time spent on delayed selection based on F1 performance (Kawadiwale et al. 2024). To better understand genetic diversity and population structure in pearl millet, 1.16 Tb of whole-genome resequencing (WGRS) was performed on PMiGAP lines, along with 116 Gb of WGRS on the crossing parents. Additionally, genotyping by sequencing (GBS) generated 78.9 Gb of data (Elshire et al. 2011), and RAD sequencing was employed to generate R and B lines (Miller et al. 2007). In the genome of pearl millet, 74,891 SSR primers were designed from 88,256 identified SSR (simple sequence repeats) using the different microsatellites (Thiel et al. 2003).

Furthermore, in PMiGAP lines, 29,542,173 SNPs (single-nucleotide polymorphisms) were identified, including 3,844,446 deletions and insertions and 423,118 structural variations covering the whole genome (Varshney et al. 2017).

The different germplasm accessions of pearl millet exhibit high levels of heterogeneity and heterozygosity, complicating association mapping and limiting strategies for genetic diversity analysis (Kannan et al. 2014). To address this, a genome-wide association study (GWAS) was conducted on three germplasm sets of pearl millet, utilizing SNP data covering the whole genome to assess LDD (linkage disequilibrium decay). This study identified approximately 1054 marker-trait associations (MTAs) using GWAS on 288 testcross progenies of PMiGAP lines for various traits: panicle yield (9), panicle diameter (1), panicle length (3), panicle harvest index (1), stover dry matter yield (5), fresh stover yield (38), tillers per plant (147), grain number per panicle (91), grains per square meter (75), plant population (68), plant height (344), panicle number (246), 1000 grain weight (10), grain yield (11), and grain harvest index (5). These MTAs collectively explained 9–27% of the total phenotypic variation (Varshney et al. 2017).

Genome-wide association studies (GWAS) have also been conducted in minor millets (Table 2). For example, a GWAS in proso millet identified 13 marker-trait associations related to key agronomic and seed traits (Boukail et al. 2021). Similarly, genotyping by sequencing (GBS) technology was utilized in foxtail millet to discover genes responsible for grain nutritional traits. The identified SNPs were associated with candidate genes linked to different metabolic processes, encoding proteins such as succinate dehydrogenase, MFS transporter, F-box domain proteins, zinc finger leucine-rich repeat proteins, X-box transcription factor-related proteins, raffinose synthase, protein-tyrosine kinase, anthocyanidin reductase, ionotropic glutamate receptor, NADH dehydrogenase flavoprotein, and PPR repeat family proteins (Jaiswal et al. 2019). Additionally, a study by Jia et al. (2013) in foxtail millet analyzed 916 diverse accessions and identified 2.58 million SNPs using GBS, pinpointing 512 loci associated with 47 agronomic traits. Another study by Yang et al. (2018) identified 149 basic helix-loop-helix (bHLH) genes through GWAS and examined the response of 8 bHLH genes under drought stress.

7 QTL Identification and MAB (Marker-Assisted Breeding) for Millet Improvement

Breeding efficiency can be improved by QTL mapping which understands the genetic basis of traits controlled by QTL on the number, mode of action, and effect basis. QTL mapping also helps in locating the genetic and physiological basis of biotic and abiotic stress tolerance along with nutritional factors which help in improving the tolerance of millets. The identification of such valuable traits linked with QTLs is the first major step in MAB which thereby produces improved millet varieties. But only a little effort has been undertaken so far in the identification of

Table 2 Details of genome-wide association studies (GWAS) conducted in millets

Name	Size of mapping population	Mapping population	Traits	Outcomings	Candidate genes identified	References
Pearl millet	222	Association mapping	Agronomic and seed traits	10 MTAs for seed-related traits and three for agronomic traits	–	Boukail et al. (2021)
	281	Inbred lines	Nutrient content	78 MTAs associated with zinc, iron, and protein	45	Pujar et al. (2020)
Finger millet	133	Association mapping	Agronomic	109 novel SNPs associated with important agro-morphological traits	05	Sharma et al. (2017)
Foxtail millet	916	Haplotype mapping	Agronomic	512 loci linked to economically important traits	17	Jia et al. (2013)
	93	Association mapping	Agronomic	81 MTAs for ten different traits across the genome	27	Jaiswal et al. (2019)
	142	Association mapping	Nutritive	74 MTAs for various nutrition-related traits	29	Jaiswal et al. (2019)
Proso millet	88	Association mapping	Anti-oxidant	218 SNPs associated with DPPH and FRAP activity	18	Yadav et al. (2021)

MTA marker trait associations

important QTLs (associated with resistance genes) in millets. Any studies so far done are majorly based on low-throughput "first-generation" molecular markers. For example, Jones et al. (2002) used RFLPs (restriction fragment length polymorphisms) markers to identify the QTL associated with downy mildew resistance in pearl millet. A list of QTLs associated with different traits in millets is given in Table 3.

In sorghum, QTL mapping has been widely conducted by various researchers to identify genes controlling plant height (Srinivas et al. 2009), maturity (Srinivas et al. 2009), grain yield and related traits (Srinivas et al. 2009), post-flowering drought tolerance (Haussmann et al. 2002), shoot fly resistance (Deshpande et al. 2003), and cold tolerance (La Borde et al. 2023). Sharma et al. (2017) identified valuable QTL linked with 50% flowering time, maturity time, ear, flag leaf, and grain yield traits through GBS using 113 accessions.

Table 3 List of QTLs controlling traits in millets. Proso millet (*Panicum miliaceum*)

Traits of QTLs studied	Number of QTLs	Mapping tool: marker	Mapping population	Map length (cM)	References
Plant height (*PH1.1*)	1	MapChart 2.2	F_2 and $F_{2:3}$ (Longmi12* Zhang778)	109-kb	Liu et al. (2022)
8 Morpho-agronomic traits	18	IciMapping v.$0.0.6.0	RIL derived from huntsman*Minsum	54.6 cM to 236 cM	Rajput et al. (2016)

QTL mapping through GBS showed the resemblance between genomic regions of candidate genes of rice are homologous with foxtail millet genes controlling plant growth, flowering, and grain yield.

8 Functional Genomics for Millet Improvement

8.1 Transcriptomics

The analysis of the transcriptome of an organism can be performed through RNA-Seq which not only helps in analyzing the transcriptome of an organism but also provides gene insights expressing at a specific time under specific conditions (Azzouz-Olden et al. 2020). Such transcriptome analysis study has been undertaken in sorghum for many traits by various researchers like drought-responsive microRNAs (Azzouz-Olden et al. 2020; Fracasso et al. 2016; Hamza et al. 2016), Anthracnose (Fu et al. 2020), heat and drought stress (Johnson et al. 2014), salinity (Sun et al. 2023) which helped in providing the gene expression profile of sorghum. Similarly, candidate genes for water stress tolerance, micronutrient accumulation using comparative transcriptome analysis were identified in barnyard millet by Jayakodi et al. (2019).

Although the draft genome sequencing of pearl millet was completed in 2017, incomplete annotations pose challenges for mapping short reads. Next-generation sequencing technologies, such as PacBio sequencing (single-molecule real-time sequencing), have been employed to generate full-length transcripts, thereby enhancing transcript recovery (Abdel-Ghany et al. 2016; Wang et al. 2018). Transcription profiling in pearl millet has identified 10 differentially expressed genes associated with drought tolerance, including calmodulin-like proteins, aspartic proteinase oryzasin, DnaJ-like protein, Rab7, glyoxalase, putative beta-1,3-glucanase, inosine-5′-monophosphate, ascorbate peroxidase, and abscisic stress ripening protein (Choudhary 2015). A list of different abiotic stress-related genes is given in Table 4.

Table 4 Details of abiotic stress tolerance genes identified in various millets

Name of crop	Type of stress	Stress-responsive gene
Finger millet	Tolerance to drought and multiple stress	*EcCIPK31-like* (Nagarjuna et al. 2016), *EcDehydrin7* (Singh et al. 2015), *MT, RISBZ4, ATFP6, TR, PP2A, ELIP, FPS and TAF6* (Parvathi and Nataraja 2017; Parvathi et al. 2013), *EcDREB1A* (Mohanty et al. 2013).
	Drought and salinity	*mtlD* (Hema et al. 2014)
Foxtail millet	Cold and salinity	*SiSET14* (Yadav et al. 2016)
	Low-potassium tolerance	*SiMYB3* (Cao et al. 2019)
	Drought and oxidative stress	*SiASR1* (Feng et al. 2016)
Pearl millet	Drought, heat, and salinity	*PgGRAS* (Jha et al. 2021)
	Cold and salt stress tolerance	*PgPmp3-1 and PgPmp3-2* (Yeshvekar et al. 2017)
	Salinity tolerance	*PgNAC21* (Shinde et al. 2019)

8.2 *Proteomics and Metabolomics*

Comprehensive proteome and metabolome studies play crucial roles in the analysis of genes involved in growth and development (transcriptomics), the components of stress response networks involving proteins (proteomics), and the metabolic pathways that enable organisms to overcome stress.

In foxtail millet, 2474 DEPs (differentially expressed proteins) were found to be involved in seedling response in drought stress which was analyzed through proteomic studies (Pan et al. 2018) showing the abundance presence of LEA and ROS scavenging enzymes which are involved actively in the defense mechanism. The authors later on, through phosphoproteomic profiling of contrasting foxtail millet varieties, reported differentially regulated phosphoproteins (DRPPs) to be involved in metabolism since they were sources of gene expression, active in signaling pathways, and directly involved in ion transport. The genes controlling biosynthesis of phenylpropanoids, phenylalanine, and its signal transduction regulate the drought stress in seedlings which was an outcome of an integrative study of transcriptomics and metabolomics in foxtail millet (Yu et al. 2020).

In finger millet, 80,602 differentially expressed genes (DEGs) and 3009 differentially expressed proteins (DEPs) were identified, associated with the drought response mechanism at transcriptomic and proteomic levels, respectively. This coordination between DEGs and DEPs was found to be crucial for enhanced drought tolerance in finger millet (Li et al. 2021). In proso millet (*Panicum miliaceum* L.), metabolomics analysis revealed 2082 differential metabolites related to primary and secondary metabolites, including anthocyanin production, energy metabolism, and photosynthesis, which are essential for drought tolerance (Cao et al. 2022).

8.3 *Epigenomics*

The inaccessibility of genomic DNA to various transcription factors, resulting from alterations in chromatin structure, modulates gene expression and is known as epigenetics. Advances in high-throughput sequencing technology have enabled the study of epigenetic mechanisms at the genomic level, enhancing our understanding of these regulatory pathways in plant genomes through genetic and biochemical approaches. Novel epigenetic insights were observed in two varieties of foxtail millet, showing different methylation patterns (Zhang et al. 2012). In pearl millet, DNA methylation and hemi-methylation play key roles in regulating defense responses and provide tolerance to salicylic acid application (Ngom et al. 2018).

9 GS (Genomic Selection) to Fasten Breeding Program for Improvement of Millets

To enhance genetic gain through selection cycles in a breeding scheme over a given time period, GS can be a promising technology which speeds up and enhances the efficiency of breeding programs. Predicting the genomic estimated breeding values (GEBV) through genome-wide distributed DNA markers is the main concept of genomic selection. The data for genomic selections is lacking in millets except for pearl millet and sorghum. In a study using the GBS approach in pearl millet inbred lines, the GEBV was predicted which showed that GBS (Liang et al. 2018) along with whole-genome resequencing (WGRS) (Varshney et al. 2017) can enable GS in pearl millet for predicting and enhancing the grain yield.

In sorghum, genomic selection led to an enhancement in grain yield and drought tolerance through multi-trait analysis (Velazco et al. 2019). Different traits in sorghum like height, brix, juice weight, leaf weight, earliness, stem weight, and grain weight were targeted for developing the Practical Haplotype Graph (PHG) training population with 250 genotypes through the Chibas sorghum breeding program concluding that PHG study can make genotyping more feasible to cost effectiveness in GS in Sorghum (Jensen et al. 2020). Apart from this, many other agronomic traits have been studied in sorghum through GS such as antioxidant content (Habyarimana and Lopez-Cruz 2019), grain yield and related components, protein, fat, fiber, and ash content (Sapkota et al. 2020).

10 Genome-Editing for Millet Improvement

The WGS data availability has enabled scope for GE approaches through the possible insertion of desirable traits. Such target precision for any gene of interest can be achieved through utilizing the currently available GE tools such as ZFNs (zinc finger nucleases), mega-nucleases, TALENs (transcriptional activator-like effector nuclease), CRISPR (clustered regularly interspaced short palindromic repeat)/Cas9 (CRISPR-associated protein 9), and their modifications as and when needed (Jaganathan et al. 2018).

Despite the availability of numerous genetic engineering tools, only a few technologies have been effectively utilized in millets, such as the CRISPR/Cas9 system. This system has been employed in millets to identify genes responsive to various abiotic stresses, including drought, salinity, heat, cold, oxidative stress, and nutrient starvation (Cao et al. 2019). The identification of genes associated with each biotic and abiotic stress can contribute to the enhancement of growth and yield in small millets under challenging environmental conditions, as reported by Lin et al. (2018). Lin et al. found a mutant phenotype (albino) when targeting the PDS (phytoene desaturase) gene in *Setaria italica* (foxtail millet) using the targeted genome editing, i.e., CRISPR/Cas9. Additionally, induction of a haploid embryo using seeds was performed to mutate the *SiMTL* gene using CRISPR/Cas9 base editor by Cheng et al. (2021). These doubled-haploid lines with an HIR (high induction rate) can be utilized by breeders for foxtail millet improvement through gene stacking approaches.

Recently, an efficient transformation method and new CRISPR/Cas9 genome-editing tools for foxtail millet have been established (Liang et al. 2022) which can prove a model plant in applying CRISPR/Cas9 genome-editing tools for millet improvement. Further, reports of successful Agrobacterium-mediated transformation with ~27% transformation efficiency (Sood et al. 2020) and 76% transformation efficiency (Zhang et al. 2021) have been reported in foxtail millet.

The reports of successful embryogenic callus cultures by Vishnoi and Kothari (1996) along with direct somatic embryos have been established in vitro for foxtail millet. Despite such reports, very little, i.e., only one successful regeneration of a transgenic plant using *Agrobacterium,* has been reported with an efficiency of 6.6% (Liu et al. 2005). Further, authors reported in 2007 that plant genotype, explant source, time of inoculation, and length of cocultivation influenced gene transfer using *Agrobacterium* (Liu et al. 2007).

10.1 Potential Use of NGS for Millet Improvement

A new pace in millet improvement through novel traits can be achieved by applying advanced biotechnological tools to develop millet cultivars. NGS data can be utilized in the following ways (with practical examples given) to incorporate it into the breeding pipeline:

(a) **Marker-Assisted selection**

Using MAS, QTL for water stress tolerance was identified in the most popular cultivar HHB 226 of pearl millet (Bidinger et al. 2005).

(b) **Mutation breeding**

Mutants developed in finger millet for late emergence of seedling, high yield, and earliness characters in proso millet and for longer seed viability in kodo millet (Ambavane et al. 2014; Bhave et al. 2016; Jency et al. 2016) have already been successfully reported. NGS opens the gate to mutation-based technologies like MutMap, MutMap+, SHOREmap, Mutmap-Gap, and QTL-Seq.

(c) **Transgenics**

The development of transgenic finger millet varieties resistant to leaf blast disease has been reported. This was achieved by incorporating the PIN (antifungal protein) gene from prawn (Latha et al. 2005) and the *Chi11* (*Chitinase11*) gene from the genus *Oryza* using *Agrobacterium*-mediated transformation (Antony Ceasar and Ignacimuthu 2011). Additionally, the development of finger millet varieties tolerant to salinity stress has been successfully reported. This was accomplished by introducing the gene from *Porteresia coarctata* (*PcSrp*, a serine-rich protein) gene from (Mahalakshmi et al. 2006).

(d) **Genome editing**

The least success of GE tools like the CRISPR/Cas9 system has been reported in sorghum for genes like *PDS* (phytoene desaturase) and *CAD* (cinnamyl alcohol dehydrogenase) (Liu et al. 2008). There are a few reports of successful applications of genome editing tools in millets. In one, the *PDS* gene was targeted through the CRISPR-Cas 9 system, whereas in the second, the *SiMTL* gene for double haploid induction was targeted in *S. italica* Matrilineal (both in Foxtail millet) (Lin et al. 2018; Cheng et al. 2021).

10.2 Future Prospects

Millets, including major and minor millets, have always been proven to be climate-smart nutria-cereals necessary for covering the food and biofortification of nutrition security in developing countries like India. Two essential uses of NGS, de novo sequencing and resequencing, are appealing choices for analyzing vast and complicated millet genomes, pre-breeding activities, and detecting valuable genetic variants in millets. Reduction of antinutritional factors through using novel promoters can be undertaken for good grain quality and the presence of a significant amount of nutrients having essential amino acids, vitamins, and minerals. Tools like RNAi and genome editing through CRISPR/CAS9 need to be employed to knockout genes controlling antinutritional factors. In the future, CRISPR-based genome editing systems, such as base editing, prime editing, and other improved CRISPR/CAS9 systems, will undoubtedly provide several opportunities to improve millets and overcome the negative impacts of biotic and abiotic stresses. It may help to improve millet production and strengthen food security in developing countries.

Advanced genetic transformation techniques utilizing *Agrobacterium* or biolistic methods, along with genome editing tools, offer opportunities for developing metabolic pathways associated with stress tolerance genes. This approach enhances our understanding of developmental, physiological, and stress-related biology in millet crops. The development and complete annotation of the foxtail millet genome, along with establishing synteny with several important cereal crops, serve as valuable resources for investigating C4 grasses. These grasses are particularly relevant as potential biomass sources for biofuel production. Foxtail millet has emerged as a target model system due to its phylogenetic position and well-established photosynthetic pathway, facilitated by its ease of cultivation and small genome size. Such choices are expected to become more prevalent in the future as sequencing costs decrease. Model systems are selected based on their capacity to address specific questions or represent particular phylogenetic groups. With many sequenced genomes and the simplicity of comparative genomics among grasses with largely colinear chromosomes, new genomic sequences can be successfully annotated and connected with genetic data from other grasses.

References

Abdel-Ghany SE, Hamilton M, Jacobi JL, Ngam P, Devitt N, Schilkey F, Ben-Hur A, Reddy AS (2016) A survey of the sorghum transcriptome using single-molecule long reads. Nat Commun 7(1):11706

Adeyanju A, Little C, Yu J, Tesso T (2015) Genome-wide association study on resistance to stalk rot diseases in grain sorghum. G3: Genes, Genomes Genetics 5(6):1165–1175

Ajeesh Krishna TP, Maharajan T, Ceasar SA (2022) Improvement of millets in the post-genomic era. Physiol Mol Biol Plants 28(3):669–685

Ambavane AR, Sawardekar SV, Gokhale NB, Desai SS, Sawant SS, Bhave SG, Devmore JP (2014) Studies on mutagenic effectiveness and efficiency of finger millet [*Eleucina coracana* (L.) Gaertn] in M1 generation and effect of gamma rays on its quantitative traits during M2 generation

Antony Ceasar S, Ignacimuthu S (2011) Agrobacterium-mediated transformation of finger millet (*Eleusine coracana* (L.) Gaertn.) using shoot apex explants. Plant Cell Rep 30:1759–1770

Azzouz-Olden F, Hunt AG, Dinkins R (2020) Transcriptome analysis of drought-tolerant sorghum genotype SC56 in response to water stress reveals an oxidative stress defense strategy. Mol Biol Rep 47(5):3291–3303

Babu RN, Jyothi MN, Sharadamma N, Sahu S, Rai DV, Devaraj VR (2013) Computational identification of conserved micro RNAs from kodo millet (*Paspalum scrobiculatum*). Afr Crop Sci J 21(1):75–83

Bajwa AA, Jabran K, Shahid M, Ali HH, Chauhan BS (2015) Eco-biology and management of Echinochloa crus-galli. Crop Prot 75:151–162

Bennetzen JL, Schmutz J, Wang H, Percifield R, Hawkins J, Pontaroli AC, Estep M, Feng L, Vaughn JN, Grimwood J, Jenkins J (2012) Reference genome sequence of the model plant Setaria. Nat Biotechnol 30(6):555–561

Bhatt R, Asopa PP, Jain R, Kothari-Chajer A, Kothari SL, Kachhwaha S (2021) Optimization of agrobacterium mediated genetic transformation in *Paspalum scrobiculatum* L. (Kodo millet). Agronomy 11(6):1104

Bhave KG, Dalvi VV, Thaware BL, Mahadik SG, Kasture MC, Desai SS (2016) Mutagenesis in proso millet (*Panicum miliaceum* L.). Int J Sci Res 5(3):1635–1638

Bidinger FR, Serraj R, Rizvi SMH, Howarth C, Yadav RS, Hash CT (2005) Field evaluation of drought tolerance QTL effects on phenotype and adaptation in pearl millet [*Pennisetum glaucum* (L.) R. Br.] topcross hybrids. Field Crop Res 94(1):14–32

Blench RM (1997) Neglected species, livelihoods and biodiversity in difficult areas: how should the public sector respond. ODI Natural Resources Paper, 23

Boatwright JL, Sapkota S, Kresovich S (2023) Functional genomic effects of indels using Bayesian genome-phenome wide association studies in sorghum. Front Genet 14:1143395

Bolger ME, Weisshaar B, Scholz U, Stein N, Usadel B, Mayer KF (2014) Plant genome sequencing – applications for crop improvement. Curr Opin Biotechnol 26:31–37

Boukail S, Macharia M, Miculan M, Masoni A, Calamai A, Palchetti E, Dell'Acqua M (2021) Genome wide association study of agronomic and seed traits in a world collection of proso millet (*Panicum miliaceum* L.). BMC Plant Biol 21:1–12

Brutnell TP, Wang L, Swartwood K, Goldschmidt A, Jackson D, Zhu XG, Kellogg E, Van Eck J (2010) Setaria viridis: a model for C4 photosynthesis. Plant Cell 22(8):2537–2544

Cao X, Hu L, Chen X, Zhang R, Cheng D, Li H, Xu Z, Li L, Zhou Y, Liu A, Song J (2019) Genome-wide analysis and identification of the low potassium stress responsive gene SiMYB3 in foxtail millet (Setaria italica L.). BMC Genomics 20:1–13

Cao X, Hu Y, Song J, Feng H, Wang J, Chen L, Wang L, Diao X, Wan Y, Liu S, Qiao Z (2022) Transcriptome sequencing and metabolome analysis reveals the molecular mechanism of drought stress in millet. Int J Mol Sci 23(18):10792

Ceasar SA (2019) Genome-wide identification and in silico analysis of PHT1 family genes and proteins in Setaria viridis: the best model to study nutrient transport in millets. Plant Genome 12(1):180019

Ceasar SA, Ignacimuthu S (2010) Effects of cytokinins, carbohydrates and amino acids on induction and maturation of somatic embryos in kodo millet (*Paspalum scorbiculatum* Linn.). PCTOC 102:153–162

Chandrasekara A, Shahidi F (2011) Bioactivities and antiradical properties of millet grains and hulls. JAFC 59(17): 9563–9571

Cheng Z, Sun Y, Yang S, Zhi H, Yin T, Ma X, Zhang H, Diao X, Guo Y, Li X, Wu C (2021) Establishing in planta haploid inducer line by edited SiMTL in foxtail millet (*Setaria italica*). Plant Biotechnol J 19(6):1089

Chhapekar SS, Gaur R, Kumar A, Ramchiary N (2016) Reaping the benefits of next-generation sequencing technologies for crop improvement – solanaceae. In: Next generation sequencing-advances, applications and challenges. InTech, London, pp 247–285

Choudhary M et al (2015) Transcriptional profiling in pearl millet (*Pennisetum glaucum* LR Br.) for identification of differentially expressed drought responsive genes. Physiol Mol Biol Plants 21:187–196

De Wet JMJ (1992) The three phases of cereal domestication. In: Grass evolution and domestication, pp 176–198

De Wet JMJ, Prasada Rao KE, Brink DE (1983) Systematics and domestication of *Panicum sumatrense* (Graminae). JATBA 30(2):159–168

Deshpande VP, Kamatar MY, Kathnalli DS, Malleshappa SM, Nayakar NY (2003) Screening of sorghum genotypes against shootfly, Atherigona soccata (Rondani). Indian J Plant Protect 31(1):90–93

Deshpande SS, Mohapatra D, Tripathi MK, Sadvatha RH (2015) Kodo millet-nutritional value and utilization in Indian foods. J Grain Process Storage 2(2):16–23

Devi KU, Rao MK, Croker SJ, Hedden P, Rao SA (1994) Coleoptile length, gibberellin sensitivity and concentrations in five non-allelic dwarf mutants of pearl millet – *Pennisetum glaucum* (L.) R Br. Plant Growth Regul 15:215–221

Devos KM, Pittaway TS, Reynolds A, Gale MD (2000) Comparative mapping reveals a complex relationship between the pearl millet genome and those of foxtail millet and rice. Theor Appl Genet 100:190–198

Dida MM, Wanyera N, Harrison Dunn ML, Bennetzen JL, Devos KM (2008) Population structure and diversity in finger millet (*Eleusine coracana*) germplasm. Trop Plant Biol 1:131–141

Doust AN, Mauro-Herrera M, Hodge JG, Stromski J (2017) The C4 model grass Setaria is a short day plant with secondary long day genetic regulation. Front Plant Sci 8:241502

Dwivedi SL, Upadhyaya HD, Senthilvel S, Hash CT, Fukunaga K, Diao X, Santra D, Baltensperge D, Prasad M (2012) Millets: genetic and genomic resources

Elshire RJ, Glaubitz JC, Sun Q, Poland JA, Kawamoto K, Buckler ES, Mitchell SE (2011) A robust, simple genotyping-by-sequencing (GBS) approach for high diversity species. PLoS One 6(5):e19379

Feng ZJ, Xu ZS, Sun J, Li LC, Chen M, Yang GX, He GY, Ma YZ (2016) Investigation of the ASR family in foxtail millet and the role of ASR1 in drought/oxidative stress tolerance. Plant Cell Rep 35:115–128

Fracasso A, Trindade LM, Amaducci S (2016) Drought stress tolerance strategies revealed by RNA-Seq in two sorghum genotypes with contrasting WUE. BMC Plant Biol 16:1–18

Fu F, Girma G, Mengiste T (2020) Global mRNA and microRNA expression dynamics in response to anthracnose infection in sorghum. BMC Genomics 21:1–16

Geervani P, Eggum BO (1989) Nutrient composition and protein quality of minor millets. Plant Foods Hum Nutr 39:201–208

Goron TL, Raizada MN (2015) Genetic diversity and genomic resources available for the small millet crops to accelerate a new green revolution. Front Plant Sci 6:132042

Guo L, Qiu, J, Ye C, Jin G, Mao L, Zhang H, Fan L (2017) Echinochloa crus-galli genome analysis provides insight into its adaptation and invasiveness as a weed. Nat Commun 8(1):1031

Gupta SM, Arora S, Mirza N, Pande A, Lata C, Puranik S, Kumar J, Kumar A (2017) Finger millet: a "certain" crop for an "uncertain" future and a solution to food insecurity and hidden hunger under stressful environments. Front Plant Sci 8:643

Habyarimana E, Lopez-Cruz M (2019) Genomic selection for antioxidant production in a panel of Sorghum bicolor and S. Bicolor× S. Halepense lines. Genes 10(11):841

Hamza NB, Sharma N, Tripathi A, Sanan-Mishra N (2016) MicroRNA expression profiles in response to drought stress in Sorghum bicolor. Gene Expr Patterns 20(2):88–98

Hatakeyama M, Aluri S, Balachadran MT, Sivarajan SR, Patrignani A, Grüter S, Poveda L, Shimizu-Inatsugi R, Baeten J, Francoijs KJ, Nataraja KN (2018) Multiple hybrid de novo genome assembly of finger millet, an orphan allotetraploid crop. DNA Res 25(1):39–47

Haussmann B, Mahalakshmi V, Reddy B, Seetharama N, Hash C, Geiger H (2002) QTL mapping of stay-green in two sorghum recombinant inbred populations. Theor Appl Genet 106:133–142

Hema R, Vemanna RS, Sreeramulu S, Reddy CP, Senthil-Kumar M, Udayakumar M (2014) Stable expression of mtlD gene imparts multiple stress tolerance in finger millet. PLoS One 9(6):e99110

Hilu KW (1994) Evidence from RAPD markers in the evolution of Echinochloa millets (Poaceae). Plant Syst Evol 189:247–257

Hittalmani S, Mahesh HB, Shirke MD, Biradar H, Uday G, Aruna YR, Lohithaswa HC, Mohanrao A (2017) Genome and transcriptome sequence of finger millet (*Eleusine coracana* (L.) Gaertn.) provides insights into drought tolerance and nutraceutical properties. BMC Genomics 18:1–16

Hu Z, Mbacké B, Perumal R, Guèye MC, Sy O, Bouchet S, Prasad PV, Morris GP (2015) Population genomics of pearl millet (Pennisetum glaucum (L.) R. Br.): comparative analysis of global accessions and Senegalese landraces. BMC Genomics 16:1–12

Hu Z, Olatoye MO, Marla S, Morris GP (2019) An integrated genotyping-by-sequencing polymorphism map for over 10,000 sorghum genotypes. Plant Genome 12(1):180044

Jaiswal V, Gupta S, Gahlaut V, Muthamilarasan M, Bandyopadhyay T, Ramchiary N, Prasad M (2019) Genome-wide association study of major agronomic traits in foxtail millet (Setaria italica L.) using ddRAD sequencing. Sci rep 9(1):5020

Jaganathan D, Ramasamy K, Sellamuthu G, Jayabalan S, Venkataraman G (2018) CRISPR for crop improvement: an update review. Front Plant Sci 9:364675

Jayakodi M, Madheswaran M, Adhimoolam K, Perumal S, Manickam D, Kandasamy T, Yang TJ, Natesan S (2019) Transcriptomes of Indian barnyard millet and barnyard grass reveal putative genes involved in drought adaptation and micronutrient accumulation. Acta Physiol Plant 41:1–11

Jency JP, Ravikesavan R, Sumathi P, Raveendran M (2016) Determination of lethal dose and effect of physical mutagen on germination percentage and seedling parameters in kodomillet variety CO3. Electron J Plant Breed 7(4):1122–1126

Jensen SE, Charles JR, Muleta K, Bradbury PJ, Casstevens T, Deshpande SP, Gore MA, Gupta R, Ilut DC, Johnson L, Lozano R (2020) A sorghum practical haplotype graph facilitates genome-wide imputation and cost-effective genomic prediction. Plant Genome 13(1):e20009

Jha DK, Chanwala J, Sandeep IS, Dey N (2021) Comprehensive identification and expression analysis of GRAS gene family under abiotic stress and phytohormone treatments in pearl millet. Funct Plant Biol 48(10):1039–1052

Jia G, Huang X, Zhi H, Zhao Y, Zhao Q, Li W, .Han B (2013) A haplotype map of genomic variations and genome-wide association studies of agronomic traits in foxtail millet (Setaria italica). Nat genet 45(8):957–961

Johnson SM, Lim FL, Finkler A, Fromm H, Slabas AR, Knight MR (2014) Transcriptomic analysis of Sorghum bicolor responding to combined heat and drought stress. BMC Genomics 15:1–19

Kannan B, Senapathy S, Bhasker Raj AG, Chandra S, Muthiah A, Dhanapal AP, Hash CT (2014) Association analysis of SSR markers with phenology, grain, and Stover-yield related traits in pearl millet (*Pennisetum glaucum* (L.) R. Br.). Sci World J 2014:1

Kawadiwale AK, Chauhan RM, Patel MS, Bhadauria HS, Tiwari KK (2024) Identification of diversity pools of pearl millet [*Pennisetum glaucum* (L.) R. Br.] restorer lines based on phenotypic and genotypic data. Vegetos 38:1–9

Khound R, Mathivanan RK, Santra DK (2023) Proso millet Nutraceutomics for human health and nutritional security. In: Compendium of crop genome designing for nutraceuticals, pp 1–31

Kimani W, Zhang LM, Wu XY, Hao HQ, Jing HC (2020) Genome-wide association study reveals that different pathways contribute to grain quality variation in sorghum (*Sorghum bicolor*). BMC Genomics 21:1–19

Krishnamurthy L, Upadhyaya HD, Gowda CLL, Kashiwagi J, Purushothaman R, Singh S, Vadez V (2014) Large variation for salinity tolerance in the core collection of foxtail millet (*Setaria italica* (L.) P. Beauv.) germplasm. Crop Past Sci 65(4):353–361

Kulkarni LR, Naik RK (2000) Nutritive value, protein quality and organoleptic quality of kodo millet (*Paspalum scrobiculatum*). Karnataka J Agric Sci 13(1):125–129

Kumari K, Muthamilarasan M, Misra G, Gupta S, Subramanian A, Parida SK, Chattopadhyay D, Prasad M (2013) Development of eSSR-markers in Setaria italica and their applicability in studying genetic diversity, cross-transferability and comparative mapping in millet and non-millet species. PLoS One 8(6):e67742

Kushwaha H, Jillo KW, Singh VK, Kumar A, Yadav D (2015) Assessment of genetic diversity among cereals and millets based on PCR amplification using Dof (DNA binding with one finger) transcription factor gene-specific primers. Plant Syst Evol 301:833–840

La Borde N, Rajewski J, Dweikat I (2023) Novel QTL for chilling tolerance at germination and early seedling stages in sorghum. Front Genet 14:1129460

Latha AM, Rao KV, Reddy VD (2005) Production of transgenic plants resistant to leaf blast disease in finger millet (*Eleusine coracana* (L.) Gaertn.). Plant Sci 169(4):657–667

Le Thierry d'Ennequin M, Panaud O, Toupance B, Sarr A (2000) Assessment of genetic relationships between Setaria italica and its wild relative S. viridis using AFLP markers. Theor Appl Genet 100:1061–1066

Li J, Wang Y, Wang L, Zhu J, Deng J, Tang R, Chen G (2021) Integration of transcriptomic and proteomic analyses for finger millet [*Eleusine coracana* (L.) Gaertn.] in response to drought stress. PLoS One 16(2):e0247181

Liang Z, Gupta SK, Yeh CT, Zhang Y, Ngu DW, Kumar R, Patil HT, Mungra KD, Yadav DV, Rathore A, Srivastava RK (2018) Phenotypic data from inbred parents can improve genomic prediction in pearl millet hybrids. G3: Genes, Genomes Genetics 8(7):2513–2522

Liang Z, Wu Y, Ma L, Guo Y, Ran Y (2022) Efficient genome editing in Setaria italica using CRISPR/Cas9 and base editors. Front Plant Sci 12:815946

Lin CS, Hsu CT, Yang LH, Lee LY, Fu JY, Cheng QW, Wu FH, Hsiao HCW, Zhang Y, Zhang R, Chang WJ (2018) Application of protoplast technology to CRISPR/Cas9 mutagenesis: from single-cell mutation detection to mutant plant regeneration. Plant Biotechnol J 16(7):1295–1310
Liu CJ, Witcombe JR, Pittaway TS, Nash M, Hash CT, Busso CS, Gale MD (1994) An RFLP-based genetic map of pearl millet (*Pennisetum glaucum*). Theor Appl Genet 89:481–487
Liu Z, Park BJ, Kanno A, Kameya T (2005) The novel use of a combination of sonication and vacuum infiltration in agrobacterium-mediated transformation of kidney bean (*Phaseolus vulgaris* L.) with lea gene. Mol Breed 16:189–197
Liu SJ, Wei ZM, Huang JQ (2008) The effect of co-cultivation and selection parameters on agrobacterium-mediated transformation of Chinese soybean varieties. Plant Cell Rep 27:489–498
Liu T, Liu X, He J, Dong K, Pan W, Zhang L, Ren R, Zhang Z, Yang T (2022) Identification and fine-mapping of a major QTL (PH1. 1) conferring plant height in broomcorn millet (*Panicum miliaceum*). Front Plant Sci 13:1010057
Mahalakshmi S, Christopher GSB, Reddy TP, Rao KV, Reddy VD (2006) Isolation of a cDNA clone (PcSrp) encoding serine-rich-protein from Porteresia coarctata T. and its expression in yeast and finger millet (*Eleusine coracana* L.) affording salt tolerance. Planta 224:347–359
Maxted N, Kell SP (2009) Establishment of a global network for the in situ conservation of crop wild relatives: status and needs. FAO Commission on Genetic Resources for Food and Agriculture, Rome, Italy, p 266
McSweeney MB, Seetharaman K, Dan Ramdath D, Duizer LM (2017) Chemical and physical characteristics of proso millet (*Panicum miliaceum*)-based products. Cereal Chem 94(2):357–362
Miller MR, Dunham JP, Amores A, Cresko WA, Johnson EA (2007) Rapid and cost-effective polymorphism identification and genotyping using restriction site associated DNA (RAD) markers. Genome Res 17(2):240–248
Mohanty IC, Gangasagar PS, Rath SN (2013) Amplification and molecular characterization of DREB1A transcription factor fragment from finger millet [(Eleusine coracana (L.) Gaertn)]. J Agric Sci 5(8):37–49
Morris GP, Ramu P, Deshpande SP, Hash CT, Shah T, Upadhyaya HD, Riera-Lizarazu O, Brown PJ, Acharya CB, Mitchell SE, Harriman J (2013) Population genomic and genome-wide association studies of agroclimatic traits in sorghum. Proc Natl Acad Sci 110(2):453–458
M'Ribu HK, Hilu KW (1996) Application of random amplified polymorphic DNA to study genetic diversity in *Paspalum scrobiculatum* L. (Kodo millet, Poaceae). Genet Resour Crop Evol 43:203–210
Muthamilarasan MEHANATHAN, Venkata Suresh B, Pandey GARIMA, Kumari KAJAL, Parida SK, Prasad MANOJ (2014) Development of 5123 intron-length polymorphic markers for large-scale genotyping applications in foxtail millet. DNA Res 21(1):41–52
Muthamilarasan M, Shweta S, Prasad M (2017) Foxtail millet genome sequencing, assembly, annotation, and application. In: The foxtail millet genome, pp 11–22
Nagarjuna KN, Parvathi MS, Sajeevan RS, Pruthvi V, Mamrutha HM, Nataraja KN (2016) Full-length cloning and characterization of abiotic stress responsive CIPK31-like gene from finger millet, a drought-tolerant crop. Curr Sci 111:890–894
Ngom B, Mamati E, Goudiaby MF, Kimatu J, Sarr I, Diouf D, Kane NA (2018) Methylation analysis revealed salicylic acid affects pearl millet defense through external cytosine DNA demethylation. J Plant Interact 13(1):288–293
Pan J, Li Z, Wang Q, Garrell AK, Liu M, Guan Y, Zhou W, Liu W (2018) Comparative proteomic investigation of drought responses in foxtail millet. BMC Plant Biol 18:1–19
Pandey G, Misra G, Kumari K, Gupta SARIKA, Parida SK, Chattopadhyay DEBASIS, Prasad M (2013) Genome-wide development and use of microsatellite markers for large-scale genotyping applications in foxtail millet [*Setaria italica* (L.)]. DNA Res 20(2):197–207
Parvathi MS, Nataraja KN (2017) Discovery of stress responsive TATA-box binding protein associated Factor6 (TAF6) from finger millet (Eleusine coracana (L.) Gaertn). J Plant Biol 60:335–342

Parvathi MS, Nataraja KN, Yashoda BK, Ramegowda HV, Mamrutha HM, Rama N (2013) Expression analysis of stress responsive pathway genes linked to drought hardiness in an adapted crop, finger millet (Eleusine coracana). J Plant Biochem Biotechnol 22:193–201

Paterson AH, Bowers JE, Bruggmann R, Dubchak I, Grimwood J, Gundlach H, Haberer G, Hellsten U, Mitros T, Poliakov A, Schmutz J (2009) The Sorghum bicolor genome and the diversification of grasses. Nature 457(7229):551–556

Poland JA, Rife TW (2012) Genotyping-by-sequencing for plant breeding and genetics. Plant Genome 5(3)

Puranik S, Sahu PP, Beynon S, Srivastava RK, Sehgal D, Ojulong H, Yadav R (2020) Genome-wide association mapping and comparative genomics identifies genomic regions governing grain nutritional traits in finger millet (*Eleusine coracana* L Gaertn.). Plants, People, Planet 2(6):649–662

Rajaram V, Nepolean T, Senthilvel S, Varshney RK, Vadez V, Srivastava RK, Shah TM, Supriya A, Kumar S, Ramana Kumari B, Bhanuprakash A (2013) Pearl millet [*Pennisetum glaucum* (L.) R. Br.] consensus linkage map constructed using four RIL mapping populations and newly developed EST-SSRs. BMC Genomics 14:1–16

Rajput SG, Santra DK, Schnable J (2016) Mapping QTLs for morpho-agronomic traits in proso millet (*Panicum miliaceum* L.). Mol Breed 36:1–18

Rao BL, Husain A (1985) Presence of cyclopiazonic acid in kodo millet (*Paspalum scrobiculatum*) causing 'kodua poisoning' in man and its production by associated fungi. Mycopathologia 89(3):177–180

Reddy VG, Upadhyaya HD, Gowda CLL (2006) Characterization of world's foxtail millet germplasm collections for morphological traits. Int Sorghum Millets Newsl 47:107–109

Reddy VG, Upadhyaya HD, Gowda CLL (2007) Morphological characterization of world's proso millet germplasm collection. J SAT Agric Res 3:4

Renganathan VG, Vanniarajan C, Karthikeyan A, Ramalingam J (2020) Barnyard millet for food and nutritional security: current status and future research direction. Front Genet 11:497319

Roche GS FLX + System (n.d.) Sanger-like read lengths – the power of next-gen through- put. Roche Diagnostics GmbH; Available from: http://454.com/downloads/GSFLXApplicationFlyer_FINALv2.pdf

Ruperao P, Thirunavukkarasu N, Gandham P, Selvanayagam S, Govindaraj M, Nebie B, Manyasa E, Gupta R, Das RR, Odeny DA, Gandhi H (2021) Sorghum pan-genome explores the functional utility for genomic-assisted breeding to accelerate the genetic gain. Front Plant Sci 12:666342

Sachdev N, Goomer S, Singh LR (2021) Foxtail millet: a potential crop to meet future demand scenario for alternative sustainable protein. J Sci Food Agric 101(3):831–842

Saha D, Gowda MC, Arya L, Verma M, Bansal KC (2016) Genetic and genomic resources of small millets. Crit Rev Plant Sci 35(1):56–79

Santra DK, Khound R, Das S (2019) Proso millet (*Panicum miliaceum* L.) breeding: progress, challenges and opportunities. In: Advances in plant breeding strategies: cereals, vol 5, pp 223–257

Sapkota S, Boyles R, Cooper E, Brenton Z, Myers M, Kresovich S (2020) Impact of sorghum racial structure and diversity on genomic prediction of grain yield components. Crop Sci 60(1):132–148

Schnable PS, Ware D, Fulton RS, Stein JC, Wei F, Pasternak S, Liang C, Zhang J, Fulton L, Graves TA, Minx P (2009) The B73 maize genome: complexity, diversity, and dynamics. Science 326(5956):1112–1115

Sehgal D, Rajaram V, Armstead IP, Vadez V, Yadav YP, Hash CT, Yadav RS (2012) Integration of gene-based markers in a pearl millet genetic map for identification of candidate genes underlying drought tolerance quantitative trait loci. BMC Plant Biol 12:1–13

Serba DD, Yadav RS (2016) Genomic tools in pearl millet breeding for drought tolerance: status and prospects. Front Plant Sci 7:220903

Serba DD, Yadav RS, Varshney RK, Gupta SK, Mahalingam G, Srivastava RK, Gupta R, Perumal R, Tesso TT (2020) Genomic designing of pearl millet: a resilient crop for arid and semi-arid environments. In: Genomic designing of climate-smart cereal crops, pp 221–286

Shan Z, Jiang Y, Li H, Guo J, Dong M, Zhang J, Liu G (2020) Genome-wide analysis of the NAC transcription factor family in broomcorn millet (*Panicum miliaceum* L.) and expression analysis under drought stress. BMC Genomics 21:1–13

Sharma D, Jamra G, Singh UM, Sood S, Kumar A (2017) Calcium biofortification: three pronged molecular approaches for dissecting complex trait of calcium nutrition in finger millet (*Eleusine coracana*) for devising strategies of enrichment of food crops. Front Plant Sci 7:228592

Shi J, Ma X, Zhang J, Zhou Y, Liu M, Huang L, Sun S, Zhang X, Gao X, Zhan W, Li P (2019) Chromosome conformation capture resolved near complete genome assembly of broomcorn millet. Nat Commun 10(1):464

Shinde H, Dudhate A, Tsugama D, Gupta SK, Liu S, Takano T (2019) Pearl millet stress-responsive NAC transcription factor PgNAC21 enhances salinity stress tolerance in Arabidopsis. Plant Physiol Biochem 135:546–553

Singh RK, Singh VK, Raghavendrarao S, Phanindra MLV, Venkat Raman K, Solanke AU, Kumar PA, Sharma TR (2015) Expression of finger millet EcDehydrin7 in transgenic tobacco confers tolerance to drought stress. Appl Biochem Biotechnol 177:207–216

Sood P, Singh RK, Prasad M (2020) An efficient agrobacterium-mediated genetic transformation method for foxtail millet (*Setaria italica* L.). Plant Cell Rep 39:511–525

Srinivas G, Satish K, Madhusudhana R, Nagaraja Reddy R, Murali Mohan S, Seetharama N (2009) Identification of quantitative trait loci for agronomically important traits and their association with genic-microsatellite markers in sorghum. Theor Appl Genet 118(8):1439–1454

Subramanian A, Nirmalakumari A, Veerabadhiran P (2010) Trait based selection of superior kodo millet (*Paspalum scrobiculatum* L.) genotypes. Electron J Plant Breed 1(4):852–855

Sun X, Zheng HX, Li S, Gao Y, Dang Y, Chen Z, Sui N (2023) MicroRNAs balance growth and salt stress responses in sweet sorghum. The Plant Journal 113(4):677–697

Swaminaidu N, Ghosh S, Mallikarjuna K (2015) Millets: the miracle grain. Int J Pharm Bio Sci 6(4):440–446

Tao Y, Luo H, Xu J, Cruickshank A, Zhao X, Teng F, Hathorn A, Wu X, Liu Y, Shatte T, Jordan D (2021) Extensive variation within the pan-genome of cultivated and wild sorghum. Nat Plants 7(6):766–773

Thiel T, Michalek W, Varshney R, Graner A (2003) Exploiting EST databases for the development and characterization of gene-derived SSR-markers in barley (*Hordeum vulgare* L.). Theor Appl Genet 106:411–422

Tuberosa R (2012) Phenotyping for drought tolerance of crops in the genomics era. Front Physiol 3:28919

Upadhyaya HD, Gowda CLL, Reddy KN, Singh S (2009) Augmenting the pearl millet core collection for enhancing germplasm utilization in crop improvement. Crop Sci 49(2):573–580

Upadhyaya HD, Yadav D, Reddy KN, Gowda CLL, Singh S (2011) Development of pearl millet minicore collection for enhanced utilization of germplasm. Crop Sci 51(1):217–223

Upadhyaya HD, Dwivedi SL, Singh SK, Singh S, Vetriventhan M, Sharma S (2014a) Forming core collections in barnyard, kodo, and little millets using morphoagronomic descriptors. Crop Sci 54(6):2673–2682

Upadhyaya HD, Sharma S, Dwivedi SL, Singh SK (2014b) Sorghum genetic resources: conservation and diversity assessment for enhanced utilization in sorghum improvement

Varshney RK, Shi C, Thudi M, Mariac C, Wallace J, Qi P, Zhang H, Zhao Y, Wang X, Rathore A, Srivastava RK (2017) Pearl millet genome sequence provides a resource to improve agronomic traits in arid environments. Nat Biotechnol 35(10):969–976

Velazco JG, Jordan DR, Mace ES, Hunt CH, Malosetti M, Van Eeuwijk FA (2019) Genomic prediction of grain yield and drought-adaptation capacity in sorghum is enhanced by multi-trait analysis. Front Plant Sci 10:997

Vishnoi RK, Kothari SL (1996) Somatic embryogenesis and efficient plant regeneration in immature inflorescence culture of Setaria italica (L.) Beauv. Cereal Res Commun:291–297

Wang M, Wang P, Liang F, Ye Z, Li J, Shen C, Pei L, Wang F, Hu J, Tu L, Lindsey K (2018) A global survey of alternative splicing in allopolyploid cotton: landscape, complexity and regulation. New Phytol 217(1):163–178

Wallace JG, Upadhyaya HD, Vetriventhan M, Buckler ES, Tom Hash C, Ramu P (2015) The genetic makeup of a global barnyard millet germplasm collection. The Plant Genome 8(1). https://doi.org/10.3835/plantgenome2014.10.0067

Yadav OP, Rai KN (2013) Genetic improvement of pearl millet in India. Agric Res 2:275–292

Yadav CB, Muthamilarasan M, Pandey G, Khan Y, Prasad M (2014) Development of novel microRNA-based genetic markers in foxtail millet for genotyping applications in related grass species. Mol Breed 34:2219–2224

Yadav A, Khan Y, Prasad M (2016) Dehydration-responsive miRNAs in foxtail millet: genome-wide identification, characterization and expression profiling. Planta 243:749–766

Yadav OP, Upadhyaya HD, Reddy KN, Jukanti AK, Pandey S, Tyagi RK (2017) Genetic resources of pearl millet: status and utilization. Indian J Plant Genetic Resour 30(1):31–47

Yadav OP, Singh DV, Dhillon BS, Mohapatra T (2019) India's evergreen revolution in cereals. Curr Sci 116(11):1805–1808

Yan Q, Wu F, Xu P, Sun Z, Li J, Gao L, Lu L, Chen D, Muktar M, Jones C, Yi X (2021) The elephant grass (*Cenchrus purpureus*) genome provides insights into anthocyanidin accumulation and fast growth. Mol Ecol Resour 21(2):526–542

Yang Y, Wang L, Li L, Li W, Zhang Y, Chang H, Xiao X, Li M, Lv L (2018) Genetic association and meta-analysis of a schizophrenia GWAS variant rs10489202 in east Asian populations. Transl Psychiatry 8(1):144

Yeshvekar RK, Nitnavare RB, Chakradhar T, Bhatnagar-Mathur P, Reddy MK, Reddy PS (2017) Molecular characterization and expression analysis of pearl millet plasma membrane proteolipid 3 (Pmp3) genes in response to abiotic stress conditions. Plant Gene 10:37–44

Yu A, Zhao J, Wang Z, Cheng K, Zhang P, Tian G, Liu X, Guo E, Du Y, Wang Y (2020) Transcriptome and metabolite analysis reveal the drought tolerance of foxtail millet significantly correlated with phenylpropanoids-related pathways during germination process under PEG stress. BMC Plant Biol 20:1–17

Yue H, Wang L, Liu H, Yue W, Du X, Song W, Nie X (2016) De novo assembly and characterization of the transcriptome of broomcorn millet (*Panicum miliaceum* L.) for gene discovery and marker development. Front Plant Sci 7:1083

Zhang G, Liu X, Quan Z, Cheng S, Xu X, Pan S, Xie M, Zeng P, Yue Z, Wang W, Tao Y (2012) Genome sequence of foxtail millet (*Setaria italica*) provides insights into grass evolution and biofuel potential. Nat Biotechnol 30(6):549–554

Zhang X, Xu G, Cheng C, Lei L, Sun J, Xu Y, Deng C, Dai Z, Yang Z, Chen X, Liu C (2021) Establishment of an agrobacterium-mediated genetic transformation and CRISPR/Cas9-mediated targeted mutagenesis in hemp (*Cannabis Sativa* L.). Plant Biotechnol J 19(10):1979–1987

Zou C, Li L, Miki D, Li D, Tang Q, Xiao L, Rajput S, Deng P, Peng LI, Jia W, Huang R (2019) The genome of broomcorn millet. Nat Commun 10(1):436

Accelerated Breeding Approaches for Improved Productivity and Quality in Dual-Purpose Oats

Parichita Priyadarshini, Jyotsana Tilgam, Indu, Subhash Chand, Vinita Ramtekey, Deepanshu Jaiswal, Maneet Rana, Krishna Kumar Dwivedi, and Shahid Ahmed

1 Introduction

Oat (*Avena sativa* L.) is an important cereal crop because it has a wealth of nutrients that are useful for animal feed, human diet, health care, and cosmetics (Butt et al. 2008; Varma et al. 2016). It is among the most ancient crops ever discovered by humankind, cultivated annually for over two millennia across many regions of the globe (Lásztity 1998; Sang and Chu 2017). This cereal is a good source of fats, various phenolic compounds, vitamins, minerals, balanced protein, dietary soluble fiber, and carbohydrates (Joyce et al. 2019). Oats have drawn more attention from scientific academics and companies as people become more conscious of the benefits of healthy eating practices. Oat beta-glucan (OBG) is a viscous polysaccharide consisting of a linear branched chain of D-glucose monosaccharides linked by mixed β $(1 \rightarrow 3)$ and β $(1 \rightarrow 4)$ links. It is one of the main constituents of soluble fiber and is found in the endosperm cell wall of the kernel (Paudel 2018). It is thought to be the primary active ingredient in oat and has a variety of functional and nutritional qualities, most notably the ability to decrease cholesterol (Whitehead et al. 2014) and antidiabetic benefits (Andrade et al. 2015). As per reports, OBG content ranges from 1.8% to 7% (Martínez-Villaluenga and Peñas 2017). The

P. Priyadarshini (✉) · S. Chand · M. Rana · K. K. Dwivedi · S. Ahmed
ICAR-Indian Grassland and Fodder Research Institute, Jhansi, Uttar Pradesh, India

J. Tilgam
ICAR-National Bureau of Agriculturally Important Microorganisms, Mau, Uttar Pradesh, India

Indu
ICAR-Indian Institute of Sugarcane Research, Lucknow, Uttar Pradesh, India

V. Ramtekey · D. Jaiswal
ICAR-Indian Institute of Seed Science, Mau, Uttar Pradesh, India

S. H. Wani, S. S. Gosal (eds.), *Accelerated Plant Breeding, Volume 5*,
https://doi.org/10.1007/978-3-032-04527-0_10

protein content in oat groats varies from 13% to 20% (Zhou et al. 2019). Antioxidant substances such as tocols, phenolic compounds, and sterols are other minor components of oats that are linked to positive health outcomes.

Oat is also grown as an annual winter feed crop. Livestock productivity is low during the winter months since these animals—cattle, buffalo, goats, etc.—mainly rely on rice straw, byproducts, and semi-dried grasses found in arid areas. Therefore, oat fodder may be successfully grown in rice fallow fields in both plain and hilly places to help alleviate the acute scarcity of green fodder during the lean period of the winter season (Luikham et al. 2015). The crop shows promise for multi-cutting and works well with farming systems to supply both quantity and quality fodder during the winter feed shortage (December to April). According to FAO estimates from 2010 to 2011, oat is placed first globally in terms of fodder production figures and sixth globally among food crops, including wheat, maize, rice, barley, and sorghum. Although oat is typically offered as green fodder, excess is sometimes turned into hay or silage for use in times when there is a shortage of fodder (Suttie and Reynolds 2004). According to Suttie and Reynolds (2004), oats grow quickly and are tasty, succulent, and nutritious. When supplemented with other winter fodder legumes like berseem, lucerne, pea, and vetch, they make a great combination. According to Stevens et al. (2004), oat grows well in a variety of soil types, but perform better in acidic soils. The crop is highly vulnerable to hot and dry conditions, especially during the time of head emergence until maturity (Suttie and Reynolds 2004). Around 75% of the world's supply of grain, seed, and industrial-grade oat comes from Russia, the former Soviet Union, the USA, Canada, Germany, and Poland. The percentage of oats used for feed has decreased in the USA and Canada since the 1960s, stayed constant in Poland and the former Soviet Union, and somewhat rose in Germany (Stevens et al. 2004).

According to Ingalls et al. (1965), the quality of fodder indicates the quantity and chemical composition of nutrients, as well as their palatability, intake, digestibility, and antinutritional variables that are related to cattle performance. Certain critical aspects that may impact the chemical makeup and nutritional value of the product include cultivar, harvest maturity stage, and storage techniques. In addition to these, environmental elements including the temperature during plant growth, day duration, and soil fertility and types play a significant role in influencing quality. Similarly, Jung and Allen (1995) found that digestibility depends on the availability and composition of the cell wall (NDF), whereas forage intake depends on the cell wall content (ADF).

One environmental element that may have a direct impact on the quality of the fodder is the temperature. According to reports, a rise in temperature increases the components of the cell wall, promotes lignification, lowers the concentration of soluble carbohydrates, and reduces digestibility (Pearson and Ison 1997). Additionally, because the stems are less digestible than the leaves, it lowers the forage's leaf-to-stem ratio, which has a direct impact on the dry matter digestibility of the forage (Buxton et al. 1995).

Accelerated breeding techniques are essential in the context of a changing climate to improve the quality and productivity of dual-purpose oat, which is grown for both grain and fodder (Bazzer et al. 2025). These methods seek to ensure the generation of improved oat varieties with features relevant to grain and fodder production, while also accelerating the breeding process. Numerous rapid breeding techniques, including as double haploid (DH) technology, genome editing, speed breeding, high throughput phenotyping (HTP), marker-assisted selection (MAS), genomic selection (GS), and speed breeding, have been implemented to improve crop yields.

This chapter describes the different types of accelerated breeding approaches and their application for the advancement of the oat crop in terms of productivity and nutritional quality.

2 Oat as a Dual-Purpose Crop

In many parts of the world, oat is grown for grain as well as a fodder crop. Livestock grain feed is still the primary use of oat crops. Compared to other cereal crops, oat is better adapted to variable soil types. The protein of rolled (flakes) oats is usually greater than that found in other cereal grains. Food uses for oats include oatmeal, oat flour, and oat flakes as breakfast cereals and ingredients in other food items (Barsila 2018; Ahmad et al. 2014; Pathan et al. 2020). Oat bran and germ contain many vitamins and minerals. In India, particularly western and northwestern regions, oat has wider adaptability due to its excellent growth habit with superior nutritional quality and quick regeneration habit. As far as nutrition is concerned, oats are one of the foremost and demanding nutritious grain cereals with high protein and fiber (Ahmad et al. 2014; Pathan et al. 2020).

2.1 Oat as a Fodder Crop

The farmers face fodder deficiency in winter as they have only dry stalks of summer cereal fodders or dry summer grasses. Being a winter-hardy crop, oat is best suited as a forage crop during the *rabi* season and is also preferred feed for animals (Barsila 2018; Gadisovich 2017). Oats are a crucial winter fodder, mostly fed as green, but surplus is converted into silage or hay to use during fodder deficit periods. Oat grain has additional nutritional value which makes it a valuable feed for horses and poultry. Interest in oat hay for the dairy, feed-stuff, and horse industries has grown in recent years (Kaur et al. 2018; Pathan et al. 2020). This is often mainly an improvement in oat hay quality caused by higher quality standards demanded by the export hay market. The forage oat varieties with higher productivity, better quality, and tolerance to abiotic stress are the need of the hour to surpass the gap between demand and supply of green fodder (Singh et al. 2018). Understanding the gene action and combining ability may help in selection for the suitable breeding method for forage oat crop improvement. In agreement with the aforesaid statement, there is an urgent need to exploit new research technologies to enhance forage yield in terms of higher yield of green fodder and dry matter per unit area. The competition for utilization of land for food grains and fodder necessitates intensified efforts towards more efficient forage production. (Ahmad et al. 2014; Pathan et al. 2020).

2.2 *Oat as a Grain Crop*

The human consumption of oat grain has increased progressively, owing to its nutritional values and better dietary benefits. Oat grain contains high dietary fiber and β-glucan content. The protein content of the hull-less oat kernel (groat) ranges from 13% to 20% (Zhou et al. 2019), the very best among cereals. Oats contain soluble fibers, well-balanced proteins, and many other vitamins and minerals which are essential for human health (Esposito et al. 2005; Gadisovich 2017). Oats contain a high amount of lipids with a considerable level of essential linoleic acid. The popularity of oat meal and oat-based products increased after January 1998, when Food and Drug Administration (FDA) issued a final permit to the food companies to make health claims on food labels of foods having soluble fiber from whole oats and daily consumption of 3.0 g of soluble fiber from these foods may reduce the heart disease risks (Kriger et al. 2018; Rauf et al. 2019). Furthermore, oats are a rich source of several natural antioxidants, like tocopherol, alkylresorcinol, and phenolic acids and their derivatives, and the single source of avenanthramides (cinnamoylanthranilate alkaloids) and avenalumic acids (ethylenic homologs of cinnamic acids). All of those phenolic compounds possess potential health-promoting properties due to their antioxidant activities and/or membrane-modulating effects (alk(en)ylresorcinols) (Kouřimská et al. 2018; Rauf et al. 2019). Moreover, β-glucans, which also exhibit an antioxidant property, are included within the soluble dietary fiber fractions of oats that participate in glucoregulation and cause a reduction in serum cholesterol levels in humans. Therefore, consumption of oats is an essential component of a hypercholesterolemic patient's diet (Ahmad et al. 2014; Gorash et al. 2017; Kaur et al. 2018).

β-Glucan is a soluble fiber readily available from oat grain, which has been gaining interest owing to its multiple functional and bioactive properties (Szpicer et al. 2020). Its beneficial role in insulin resistance, dyslipidemia, hypertension, and obesity is being continuously documented. Consequently, the applicability of β-glucan as a food ingredient is being widely considered with the twin purposes of accelerating the fiber content of food products and enhancing their health properties (Mäkelä et al. 2021). β-Glucan has viscosity properties because oats can be used in the bakery industry for products like biscuits and pasta. Its use could also include frozen desserts, breakfast foods, beverages, meats, on-dairy creamers, and canned soups, especially as fat replacers (Ames et al. 2021). The cheese industry can benefit from β-glucan which optimizes the process of raw material and improves the cheese structure. The combination of insulin and β-glucan also has very interesting properties that can be put to use as a fat-replacer in products such as low-fat ice-creams.

This has been proven to be a very successful dietary source of the soluble fiber β-glucan and has been related to a reduced risk for many diseases (Ahmad et al. 2014).

Oat milk is also becoming popular as it is a nutritious and cheaper alternative to dairy milk. Western herbalists regard oats as a tonic for the systema nervosum (Demír et al. 2021). It has high fiber, zero fat, vitamin E, vitamin B, phytochemicals (β-carotene), cholesterol, and lactose-free (Syed et al. 2020).

3 Conventional Breeding Efforts for Improved Productivity in Dual-Purpose Oats

Oats are susceptible to various pathogens and environmental stresses which perpetually decrease oat yield and affect the composition, hence the availability of good-quality oats for consumption. Almost all beneficial oat components, namely, β-glucan, proteins, oils, starch, and phytochemicals, are influenced by environmental factors (Stewart and McDougall 2014). Currently, the changing climate is aggravating the negative effects of abiotic stress on oat cultivation. A study conducted by Tamm (2003) has presented the influence of high temperature, drought, heavy rain, and wind on oat yield realization. Similarly, pathogens like bacteria, fungi, and viruses limit oat yield by causing direct and indirect damages. In oats, diseases cause about 20–30% of losses, and wherein serious disease epidemics results in complete yield loss (Saini et al. 2019). The diseases that directly compromise the oat yield include stem rust, crown rust, and leaf blotch caused by *Pyrenophora* spp., Barley yellow dwarf virus, and *Septoria* spp., respectively (Saini et al. 2019). Infection by fungi of the *Fusarium* genus leads to indirect damage by affecting the grain quality. The fungi produce toxins in the grain which are unfit for consumption as food or feed (Li et al. 2011; Saini et al. 2019). Table 1 includes the list of commonly found oat diseases. Among the environmental stresses, temperature stress has an important role in delineating the area of adaptation of oat. Generally, winter oat is cultivated in the area having relatively mild winters and warm summers whereas spring oat is cultivated in the area having relatively colder winters and cooler summers. According to Marshall (1992), oat cultivation in the winter-spring transit zone is speculative because winter-sown oat may suffer from chilling injury caused by low-temperature stress in the winter, whereas high-temperature stress in summer may decrease the grain quality and yield of spring-sown oats. Heat stress brings drought events more frequently, and, in most cases, drought and heat effects are confounded, and specific damages are difficult to evaluate. Drought and heat stress cause grain abortion in oats that results in empty white spikelets (Sánchez-Martín et al. 2017). Hence, the quality of grain is compromised because of poor grain filling and a higher hull percentage. Heavy rain and wind also limit oat production by causing lodging of the plants. It is reported that lodging may reduce yield up to 50% in oat (Bailey-Serres et al. 2012). Plant lodging mainly hampers the grain

Table 1 The list of commonly found oat diseases and their causal organisms

Sl. no	Type of disease	Causal organism	Oat diseases
1.	Viral	*Barley yellow dwarf virus*	Barley yellow dwarf
		Oat mosaic virus	Soil-borne oat mosaic
2.	Bacterial	*Pseudomonas coronafaciens*	Halo blight
		Bipolaris sorokiniana, Fusarium, and Pythium spp.	Seed and seedling diseases
3.	Fungal	*Puccinia coronata* f. sp. *avenae*	Crown rust
		Puccinia graminis f. sp. *avenae*	Stem rust
		Drechslera avenae	Helminthosporium leaf blotch
		Erysiphe graminis f. sp. *avenae*	Powdery mildew
		Septoria avenae	Septoria leaf blotch
		Ustilago avenae	Loose smut
		Colletotrichum graminicola	Anthracnose
		Fusarium graminearum	Fusarium head blight (Scab)

filling if it happens before physiological maturity, as well as causes grain loss at the time of harvesting. Moreover, the possibility of lodging rises with the soil fertility level, and cultivators suffer more damage due to the use of suboptimal levels of oat fertilization.

3.1 Resistance to Various Biotic and Abiotic Stresses

3.1.1 Biotic Stresses

The practice of oat improvement started when the species was first cultivated. Through oat plant selection, famous cultivars 'Potato' and 'Sandy' were made in the years 1788 and 1852, respectively (Lawson and Son 1852; Hunter 1924). The breeding approaches for oat improvements include introduction, selection, hybridization (interspecific and inter-varietal), and then bulk, backcross, pedigree, and single-seed descent methods (Choubey and Roy 2005). The recurrent selection was also found beneficial in some cases. For oat fodder, few modified methods were attempted, such as the use of North Carolina designs I and II. Biparental matings among genotypes within and between F2 progenies have also been made to select superior plant populations (Pant et al. 1992). In oat breeding for disease resistance, there have been several breeding strategies proposed: incorporation of major race-specific genes, selection for partial resistance (non-race-specific or adult plant resistance), gene pyramiding, and use of multiline cultivars (Ohm and Shaner 1992).

All the varieties that developed before the 1930s were made by plant selection from the introduced landraces. The resistance trait of these varieties, possibly governed by race-specific resistance, single genes, and selection, was conducted for maximum levels of resistance that were readily visible. From the early 1920s, the selection following hybridization had become a common practice which also brought various variations of pedigree and bulk methods of breeding. Race-specific resistance was incorporated in new cultivars because of easy selection, dramatic effect, and simple Mendelian inheritance (Ohm and Shaner 1992).

Initially, new resistant cultivars were produced from single crosses between resistant and existing susceptible parents. The concept of hybridization for resistance was first accepted by Europe rather than the USA, and several rust-resistant varieties, including Red Rustproof and Kherson, were introduced. In 1927, crown rust- and smut-resistant oat variety Victoria (*Avena byzantina*) was introduced in the USA which was used to develop new resistant varieties. Before introducing Victoria variety in the USA, White cross and Lee were selected from the progeny of crosses (Coffman et al. 1961). Besides Victoria, another smut-resistant variety, Bond, was utilized in breeding programs to derive disease-resistant oat varieties in the USA. Bond was developed by crossing wild red oat (*Avena sterilis*) and cultivated Golden Rain (*A. sativa*), a good example of oats through introgression of resistance genes from wild progenitors. Bond-derived cultivars are superior to Victoria-derived cultivars in terms of higher yield, test weight, and resistance to diseases. The Clinton variety derived from the cross between Iowa D69 and Bond showed high agronomic performance such as high yield, quality, disease, and lodging resistance. Benton, a sister selection of Clinton, also had a great rust resistance trait. Victoria-derived cultivars faced major damage in 1946 from *Helminthosporium* blight, whereas Bond-derived crosses showed resistance. The cross of Columbia/Victoria/Richland was found to be resistant to blight (Poehlman and Kingsolver 1950). In general, breeding for Barley yellow dwarf virus (BYDV) tolerance has included aggregation of genes for tolerance via recurrent selection. The oat strains such as 'Kherson', 'Red Rustproof', and their derivatives were found as sources of BYDV tolerance in hexaploid oats (Endo and Brown 1964). In 1973, Brown and Jedeinski released Otee, a transgressive segregant that showed a moderate level of BYDV tolerance along with loose smut (*Ustilago avenae*) and covered smut (*U. kolleri*) resistance traits. Various germplasm lines developed and released as oat cultivars, having a high level of BYDV tolerance (Brown and Jedlinski 1978). Varieties Larry, Ogle, Don, and Hazel developed by single-seed descent procedure had good BYDV tolerance (Mcmullen and Patterson 1992). Soilborne mosaic caused by oat mosaic virus (OMV), controlled through the release of resistant or tolerant varieties such as Arlington 23, Century, Sumter, Sumter 3, and Bruce. Sumter, Sumter 3, and Bruce also have resistance to culm rot, Victoria blight (Mcmullen and Patterson 1992). Through breeding, fungal disease, powdery mildew, was controlled by the use of resistance genes as well as the identification and introgression of the new resistance gene. Resistance genes against powdery mildew are designated as oat mildew resistance (OMR) groups in oat plants (Jones and Jones 1978). The *A. sativa* and *A. sterilis* have contributed major genes for resistance in cultivars (Frey 1986). Herrmann and Roderick (1996) reported that some wild oats such as *A. strigose*, *A. occidentalis*, *A. fatua*, and *A. sterilis* also show resistance against powdery mildew.

Adult plant resistance was first identified by Peturson (1944), and he suggested that hybrid lines should not be discarded in the seedling stages as resistance could progressively develop in adult plants (Hite et al. 1977). Many tetraploid and diploid oat cultivars and accessions have been reported with durable APR resistance against crown rust (Cabral et al. 2011). In oat cultivars, pyramiding of race-specific genes was also employed to obtain durable resistance against new rust races (Patterson and Schafer 1978). Browning and Frey (1969) proposed the role of multiline cultivars to control disease damage in self-pollinated crops like oats. A multiline cultivar is a mixture of isogenic lines in which each one has a different gene for specific resistance. Multiline cultivars effectively control disease when a disease is caused by genetically variable organisms, e.g., crown rust. Multiline cultivars can maintain simple pathogen races (having few virulence genes) and delayed inoculum build-up and rate of disease spread in susceptible plants (Browning and Frey 1969). Multiline cultivars M68, E68, M69, E69, E70, and M70 were released for crown rust resistance. Race-nonspecific resistance is quantitative and governed by many genes. In the breeding program, expression of race-nonspecific resistance was increased by pedigree methods, recurrent selection, and the backcrossing method. The higher level of race-nonspecific resistance achieved against pathogen *P. coronata*, *E. graminis*, and BYDV (Simons 1972).

Nishiyama (1929) has investigated the cytogenetics of oats and thus established the basis for the application of interspecific hybrids for the introduction of desirable genes from wild to cultivated oats. Landraces and wild relatives offer plenty of genetic resources for disease resistance and abiotic stress tolerance and can lend these traits to cultivated varieties upon hybridization. Through a chromosomal manipulation approach, resistance from the tetraploid wild species, *A. barbata*, has been transferred into *A. sativa* (Thomas et al. 1980). The establishment of Germplasm collection and its evaluation for genetic diversity and identification of key agronomic traits were the prime work of breeders in the early stage of the oat improvement program (Krull and Borlaug 1970; Diederichsen 2008; Kaur et al. 2018). Frequent emergence of new aggressive races, by recombination and mutation, parasitize the cultivars which were earlier regarded as resistant. Such pathogenic variability was addressed by utilizing resistant genes of wild relatives. For instance, crown rust-resistant varieties derived from *Avena byzantina* introduction were subjected to new races of rust fungus in due time. The new source of resistance for virulent races was found in the introduction of P. I. 193,958 of tetraploid species *A. abyssinica*, hexaploid species *A. sterilis*, various varieties of diploid species *A. strigosa* (Dherawattana and Sadanaga 1973).

Stadler (1929) introduced the first induced mutation in oat and noted that the mutation rates in polyploid oat were much less than those in diploid oat by exposing the same level of mutagenic treatment. Induced mutation breeding was reported for the introduction of rust resistance in oats. The mutation breeding was employed against *Puccinia graminis avenae* due to inadequacy of resistance in cultivated and weedy species of oat (McKenzie et al. 1971) and for resistance to *Puccinia coronata* due to the emergence of new virulent races (Simons et al. 1962). However, the limited success of mutation breeding for the introduction of rust resistance in oats and the identification of various rust resistance genes in *Avena sterilis* has led to the withdrawal of mutagenesis work in oats.

3.1.2 Abiotic Stress

Oat breeding for abiotic stress tolerance is difficult to achieve owing to the complex genetic and physiological associations. In addition, the development of a genotype with complex gene combinations for stress tolerance with all essential agronomic traits governing genes in a useful variety is a challenging process. Very few successful oat breeding attempts have been observed for abiotic stress tolerance.

The early success of winter hardiness was achieved mainly from selection within Red Rustproof and its derivatives. Two Red Rustproof (*A. byzantina*) selections, namely, Fulghum and Culbersons, contributed majorly to the winter hardiness of derivatives (Coffman et al. 1961). The occurrence of transgressive segregation for winter hardiness is well documented. In 1932, Wintok was derived from a cross between Hairy Culberson x Winter Fulghum. Wintok exceeds its parents in freezing resistance and has become a winter hardiness standard in the USA. Many winter-hardy cultivars better than Wintok were developed by breeders in the later years. Besides winter oat germplasm, domestic spring oat germplasm also served source of genetic diversity for freezing resistance. Winter oat cultivar Dubois (Clinton/Forkedeer) is more winter hardy than its winter oat parent (Forkedeer). Hexaploid species of wild oat were also considered as a gene source for winter hardiness. Few genotypes of *A. fatua* show a similar level of winter hardiness to Dubois. Freezing resistance is also provided by accessions of *A. sterilis*, *A. sterilis* var. *ludoviciana.* Marshall (1992) has produced bulk composites by hybridizing hexaploid wild oat species with cultivated winter cultivars and maintaining them for natural selection for winter hardiness.

The conventional breeding efforts were not very successful in improving drought and heat stress tolerance in oat. The simultaneous occurrences of both stresses are common and usually difficult to distinguish. There was a lack of information about the genetic and physiological basis of resistance, the range of genetic diversity for resistance, lab and field testing, effective plant tissue, and the growth stage to screen. In the oat gene pool, very few drought- and heat-resistant cultivars and lines have been found. For the improvement of spring oat, the gene pool of *A. byzantina* was identified as a potential source of effective genetic diversity for tolerance to both drought and heat (Marshall 1992). *A. byzantina*-based cultivars were more tolerant than *A. sativa* cultivars. Some genotypes of wild oat, *A. sterilis*, exhibit notable traits that were utilized for the improvement of drought resistance of cultivated oat. Certain genotypes of *A. sterilis* isogenic lines are able to maintain more turgidity and are in a viable condition in drought stress (Brinkman and Frey 1977). Helsel and Frey (1978) observed delayed leaf senescence traits in high-yielding elite lines derived from 'Clintford' *41/*A. sterilis*. *A. sterilis*-derived lines also possess a larger root mass than *A. sativa* cultivars, certainly to overcome drought stress (Marshall 1992). To escape stress, oat breeders tried to incorporate early maturity traits in cultivars; unfortunately, that results in a loss of yield potential during non-stress periods. Early maturing varieties, Rapida and Montezuma, were released for dry and hot climates were originated from *A. byzantina/A. fatua.*

Suitable conventional breeding methods to improve lodging resistance have not been identified due to a lack of basic studies. Crossing between elite lodging-resistant and non-lodging-resistant parents for more cycles is required to accumulate essential genetic variation for different characteristics in the breeding population. The probability of retrieving the lodging resistance is enhanced through sib-mating-derived individuals, which show better resistance, or via backcrossing such lines to the resistant parent. The lodging-resistant cultivars were released in the USA by using stiff-strawed from Wales or Scotland germplasm. Clintford cultivar originated from the hybridization of Welsh cv. Milford used as one parent. Other Welsh germplasm-derived cultivars were Compact, Walken, and Kenoat that contain excellent lodging resistance as well as winter-hardiness similar to Wintok (Marshall 1992). Egdolon 23 and Egdolon 26, derived from Astro /PI 193027, were widely utilized as parental lines by oat breeders for lodging resistance.

3.2 Yield-Related Traits

Oats belong to the family Gramineae with a 7 basic chromosome number and diploid, tetraploid, and hexaploid ploidy. The elite cultivar of oat is hexaploid (AA, CC, DD) with the 22 million tons annual production worldwide (Loskutov 2008; Ralla et al. 2018). Oat (*Avena sativa*) is being used as feed and food like other food grains. It is considered a nutritious source of major biomolecules along with fibers, β-glucan, tocols, and avenanthramides, which are essential for the growth and development of animals and humans (Peterson 2004). The protein, fat, and beta-glucan content in Oat are 12–17%/dm, 6–10%/dm, and 3.6–5%/dm, respectively. The presence of variable amounts of vitamin E in oat grains adds to the value of the grain for feed and food (Hampshire 2003). The Gramineae family has a diverse range of wild germplasm containing various alleles for biotic and abiotic resistance. Due course of time, breeders selected the elite cultivar to get better quality yield, resulting in many essential gene combinations being left in wild germplasms. These wild accessions are conserved as plant genetic resources which can be utilized as per requirement to boost the Oat yield. The available germplasm collection, conservation, and searching of crossable genetic resources for Oat yield improvement are the major tasks for breeders (Loskutov 2001; Loskutov 2004; Ahmad et al. 2020). The available landraces can be characterized morphologically and with molecular markers. The application of molecular markers to find the allelic diversity and further its use in the selection of a crossable gene pool or single plant selection is highly reliable (Fu et al. 2003; Katsiotis et al. 2004). Breeders used the hexaploid wild oats for the introgression of desirable genes into elite oats cultivars, whereas exploiting tetraploid and diploid oats wild species as a source of genes through breeding is a difficult affair due to the problem of stable recovery of recombinants

(Fu 2018). Based on phenotypic traits, selection of wild cultivar and crossing wild to elite cultivar through a conventional breeding program are less viable due to problems of interspecific hybrids generation and time-consuming.

Even though the yield of oats was increased up to 7% by using wild germplasm Avena sterilis (Frey et al. 1984). The yield enhancement up to 7% is not enough due to oats dual-purpose uses and high demand. Conventional breeding methods have some limitations such as insufficient information regarding the location of loci on chromosomes, linkage, epistasis, dominance, and codominance (Swamy and Sarla 2008).

Oats are usually known for their good health benefits for humans and animals (poultry and ruminants). Therefore, being a food and feed crop, its yield enhancement (number of panicles, number of seeds per panicle, and seed weight) is the need of the hour.

For many years, the traditional breeding for crop improvement involved introduction, selection, and hybridization with elite cultivars to introgress the desirable trait. To get a homozygous and immortal plant population like near near-isogenic lines and recombinant inbred lines were developed. The genotype of oats given in Table 2 could be used to introgress desirable traits into cultivars through conventional breeding. Further, developed cultivars could be used for performance analysis, and the best one will be selected for higher fodder yield. There are various studies have been conducted to compare the elite cultivar, and the selection of the best-performing variety has been recommended to farmers.

There are a few oat species which are a good source of desirable traits to enhance the fodder yield.

The oat varieties developed through a conventional breeding approach should be screened for their performance at different locations, and finding few of the best varieties should be recommended for cultivation. In this respect, several studies

Table 2 Sources of desirable traits for fodder yield enhancement

Sl. no.	Oat species	Desirable traits
1.	*Avena nuda*	Leaf size
2.	*A. strigosa*	Plant height
3.	*A. barbata*	Longer growth duration
4.	*A. canariensis*	Earliness
5.	*A. hirtula*	Tillering
6.	*A. prostrata*	Tillering
7.	*A. abyssinica*	Earliness, foliage protein
8.	*A. maroccana*	Foliage protein, longer
9.	*A. sativa*	Plant height, leaf size, longer
10.	*A. sterilis*	Plant height, tillering, foliage protein, forage yield

have been done to find the highest-yielding variety by various authors. Lodhi et al. (2009) in their study took fodder yield, plant height, and leaf size as the parameters of yield traits. Among nine improved varieties of oats, viz. No.725 (V1), X 5229–1(V2), CK1(V3), AVONXEM(V4), SCOTT(V5), No. 97081(V6), PD2LV65(V7), UGG(V8), and No. 75525(V9), highest fodder yield (47.6 ton/hectare), plant height (118.02 cm), and leaf size (124.24 cm^2) was found in Variety 1 that is No. 725 and lowest fodder yield (33.3 ton/hectare) and plant height (92.6 cm) in CK1(V3). The lowest leaf area was 88.99 cm^2 in PD2LV65 (V7). The highest yield was due to higher plant height, leaves per tiller, and more leaf area, which provided more surface area for photosynthetic accumulation (Hussain et al. 1993; Amanullah et al. 2004).

In the same manner, 17 varieties of oat, No. 663, No. 646, reil, nile, steel, jasper, foothill, valley, murray, superlate, hackey, winjardo, S-99, PD2LV65, sheikuhpura, and marlu have been evaluated for green fodder yield. Based on the parameters, plant height, green fodder yield, and leaf area were measured. The oat variety Reil had the highest yield potential, i.e., 83.9 ton/hectare (Naeem et al. 2002). The experiment conducted in Ethiopia, to find the best forage variety based on phenotypic data on plant height, green fodder yield, tiller number/plant, tillers/m2, and leaf number/tiller, Beyene et al. (2015) reported that oat variety Lampton is the best performing variety among seven oat varieties.

In Ethiopia, SRCPX80Ab2291, Grayalegris, Jasari, SRCPX80Ab2806, CI-8251, CI-8235, 79Ab384 (TX) 80SA94, Lampton, 79Ab384 (TX) 80SA95, and CI-8237 yield performance of different oat accessions was analyzed by Semman et al. (2018). Based on biomass yield, plant height, susceptibility to disease, seed yield, no lodging, and harvesting index (HI), CI-8251 was found to be superior to other tested accessions. These accessions can be used in the varietal development through the oat breeding program.

In India, various oat varieties were introduced in the 1970s. Since then, an intensive breeding program at various organizations has been organized resulting in different multi-cut oat varieties being released. The notified fodder variety JHO-851 has a high tillering capacity with multi-cut quality, becoming the popular one. Being a forage cum grain crop, it is essential to identify the genotypes or varieties that could be used for both feed and food. However, some oat genotypes, such as (JHO 802 x HFO 267) x *A. sterilis*-3 and (HFO 267 x IGO 320) x *A. sterilis*-7A, are potential oats to be used as dual-purpose crops (Choubey and Roy 2005). Wild germplasms of oats containing different allelic variations are potential candidates for introgression breeding, which incorporates the transfer of desirable traits into elite cultivars of oats (Frey et al. 1984). At ICAR-IGFRI, India, the wild germplasms *A. macrocana* ($2n = 4x = 28$) and *A. sterilis* ($2n = 6x = 42$) have been utilized to introgress traits such as biomass, high growth, and tillering capacity into the *Avena sativa* background. The ICAR-IGFRI has also developed many fodder oats varieties, such as Bundel Jai-822, Bundel sheet Jai 810 and 829, Bundel Jai 992, and various polysomic lines (Choubey and Roy 2005).

The grain yield of oats does not solely depend on the genotype. Tamm (2003) in his study on 101 oat genotypes from different countries analyzed the effect of drought, salinity, heavy winds, and lodging. He concluded that the effect of these stresses also affects the yield, along with the genotype of oats.

Being a basic raw material of any crop improvement, genetic diversity among the genotypes is essential for oat genetic improvement. To date, there are many studies that have been conducted on genetic divergence and variability. These studies give an idea about the quantitative traits responsible for crop improvement. Among descriptive traits, the focus has been given to yield-related traits. Kumari and Jindal (2019) evaluated 55 oat genotypes belonging to different *Avena* sp. to identify the quantitative traits responsible for yield. Principal component analysis revealed that the first three factors (leaves per plant, height at maturity, days to maturity, tillers per plant, green fodder yield per plant, height at fodder stage, dry fodder yield per plant, leaf length, leaf width and dry matter percentage, and flag leaf length) are key traits associated with dry fodder yield per plant, while the last principal factors (PF-6, PF-7, and PF-8) mostly contribute to seed yield. Fifty genotypes were used to estimate the genetic diversity for fodder yield and yield contributing traits in oats based on principal component analysis. PC1 to PC10 were responsible for the total variation among the genotypes which was around 89.52%. The penultimate leaf area, panicle weight, axis branch number, spikelets panicle^{-1}, florets panicle^{-1}, days to 50% flowering, days to flower initiation, number of leaves plant^{-1} dry matter, plant height, axis length, days to maturity, and 1000 seed weight are the major yield-related traits for breeding dual-purpose oat (Kujur et al. 2017). Fifty genotypes of fodder oat were studied to know the direct and indirect yield-related traits by performing correlation, path, and diversity analysis. The direct traits correlated to yield were leaf-to-stem ratio, dry matter, spike length, and spikelet/panicle number, whereas indirect traits affecting fodder yield were leaf length, spike girth, leaf breath, and 1000 seed weight, which contributed to green fodder yield, with the exception of number of leaves and stem girth. This study suggested that the selection of these traits for breeding oat genotypes will lead to improvement of green fodder yield (Krishna et al. 2014). Surje and De (2014) used 15 genotypes, which were subjected to estimate the quantitative traits for improvement of dual-purpose oat, such as green fodder yield and seed yield via different methods.

One method focused on estimating green fodder yield via one cut as green fodder and left to regenerate to produce seed yield, whereas the other method focused mainly on seed yield without taking any cut. Based on a correlation study, 13 traits were found to be correlated with green fodder and seed yield. The plant height, tiller number per plant, flag leaf weight, crude protein percentage, panicle length, grain per panicle, and spikelet number per panicle exhibited significant and positive correlation with green fodder and seed yield. Other than these traits, total chlorophyll and chlorophyll B were also positively associated with yield. This study suggested that the selection of the abovementioned traits would be fruitful for oat genetic improvement. Yadav et al. (2011) performed Mahalanobis D^2 statistics to estimate the genetic divergence among oat genotypes. Based on the analysis, genotypes were grouped into 11 clusters depending on the similarity and dissimilarity of genetic variability. The genotypes belonging to clusters VI, VII, and VIII had the highest mean values related to days to 50% flowering, plant height, and green fodder yield per plant. The crossing program among these clusters would be good enough to produce desirable segregants and higher genetic gain. This would result in the development of a variety with dual purpose, i.e., green fodder yield and seed yield.

A group of 75 oat germplasm lines was used to perform correlation and path analysis to identify the most contributing traits for dual purpose. A similar finding was obtained by Kaur et al. (2018) while evaluating dual-purpose oat genotypes. Days to 50% flowering, plant height, no. of leaves plant^{-1}, leaf length, leaf width, and leaf-to-stem ratio were the direct and indirect quantitative traits associated with green fodder yield and grain yield (Ahmed et al. 2013). Iannucci et al. (2011) estimated genetic divergence among the oat germplasm by performing univariate and multivariate analysis. This study revealed that harvest index, seed weight panicle^{-1}, 1000 seed weight, number of panicles m^{-2}, and number of seeds panicle^{-1} are directly associated with seed yield. Other than these descriptive traits, one has to put emphasis on heading time, maturity time, filling period, plant height, and seed test weight to breed oat genotypes with different purposes, mainly food and forage. Dumlupinar et al. (2012) evaluated 196 Turkish oat landraces to determine the most significant traits for oat genetic improvement. According to this study, flag leaf length and width, days to maturity, vegetative period, total biomass, and germination percentage are the most promising quantitative traits for selection. Along with genotypes, the environmental factor has also played a role in genetic divergence. These traits need to be emphasized while breeding for grain and forage yield under different environmental conditions. One of the studies conducted by Kebede et al. (2020a, b) verified different oat forage varieties. These fodder varieties were high in dry matter content, crude protein content, in vitro dry matter digestibility content, and leaf-to-stem ratio. So, in the breeding program, we have to select these key traits for the development of oat forage varieties.

A group of 75 oat germplasm lines was used to analyze genetic diversity and variability among the genotypes and to determine the yield-related traits. Based on D^2 cluster analysis, 75 lines were grouped into 9 different clusters which suggest that direct selection for days to 50% flowering, leaves plant^{-1}, leaf length and width, leaf-to-stem ratio, plant height, panicle length, and spikelet panicle^{-1} would be enable helpful to increase green fodder yield and grain yield (Ahmed et al. 2011). Sahu and Tiwari (2020) evaluated 14 fodder oat genotypes to estimate genetic variability. Green fodder yield was significantly and positively correlated with leaf-to-stem ratio, crude protein content, and dry matter yield, except for days to 50% flowering and plant height. This suggests that the genotypes with desirable quantitative yield-related traits should be considered in the forage breeding program.

Deep et al. (2019) conducted a study for traits related to fodder and grain yield based on heterosis and transgressive segregants in a set of diallel crosses. The grain yield was found to be significantly and positively associated with traits such as plant height, tillers plant^{-1}, leaves plant^{-1}, leaf length and width, dry matter, seeds panicle^{-1}, harvest index, mean groat mass, and green area index in the F1 generation while days to 50% flowering, days to maturity, plant height, tillers plant^{-1}, spike length, leaves plant^{-1}, leaf width, seeds panicle^{-1}, harvest index, hectoliter grain weight, mean groat mass, and groat length were positively correlated with grain yield per plant in the F2 generation. This suggests that these are important traits which need to be selected during the breeding program of forage. The genetic diversity and variability have been decreased in cultivated genotypes of oats because of the utilization of common parents in the breeding program. This has narrowed down the population structure and allelic diversity for traits responsible for oat improvement. To overcome this issue, breeders are focusing on the historical collection of land races and traditional cultivars to broaden the genetic base of oat (Winkler et al. 2015). The utilization of this germplasm in a pre-breeding program would eventually help plant breeders to develop dual-purpose oat varieties with higher yield.

4 Conventional Breeding Efforts for Improved Quality in Dual-Purpose Oats

4.1 Grain Quality

In the past, the primary focus of oat production was the animal feed industry. Currently, around 75% of oat grain is utilized for animal feeding worldwide (Webster and Wood 2011). Nonetheless, recent decades have seen confirmation of oats' distinct nutritional qualities and benefits as a healthy cereal for human consumption. Oats have drawn more attention lately from the milling sector and from those who produce functional foods. The development of new cultivars that satisfy the needs of millers and consumers will be the primary catalyst for the advancement of scientific studies on oat breeding and the increase in oat cultivation and consumption. The principal attributes of oat grain quality are as follows: percentage of groats, test weight, protein, fat, and quantity of β-glucan. Easy dehulling, homogeneous kernels, minimal groat breakage during dehulling, and consistency in color and flavor are all significant for milling, though to a lesser extent (Chu 2013).

Protein content and groat percentage are crucial for feed and human consumption. Feed with a high fat content is more desirable because it provides more metabolizable energy. However, excessive fat content is not recommended for human consumption, as it might shorten the shelf life of oat products and lead to rancidity. Furthermore, oat products must have both a high β-glucan content and a low-fat content to make health claims. A high β-glucan level is more advantageous for human consumption than for animal consumption. It has been reported that poultry, particularly young chicks, inefficiently use high levels of β-glucan found in oat foods (Humphreys et al. 1994; Peterson 2004).

To comply with food end product labeling regulations, many industrial uses require minimum levels of protein and β-glucan levels, and maximum requirements for fat concentration. Because of this, a processor could need oat cultivars that have a particular concentration of the desired chemicals. In the industrial sector, distinct oat components are extracted, including β-glucan and antioxidants as food additives, protein and starch for cosmetics, and polar lipids for use in pharmaceutical and dietary supplements. Consequently, unique cultivars possessing improved characteristics might be created for specific uses (Humphreys and Mather 1996; Doehlert et al. 2001; Peterson 2004).

Groat percentage is mostly determined by genotype; however, there is some evidence of environmental influence (Marshall and Shaner 1992; Doehlert et al. 2001). Groat percentage is controlled by one primary gene and three modifying genes. The heritability of groat percentage has been found to be high, which can further be improved by recurrent selection. According to Doehlert et al. (2001), test weight (hectoliter weight) is largely reliant on growing conditions and environmental factors and is less heritable.

Numerous studies suggest that the heredity of protein content is relatively substantial. High protein content, however, typically has a negative correlation with grain yield (Humphreys and Mather 1996). Both environmental factors and genotype influence the amount of protein. Numerous studies demonstrate that oat kernels with nitrogen fertilization had higher protein contents. Thus, it is possible to control the protein content by choosing the right cultivar and applying the right amount of fertilizer (Doehlert et al. 2001; May et al. 2004; Fan et al. 2009).

Higher fat content is found in oats, and in some genotypes, it is even higher in maize (Doehlert et al. 2001; Menon et al. 2016). Higher-fat oat cultivars offer an advantage when it comes to animal feeding due to their higher metabolizable energy. Recurrent selection programs were used to develop oat lines containing 18% fat (Peterson and Wood 1997; Frey and Holland 1999). β-glucan and protein concentration, however, are inversely connected with fat content. As a result, oil was reduced even if the content of protein increased when nitrogen fertilization was applied (May et al. 2004; Gazal et al. 2014). However, cultivars with elevated levels of oil and protein have been developed (Youngs and Forsberg 1979).

β-glucan is a polygenic character primarily regulated by genes with additive effects. The range for the heritability of β-glucan content is 0.27 to 0.58, as reported by Holthaus et al. (1996), Humphreys and Mather (1996), and Kibite and Edney (1998). Environmental factors such as nitrogen level and precipitation amount might affect the content of β-glucan, but genotype-related variation usually outweighs the environmental effect (Dvoncova et al. 2010). β-glucan content has a negative correlation with fat concentration but a positive correlation with protein content (Kibite and Edney 1998).

The development of cultivars with high β-glucan and low-fat content for human consumption is hindered by the negative association of β-glucan with fat content. Using laboratory techniques for β-glucan determination, early generations with sufficient seed availability can select genotypes with the appropriate β-glucan content. The most popular approach is an enzymatic one, which involves quantifying the end products after the carbohydrate is specifically broken down by an enzyme.

The selection of parents with high levels of target qualities is the foremost breeding strategy used to increase nutritional quality, which is followed by hybridization cycles, pedigree selection, and/or recurrent selection. Large segregating populations are necessary to obtain substantial genetic variation and valuable transgressive segregants with the necessary gene complexes. The phenotypic selection of genotypes with the desired quality traits may be facilitated by laboratory measurements of β-glucan, protein, and fat content. One potential method for facilitating oat breeding for desirable features is molecular genetics. For instance, markers linked to the amount of β-glucan have been discovered (Orr and Molnar 2008).

Naked oats have more nutrients (protein, fat, oil, antioxidants, and other components) than those that are covered due to the absence of hulls (Valentine 1995; Givens et al. 2004). The value of oats could rise globally via breeding for naked oats and creating cultivars aimed at niche markets. It has long been believed that cultivars of covered oats have a higher yield potential than the naked oats (Peltonen-Sainio 1997). While naked oats are multiflorous, covered (husked/conventional) oats typically have two (rarely three or four) viable florets per spikelet (Valentine 1995). Because of this multiflorous spikelet, which has the ability to produce more grains per spikelet, some researchers hypothesize that the groat yield of naked oats should be higher than that of covered oats. Others disagree since, historically, the groat yield of naked oat cultivars was far lower than that of covered cultivars. According to reports (Valentine 1995; Burrows et al. 2001), there is a negative correlation between the yield potential and the floral morphology of the naked oats. Comparison of near-isogenic naked and covered oat lines revealed no appreciable variation in groat yield between the covered and naked isolines (Burrows et al. 2001). The study concluded that the potential yield of groats was unaffected by the variation in floral morphology between covered and naked oats. Therefore, if cultivars of oats that are naked produce fewer groats than cultivars that are covered, this is most likely due to external variables rather than a genetic variation governing the naked trait (Burrows et al. 2001). According to Doehlert et al. (2001), a study over 3 years in the USA revealed that the groat yield of the naked oat cultivar Paul was higher than the groat yields of 11 covered oat cultivars in 12 different environments. Moudrý et al. (2004) showed that groat yield in the Czech Republic was 23% greater in naked than in covered oat types. According to Maunsell et al. (2004), there was no discernible difference in groat production in the UK between covered and naked lines in 2001–2002.

Compared to covered oat cultivars, naked oat cultivars offer distinct end-use advantages (Burrows et al. 1993), yet the utilization of these cultivars has been limited by a number of factors, such as decreased grain production. In addition to their genetic characteristics, naked oats are limited by their biological characteristics. The lack of a hull increases the vulnerability of the seeds to threshing or storage damage and can lower germination rates (Peltonen-Sainio 1994; Valentine 1995; Kirkkari et al. 2001).

4.2 Forage Quality

In oats, forage quality attributes are water-soluble carbohydrates (WSCs), crude protein (CP), acid detergent fiber (ADF), neutral detergent fiber (NDF), acid detergent lignin (ADL), and in vitro dry matter digestively (IVDMD). Both European white oat (*A. sativa*) and red oat (*A. byzantina*) are autogamous hexaploids well-suited for hybridization methods. Modern plant breeding mainly focuses on grain production rather than fodder production in oats. Reduced investment, scientific allocation, and infrastructural development towards fodder crops are the leading causes to have very few global references in the literature (Stevens et al. 2004). A high-temperature regime coupled with drought stress decreases oat hay quality by dropping the percentage of WSCs and IVDMD, increasing ADF and NDF content (Sadras et al. 2019). The WSC plays a vital role between produce and quality under stressful environmental conditions. In oat hay, an increase in the proportion of WSCs tends to decrease crude fiber and increase digestibility, but at the cost of yield. Foliage greenness reduces the proportion of WSCs; however, it is positively associated with yield and CP but negatively with ADF and NDF (Sadras et al. 2019).

In India, concerted efforts were made separately to increase fodder oat productivity and quality using conventional breeding methods. Artificial selection from an indigenous collection (IC) or exotic collection (EC) is the primary and effective breeding method to increase fodder yield and improve quality since long back. For example, JHO-851 (a multi-cut variety; 1997) was developed by IGFRI, Jhansi, using a single plant selection from germplasm (HIUGAKAIRYOKURO) introduced from Japan and had 10–12% CP content (Roy et al. 2020). Another cultivar, NDO-1 (single cut; 2009), was developed by NDUAT Ayodhya, using the mass selection from locally accessed material from Kothi, Barabanki (Uttar Pradesh), and had increased palatability and digestibility with high nutritive value (Roy et al. 2020). The pedigree method is the most commonly used breeding method for trait improvement in self-pollinating crops. For example, JHO-2001-3 (single cut variety; 2005) was also developed by IGFRI using a three-way inter-varietal cross [(UPO-94 × IGO 320) × Akiyutaka] followed by the pedigree method and had 9.23%, 40.50%, and 60.40% of CP, ADF, and IVDMD, respectively (Roy et al. 2020). Recently, OL 1861 (single cut; 2019) variety was developed by PAU, Ludhiana, using the single cross (HJ 8 and OL 1610) followed by the pedigree method of selection and showed high CP (8–9%) and IVDMD (52.9%) at the national level (Roy et al. 2020). Another cultivar, HFO 607 (single cut; 2019), was developed by HAU Hisar using a single cross (HJ 8 and UPO-04-1) and has high CP (9.23%) and IVDMD (56.3%) (Roy et al. 2020). Thus, forage quality can be increased using conventional breeding approaches coupled with recent advances of molecular breeding such as marker-assisted selection, marker-assisted recurrent selection, and genomic selection to increase selection efficiency as environmental factors highly influence phenotypic selection.

5 Challenges and Limitations of Conventional Breeding Approaches

Conventional breeding approaches have long been used in oat crop production to improve various traits such as yield, disease resistance, and nutritional quality. However, these approaches also come with their own set of challenges and limitations:

(i) Time-consuming: Conventional breeding methods often require many years to develop new oat varieties with improved traits. This is because the breeding process involves multiple generations of crossing, selection, and evaluation before desirable traits are fixed in the new variety.
(ii) Limited genetic variability: Oat cultivars have relatively narrow genetic diversity compared to some other crops. This can restrict the pool of potential traits that breeders can select for and may limit the pace of improvement.
(iii) Complex traits: Some traits of interest in oat, such as disease resistance or tolerance to abiotic stresses like drought or heat, are complex and controlled by multiple genes. Breeding for these traits using conventional methods can be challenging and time-consuming.
(iv) Labor-intensive: Conventional breeding often involves manual pollination, selection, and evaluation of plants in field trials. This can be labor-intensive and costly, especially when dealing with large populations or multiple traits.
(v) Limited precision: Conventional breeding relies on phenotypic selection, which means selecting plants based on observable traits. This approach may not always accurately reflect the underlying genetic makeup of the plants, leading to slower progress or unintended outcomes.
(vi) Environmental constraints: Conventional breeding often requires extensive field trials conducted over multiple seasons and locations to assess the performance of new varieties. Environmental variability can influence the results and make it challenging to identify consistently superior genotypes.
(vii) Targeted traits may be linked: In some cases, the traits that breeders aim to improve may be genetically linked to undesirable traits. This can complicate the breeding process as it may require breaking undesirable linkages through additional breeding cycles.

Despite these challenges, conventional breeding remains a valuable tool in oat crop improvement. However, to overcome these limitations and accelerate the pace of improvement, complementary approaches such as marker-assisted selection (MAS), genomic selection, and genetic engineering are increasingly being integrated into oat breeding programs. These approaches offer the potential to enhance breeding efficiency, precision, and the ability to address complex traits more effectively.

6 Approaches for Accelerated Breeding in Dual-Purpose Oats

6.1 Marker-Assisted Breeding

Marker-assisted selection (MAS) studies in oat crop improvement have become increasingly prevalent in recent years, allowing breeders to expedite the breeding process by identifying and selecting for desired traits more efficiently. Here are a few examples of MAS studies in oat crop improvement.

6.1.1 Disease Resistance

Oats are susceptible to various diseases, including crown rust and powdery mildew. MAS has been used to identify molecular markers associated with resistance genes for these diseases (details in Table 3). For example, researchers have identified molecular markers like SCAR, DArT, KBioscience competitive allele-specific (KASP) markers linked to the Pc91 gene (McCartney et al. 2011), which confers crown rust resistance to oats (Gnanesh et al. 2013). Similarly, two SNP markers linked to Pc98, a seedling crown rust resistance gene, were identified by linkage analysis and KASP assays (Zhao et al. 2020). A KASP marker linked to the Pg13 gene, which confers a stem rust resistance gene, was identified by linkage analysis and association mapping studies (Kebede et al. 2020a, b). Very few linked molecular markers have been developed for oat powdery mildew (PM) and Fusarium head blight (FHB) resistance. Yu and Hermann (2006) were the first to map the PM resistance gene Eg-5 in hexaploid oat. One SSR and four AFLP-derived PCR-based markers were identified to be linked to Eg-5. A cDNA-RFLP marker was identified to be linked to Eg-3, another PM resistance gene (Mohler et al. 2012). Okon and Kowalczyk (2012) identified a SCAR marker linked to the oat mildew resistance group 2 (OMR2) gene. GWAS of oat accession identified one DArT sequence oPt-S014, which was strongly associated with PM resistance in adult plants (Montilla-Bascón et al. 2015). Few SNPs based on GWAS of oat resistance have been developed for their association with Barley Yellow Dwarf virus tolerance in spring oat (Foresman et al. 2016). SNP markers linked to FHB resistance QTL have been identified, which could be useful in MAS of resistance QTL (He et al. 2013).

6.1.2 Beta-Glucan Content

Beta-glucan is a key component of oat grain associated with health benefits such as cholesterol reduction. MAS has been employed to select for oats with higher beta-glucan content. For instance, in order to assess the diversity of β-glucan content in this crop, 1700 oat lines with mutations induced by TILLING of high-frequency mutagenesis were created for breeding purposes using molecular-based, high-precision selection methods from cv. 'Belinda' (Sweden) (Chawade et al. 2010). Following their evaluation, 10 lines had grain β-glucan concentrations greater than 6.7%, while 10 lines had β-glucan contents less than 3.6% (cv. 'Belinda' had a 4.9% β-glucan concentration). Sikora et al. (2013) found that the polysaccharide content varied between 1.8% and 7.5% at its maximum. The identification of genes involved in the biosynthesis of β-glucans in cereals, along with the creation of their initial genetic map, presents novel prospects for the genetic enhancement of grain quality indicators and their resulting food products, which are crucial for human health. A genome-wide association study (GWAS) of oat germplasm of worldwide origin from the American Gene Bank was conducted to identify QTL related to the amount of β-glucan in grain. The study's findings included the identification of three distinct markers that were closely linked to the target trait. Upon comparing these outcomes

Table 3 List of disease resistance genes and their associated molecular markers identified in oat

Resistance gene	Type of molecular marker	Name of molecular marker
Crown rust		
Pc38	RFLP	Cdo673, wg420
Pc39	RFLP	Cdo666
Pc48	RFLP	cdo337
Pc54	RFLP	cdo1435B
Pc58a, b, c	RFLP	PSR637, RZ516D
Pc59	RFLP	Cdo549B
Pc68	RAPD	ubc269
	SNP	Pc68-SNP1, PC68-SNP2
	AFLP	U8PM22, U8PM25
	SDS-PAGE	AveX, AveY, AveZ
	RGA/RFLP	Orga1
	SCAR	ubc269a SCAR
Pc71	RFLP	cdo783, cdo1502
Pc81, 82	AFLP	isu2192, OP C18
Pc83, 84, 85	STS	Agx4, Agx9, Agx7
Pc91	RFLP	UMN145
	DArT	oPT-0350
	SCAR	oPT-0350-cdc
	KASP	oPT-0350-KOM4c2
Pc92	RFLP	OG176
Pc94	AFLP	AF94a
	SCAR	SCAR94-1, SCAR94-2
	SNP	Pc94-SNP1a
Pca	RGA/RFLP	Isu2192, L7M2.2, B9-1
Pcx	RFLP, RAPD	Xcdo1385F, XpOP6(A), Xacor458A
Stem rust		
Pg3	RAPD	ACOpR-1, ACOpR-2
	SCAR/CAPS	Pg3 SCAR/CAPS
Pg4	SCAR/CAPS	Ubc254s SCAR
Pg9	Acid-PAGE	Avenin band
	RFLP, RAPD	Xcdo1385F, Xacor458A
	SCAR/CAPS	Pg9 SCAR/CAPS
Pg13	SDS-PAGE	56.6-kDa polypeptide locus
	RFLP, RAPD	Xmog12B, Xacor254C
	SCAR	Pg13SCAR

with the rice data, it was discovered that one of the identified markers, situated on rice chromosome 7, was in close proximity to the CslF gene family, which is responsible for the β-glucan synthesis in grains. With the creation of higher-density markers in the future, GWAS in oat can therefore be a fruitful QTL finding method (Newell et al. 2012).

6.1.3 Quality Traits

MAS has been utilized to improve various quality traits in oats, including protein content, oil content, and fatty acid composition. For example, to improve groat protein content (GPC), it is imperative to identify the genetic loci linked to this feature in oats. Zhu et al. (2004) used the "Ogle"/"MAM17–5" mapping population to carry out the first quantitative trait locus (QTL) analysis for oat GPC. Seventeen QTLs on 13 linkage groups were found by this approach, and they comprised 29–42% of the overall phenotypic variance over a 2-year period. Three QTLs for GPC were found in the "Terra"/"Marion" mapping population and linkage map (De Koeyer et al. 2004) and five QTLs in the "Aslak"/"Matilda" mapping population and linkage map by subsequent QTL studies employing recombinant inbred line (RIL) mapping populations (Tanhuanpää et al. 2012). In addition, one and two QTLs linked to GPC in oat were reported by Hizbai et al. (2012) and Herrmann et al. (2014). One locus among these QTLs was found to be able to account for up to 42% of the phenotypic variation (Zhu et al. 2004). A recent GWAS research using 38,313 high-quality SNPs found 27 non-redundant QTLs and 41 SNPs that were strongly correlated with GPC. With the exception of CZ20, two QTLs on chromosomes 6C (QTL16) and 4D (QTL11) were consistently found in multiple environments. Of these, QTL16 was the most significant and explained the greatest percentage of the variation in phenotype in all tested environments. Haplotype analysis revealed that favorable haplotypes for GPC are more prevalent in hulless oats. These results lay the groundwork for upcoming initiatives to clone potential QTLs, fine map, and introgress beneficial alleles into new cultivars (Yan et al. 2023).

6.1.4 Abiotic Stress Tolerance

Oats face challenges from abiotic stresses such as drought and salinity. MAS has been employed to identify genomic regions associated with tolerance to these stresses. Oats are a great species to explore the genetic underpinnings of drought tolerance since they are frequently cultivated in harsh, semiarid environments (Wang et al. 2017). Using naked oat landraces from arid environments, a selective sweep analysis was done to identify candidate genes linked to drought resistance. For instance, Nan et al. (2023) identified 43 UDP-glucosyltransferases among the potential candidate genes. According to a previous study, rice's resistance to salt and drought stresses could be increased by overexpressing UDP-glycosyltransferase 3 (*UGT3*) (Wang et al. 2021).

6.1.5 Phenology and Agronomic Traits

MAS has been used to improve agronomic traits such as flowering time, plant height, and yield components in oats. GWAS of European oat accessions, including landraces and old and modern varieties, was done to identify markers associated with lodging tolerance. The collection was genotyped by the Infinium 6 K SNP array for oat. Phenotypes for lodging severity, plant height, and growth habit were collected under natural conditions in eight European countries. Plant height showed a high correlation to lodging severity, explaining about 30% of lodging variation. GWAS analyses detected six significant associations for lodging and two for plant height (Tumino et al. 2017).

These examples demonstrate the diverse applications of marker-assisted breeding in oat crop improvement, ranging from disease resistance to nutritional quality enhancement and agronomic trait improvement. By harnessing the power of molecular markers, breeders can accelerate the development of improved oat varieties with desirable traits, ultimately benefiting oat producers and consumers.

6.2 *Double Haploid (DH) Technology*

DH Technology has caught the eyes of many plant scientists because of its ability to generate homozygous plants in a single generation. In contrast, conventional breeding programs rely on self-pollination of homozygous plants for 6–10 years. Induction and regeneration of DH have been made possible in various ways like androgenesis, gynogenesis, and interspecies or intergenic crosses, involving chromosome elimination of one parent and pollination with desired pollens. The haploid produced by these techniques is then doubled by colchicine treatment. Haploid plants are very important test organisms for molecular genetics, plant biotechnology, and breeding programs. However, many plant species till date are recalcitrant to haploid induction. There are several factors that affect haploid production in plant species. These factors are growth media, sugar concentration, hormone, pH, and genotype.

In oats, haploid induction and regeneration mainly utilize two techniques, namely, androgenesis (Rines et al. 1997; Kiviharju et al. 2005, 2017) and wide hybridization with maize (Sidhu et al. 2006; Marcińska et al. 2013; Nowakowska et al. 2015) or pearl millet and gamagrass (Matzk 1996). Like other crop species, these techniques are also affected by various factors, which are detailed below.

6.2.1 Androgenesis

It is a process in which a gametophyte turns into a sporophyte, and the resulting embryos/plants produced inherit chromosomes of the male parent. Androgenesis is attained by two methods, anther and microspore culture. In oats, seminal works on androgenesis way demonstrated by Rines et al. (1997), who isolated haploid and double haploid plants from anther culture of the 'Stout' cultivar. Efforts were made to recover plants from cultured anthers (Sun et al. 1991; Kiviharju and Pehu 1998; Kiviharju et al. 2005). Like other plant tissue culture techniques, microspore/anther culture is also affected by various factors like genotype, stage of male gamete, growth condition, media composition, and pretreatment of anthers (Warchoł et al. 2019). Studies suggest that androgenesis in oat is strongly dependent on the genotype (Kiviharju et al. 2005, 2017; Ponitka and Slusarkiewicz-Jarzina 2009). The success of a microspore to embryo highly depends on the developmental stage of the microspore. The uninucleate or early binucleate stage is the most appropriate stage for androgenesis in cereals. The structural variations in oat panicle morphology, as well as irregular microsporogenesis maturation in the panicles, are some of the stumbling blocks in identifying the appropriate stage of anther development. Therefore, cytological screening is recommended for identifying the suitable stage for oat androgenic response. Pre-treatment has also been reported to play a critical role in the oat androgenesis process; both cold and heat pretreatment have been reported for oats (Kiviharju and Pehu 1998; Sidhu and Davies 2009). The most important factor that affects embryo induction in microspore/anther culture is media composition. Various components of media, like carbon source, macronutrient, micronutrient type, and concentration of plant hormones, play a decisive role in stimulating anther/microspore-based embryogenesis (Noga et al. 2016).

6.2.2 Wide Hybridization

Besides androgenesis, oat double haploids are also produced by wide crosses with maize (Sidhu et al. 2006; Marcińska et al. 2013; Nowakowska et al. 2015) and pearl millet (Matzk 1996). The most common and effective crosses reported by several researchers are oat x maize crosses. The main demerit of wide crosses is the abortion of the embryo produced, and for successful conversion of this embryo into a plantlet, the embryos are incised and transferred to suitable growth media by a method as embryo rescue. Pioneering work on oat pollination with maize pollen was reported by Rines and Dahleen (1990). He obtained 14 haploids from 3300 flowers. Besides maize, oat crosses have also been demonstrated in pearl millet and gamagrass (Matzk 1996). However, these types of wide crosses do not eliminate the chromosome of the male parent like maize, pearl millet. Wide hybridization technique is also affected by genotype, pH, and composition of growth, CHO types and concentration, and growth hormones. The main factor that influences haploid embryo germination is CHO type and concentration. Rines et al. (1997) utilized a bilayer medium that was customized for wheat embryo; the media doubled the amount of haploid oat plant from a haploid embryo. The sucrose concentration in this particular medium was raised to 15% from the standard 3% sucrose. The most commonly used sucrose concentrations in oat haploid embryo culture are 2, 3, and 7% (Kynast and Riera-Lizarazu 2011). Moreover, substitution of maltose (3% and 9%) with sucrose was found to increase the haploid embryo production. The number of haploids produced also depends on the genotype. Marcinska et al. (2013) reported strong variation in the frequency of embryo formation between the 17 oat genotypes crossed with maize pollen. They also reported that there was no significant difference between embryo production when the plants were applied with 2,4-D or dicamba. To study the effect of time between emasculation and pollination on haploid embryo development, they pollinated the oat floret at 0,1, 2, 3, 4, and 5 days after emasculation. The study found that pollination on the second day after emasculation was the most effective date, producing 13% embryos from 100 plants. The hormonal and molecular mechanisms behind the haploidization process are very little known. So, to study the hormonal profile of the oat haploid obtained by crossing with maize. Dziurka et al. (2019) studied the hormonal profile of ovaries with embryo (OE) and ovaries without embryo (OWE) and found that the total IAA, tZ, and KN in OE were significantly higher than in OWE.

The oat haploid study reported so far suggests that oat is recalcitrant to haploidization, and the efficiency of haploid development is lower. This deficiency limits the incorporation of the oat DH production technique into breeding programs. Therefore, it is necessary to conduct more research that will aid in increasing the efficiency of DH plant production in oats.

6.3 *Speed Breeding*

Speed breeding is a novel approach that utilizes controlled environments, such as growth chambers or high-throughput phenotyping facilities, to accelerate the breeding cycle and shorten generation times. By optimizing growth conditions, photoperiod, and temperature regimes, breeders can achieve rapid generation turnover and accelerate the development of new oat varieties. Speed breeding can be particularly useful for selecting traits relevant to dual-purpose oats, allowing breeders to quickly identify lines with superior performance under specific environmental conditions.

Plant breeding programs need to increase generations per year to reduce the breeding cycle to develop novel cultivars for different desirable traits. Plant growth can be sped up by the extent of photoperiod in most plants. However, the extent of photoperiod does not suit oat owing to its sensitivity towards changes in light and temperature by inhibiting photosynthesis (Quiles 2006). The photosynthetic impairment may adversely affect cell division during the flower primordial developmental stage and reduce seed production (Slafer et al. 2002). The embryo rescue method is mostly used for fast generation cycles in oat, which is labor-intensive, costlier, and has a low success rate (Liu et al. 2016). Recently, a method already used in wheat and barley was used to increase generation cycles per year by increasing photoperiod and foliar mineral supplements in oat. Interestingly, anthesis ensued 12–18 days earlier; however, seed count and inflorescence weight were reduced by three and two-fold than the control, respectively (Heuschele et al. 2019). Abiotic stress tends to unbalance the equilibrium between antioxidants and reactive oxygen species (ROS), triggering oxidative stress. The increased ROS adversely affects cellular pathways and developmental physiology, resulting in reduced seed weight, seed count (Weng et al. 2014), and grain quality (Suriyasak et al. 2017). Therefore, it is necessary to understand the alteration between stress response and physiological response because stress response (ROS stress) can be alleviated, whereas physiological responses cannot. Thus, it would be necessary to induce photoperiod response without a change in stress response to increasing cycles per year.

Oat is the only cereal crop where ascorbic acid (acts as an antioxidant) concentration does not increase in response to the quick generation cycle (Heuschele et al. 2019). The ROS stress cannot be controlled by insufficient production of ascorbic acid and leads to disruption in regular plant metabolic activities that could impair grain yield (Gill and Tuteja 2010). The impaired grain may affect seed germination and seedling vigor in succeeding generations and decide future breeding programs (Grass and Burris 1995). Seedling phenotyping may also be affected if seeds harvested from stressed maternal plants were used due to reduced germination and vigor.

Another challenge to breeders is to increase selection efficiency during speed breeding programs and manipulate by allocating large growing areas. Genetic gain can be improved by increasing selection intensity in a large screening population. However, the limited available space in glasshouses restricts the number of lines that can be advanced per generation. Higher planting density adversely affects seed production and generation time as the initiation of stress due to inter-plant competition for space and nutrition (Sangoi 2001). The single-seed descent method would be ideal to use fast generation cycles for moving populations, where only a few seeds are required. However, it would not be essential where bulky seed count or high seed vigor are required to evaluate traits such as disease and insect screening (Heuschele et al. 2019).

Fast generation cycling could be rewarding to synchronize the flowering period between late and early flowering accessions to enable cross-pollination. However, it could reduce the existing variability for flowering amongst genotypes due to a reduction in the entire flowering time (Araus et al. 2002) and reduce seed count (Heuschele et al. 2019). Therefore, a reduction in seed count and generation time may affect the breeding targets in oat.

Thus, speed breeding would be most effective in accelerating advanced breeding populations using the single seed descent method, but would not be rewarding when seed yield is of utmost selection criterion. Therefore, further research is needed to understand the physiological mechanisms, metabolic pathways, and anatomical behavior of oats to meet the current breeding goals.

6.4 Genomic Selection

Genomic selection (GS) is a modern breeding approach that utilizes genomic information to predict the breeding value of individuals based on their entire genome (Kumar et al. 2021; Wani et al. 2022; Fernández-González and Isidro y Sánchez 2025). The first step in genomic selection is genotyping a breeding population using high-throughput genotyping technologies, such as SNP arrays or next-generation sequencing (NGS). This allows for the identification of thousands to millions of genetic markers distributed throughout the oat genome. Concurrently, the breeding population is phenotyped for target traits of interest, such as grain yield, disease resistance, nutritional quality, and agronomic performance. Phenotypic data is collected across multiple environments to capture genotype-by-environment interactions. A genomic prediction model is developed using statistical methods to estimate the relationship between genetic markers and phenotypic traits. This model learns the associations between marker genotypes and trait performance in the training population. A subset of the breeding population, known as the training population, is used to train the genomic prediction model. This subset should be genetically diverse and represent a wide range of phenotypic variation for the target traits. Once the genomic prediction model is trained, it can be used to predict the breeding value of individuals in the breeding population based solely on their genotypic data. These predictions estimate the genetic merit of individuals for the target traits, allowing breeders to prioritize selections for further breeding efforts. Breeders use the genomic predictions to make informed selection decisions, prioritizing individuals with the highest predicted breeding values for advancement in the breeding program. By selecting individuals with superior genetic merit, breeders aim to accelerate the genetic gain for target traits in subsequent breeding cycles. The accuracy of genomic predictions is validated using independent validation populations or cross-validation methods. Validation ensures that the genomic prediction model performs well across different genetic backgrounds and environments. The model may be refined and updated based on validation results to improve prediction accuracy. GS is effective in improving a variety of traits in oats for grain. For example, GS has been used to increase yield by 5–10% per breeding cycle, reduce grain protein content by 0.5–1.0% per breeding cycle, and increase resistance to oat smut by 50–70% per breeding cycle (Würschum et al. 2015). Recently, GS has also been used for predicting herbage yield in forage oats, with prediction accuracies ranging from 0.91 to 0.97 (Rocha Sr et al. 2023). Genomic selection can be integrated into traditional oat breeding programs, complementing conventional breeding methods. It facilitates the selection of elite lines at early stages of breeding, reducing the time and resources required for field evaluations and accelerating the development of improved oat varieties.

6.5 Genome Editing

Genome editing refers to the use of molecular techniques to make precise changes to the plant genome. This technology allows researchers and breeders to introduce desirable traits into crop varieties or to modify existing traits for improved yield, quality, disease resistance, or other agronomic characteristics. As of now, genome editing in oats is an emerging field and research efforts have been relatively limited compared to major crops like rice, maize, and wheat. However, there are ongoing research initiatives aiming to harness genome editing technologies to improve oat varieties.

One of the most widely used genome editing techniques is clustered regularly interspaced short palindromic repeats (CRISPR)-CRISPR associated 9 (Cas9), which is a system derived from a bacterial defense mechanism. The steps in genome editing that can be applied to the oat crop using the CRISPR-Cas9 tool include the following: Target identification, designing the CRISPR-Cas9 system, DNA repair, verification, and selection of edited plants.

The first step in genome editing is to identify the specific gene or genes responsible for the trait of interest. Once the target gene is identified, specific sequences within that gene can be targeted for modification. A guide RNA (gRNA) is designed to complement the target DNA sequence within the oat genome. The gRNA guides the Cas9 enzyme to the target site, where it introduces a double-stranded break in the DNA. After the Cas9-induced DNA break, the cell's DNA repair machinery can repair the break through two main pathways. Non-Homologous End Joining (NHEJ), which introduces small insertions or deletions (indels) at the target site, leading to gene knockout or disruption, or Homology-Directed Repair (HDR), where a repair template with the desired sequence is provided to introduce precise changes (e.g., point mutations, insertions, or replacements) into the target gene. After genome editing, it is essential to verify that the desired changes have been successfully introduced into the oat genome. This can be done through various molecular techniques, such as PCR, sequencing, and phenotypic analysis. Once verified, the edited plants can be selected for further breeding or propagation.

There are several benefits of the genome editing tool; for instance, genome editing can speed up the breeding process by directly introducing or modifying specific traits without the need for traditional breeding methods, which can be time-consuming and labor-intensive. Genome editing allows for precise modifications at the molecular level, leading to predictable and targeted changes in the oat genome. It introduces new alleles or modifies existing ones, causing expansion of genetic diversity for oat breeding, potentially leading to the development of improved varieties with enhanced agronomic traits.

However, these modern genome editing tools also possess various challenges and considerations. One of the concerns with genome editing is the potential for off-target effects, where unintended modifications occur elsewhere in the genome. Careful design and validation of gRNAs, as well as rigorous screening of edited plants, are essential to minimize off-target effects. Secondly, the regulatory status of genome-edited crops varies by country and region. It is crucial to adhere to local regulations and engage with stakeholders, including farmers, consumers, and policymakers, to ensure the acceptance and responsible deployment of genome-edited oat varieties. In conclusion, genome editing offers promising opportunities for oat improvement by enabling the precise modification of target genes associated with desirable traits. With careful design, validation, and consideration of regulatory and societal aspects, genome-edited oat varieties could contribute to sustainable agriculture and food security.

7 Conclusions

Oats are becoming increasingly popular for human consumption because of their nutritional and health benefits, which are mostly attributed to their high β-glucan and dietary fiber content. As per clinical evidence, consumption of 3 g or more of β-glucan from oats per day may reduce the risk of coronary heart disease (Thies et al. 2014). The market for whole-grain oat products and oat bran in human diets has grown as a result of consumers' growing awareness of the heart-health benefits linked to oat soluble fiber's ability to decrease cholesterol, as well as the general advantages of whole grains. Furthermore, preliminary studies on minor oat constituents, including the antioxidants avenanthramides, are demonstrating a connection between certain oat constituents and asthma, allergies, and the growth of cancer cells. Similarly, oat is best suited as a forage crop during the *rabi* season and are also preferred feed for animals. It is a crucial winter fodder, mostly fed as green, but surplus is converted into silage or hay to use during fodder deficit periods. Due to its additional nutritional value, it is also considered a valuable feed for horses and poultry. Interest in oat hay for the dairy, feed-stuff, and horse industries has also grown in recent years. Oats have a significant genetic and morphological diversity, which makes them an important resource for agriculture and plant breeding. Significant advancements in the agronomic development of oats with increased amounts of health components are being made by modern biological approaches and plant breeding skills. More focus should be paid to resistance to common diseases, including powdery mildew, crown rust, BYDV, and mycotoxin-producing organisms, particularly the latter, as the risks linked to them are not well understood. We will have the chance to improve oats' qualities, particularly their functional and health-promoting qualities, as our knowledge of their agronomic, resistant, and health-promoting benefits advances. By integrating accelerated breeding approaches, oat breeders can expedite the development of dual-purpose varieties with improved productivity and quality, meeting the demands of both forage and grain production systems. For numerous oat populations, particularly for certain economic features, molecular-marker-based linkage maps, marker-trait relationships, QTL, marker-assisted selection, and genomic selection have been developed. However, the huge size of the oat hexaploid genome and the global fall in oat hectarage have hindered oat research progress and made oats less competitive compared to other cereals for research funding. Oat researchers will continue to expand oat biotechnology by utilizing technological advancements in DNA sequencing, bioinformatic analysis, and high-throughput marker development.

The development of gene and genome databases is essential for the development of oats, as it is for several major crops and model plant species. Moreover, the identification of the genes responsible for economically significant traits will facilitate the identification of ideal markers within the key genes, allowing for high-throughput selection and genomic prediction. Collaboratively, breeding and molecular labs show promise for advancing oat biotechnology research to tackle crop difficulties in the future and expedite oat genomic studies.

References

Ahmad M, Dar ZA, Habib M (2014) A review on oat (Avena sativa L.) as a dual-purpose crop. Sci Res Essays 9(4):52–59

Ahmad M, Jehangir IA, Rizvan R, Dar SA, Iqbal S, Wani SH et al (2020) Phylogenetic relationship of oats (Avena sativa L): a guide to conservation and utilisation of genetic resources. Int J Curr Microbiol App Sci 9(11):831–845

Ahmed S, Roy AK, Majumdar AB (2011) Genetic diversity and variability analysis in oat (Avena sativa L.). Range Manag Agrofor 32(2):96–99

Ahmed S, Roy AK, Majumdar AB (2013) Correlation and path coefficient analysis for fodder and grain yield related. Ann Biol 29(1):75–78

Amanullah PS, Zada K, Perveen S (2004) Growth characters and productivity of oat varieties at Peshawar. Sarhad J Agric 20(1):5–10

Ames N, Malunga LN, Mollard R, Johnson J, Chu Y, Thandapilly SJ (2021) Effect of processing on oat β-glucan viscosity, postprandial glycemic response and subjective measures of appetite. Food Funct 12(8):3672–3679

Andrade EF, Vieira Lobato R, Vasques de Araújo T, Zangerônimo MG, Sousa RVD, Pereira LJ (2015) Efecto de los beta-glucanos en el control de los niveles de glucosa en pacientes diabéticos: revisión sistemática. Nutr Hosp 31(1):170–177

Araus JL, Slafer GA, Reynolds MP, Royo C (2002) Plant breeding and drought in C3 cereals: what should we breed for? Ann Bot 89(7):925–940

Bailey-Serres J, Lee SC, Brinton E (2012) Waterproofing crops: effective flooding survival strategies. Plant Physiol 160(4):1698–1709

Barsila SR (2018) The fodder oat (Avena sativa) mixed legume forages farming: nutritional and ecological benefits. J Agric Nat Resour 1(1):206–222

Bazzer SK, Oliveira G, Fiedler JD, Nandety RS, Jannink JL, Caffe M (2025) Genomic strategies to facilitate breeding for increased β-Glucan content in oat (Avena sativa L.). BMC Genomics 26(1):35

Beyene G, Araya A, Gebremedhin H (2015) Evaluation of different oat varieties for fodder yield and yield related traits in Debre Berhan area, central highlands of Ethiopia. Development 27(9)

Brinkman MA, Frey KJ (1977) Growth analysis of Isoline-recurrent parent grain yield differences in oats 1. Crop Sci 17(3):426–430

Brown CM, Jedlinski H (1978) Registration of 13 germplasm lines of Oats1 (Reg. No. GP l0 to GP 22). Crop Sci 18(6):1098–1098

Browning JA, Frey KJ (1969) Multiline cultivars as a means of disease control. Annu Rev Phytopathol 7(1):355–382

Burrows VD, Cave NA, Friend DW, Hamilton RMG, Morris JM (1993) Production and feeding of naked oat

Burrows VD, Molnar SJ, Tinker NA, Marder T, Butler G, Lybaert A (2001) Groat yield of naked and covered oat. Can J Plant Sci 81(4):727–729

Butt MS, Tahir-Nadeem M, Khan MKI, Shabir R, Butt MS (2008) Oat: unique among the cereals. Eur J Nutr 47:68–79

Buxton DR, Mertens DR, Moore KJ, Boyd LJ, Oldfield JE (1995) Forage quality for ruminants: plant and animal considerations. Profess Anim Sci 11(3):121–131

Cabral AL, Singh D, Park RF (2011) Identification and genetic characterisation of adult plant resistance to crown rust in diploid and tetraploid accessions of Avena. Ann Appl Biol 159(2):220–228

Chawade A, Sikora P, Bräutigam M, Larsson M, Vivekanand V, Nakash MA et al (2010) Development and characterization of an oat TILLING-population and identification of mutations in lignin and β-glucan biosynthesis genes. BMC Plant Biol 10:1–13

Choubey RN, Roy AK (2005) Forage oat breeding in India-achievements and prospects. In: Proceedings of the workshop on Fodder Oats, fodder technology packages and small farm income network, Kathmandu, pp 120–123

Chu Y (ed) (2013) Oats nutrition and technology. Wiley

Coffman FA, Murphy HC, Chapman WH (1961) Oat breeding. In: Oats and oat improvement, pp 263–329
De Koeyer DL, Tinker NA, Wight CP, Deyl J, Burrows VD et al (2004) A molecular linkage map with associated QTLs from a hulless× covered spring oat population. Theor Appl Genet 108:1285–1298
Deep A, Shweta M et al (2019) Correlation and genetic divergence analysis for seed and fodder yield and its contributing character in oat (Avena sativa L.). Int J Pure Appl Biosci 7(3):471–477. https://doi.org/10.18782/2320-7051.7246
Demír H, Simsek M, Yıldırım G (2021) Effect of oat milk pasteurization type on the characteristics of yogurt. LWT 135:110271
Dherawattana A, Sadanaga K (1973) Cytogenetics of a crown rust-resistant Hexaploid oat with 42+ 2 fragment chromosomes 1. Crop Sci 13(6):591–594
Diederichsen A (2008) Assessments of genetic diversity within a world collection of cultivated hexaploid oat (Avena sativa L.) based on qualitative morphological characters. Genet Resour Crop Evol 55(3):419–440
Doehlert DC, McMullen MS, Hammond JJ (2001) Genotypic and environmental effects on grain yield and quality of oat grown in North Dakota. Crop Sci 41(4):1066–1072
Dumlupınar Z, Dokuyucu T, Maral H, Kara R, Akkaya A (2012) Evaluation of Turkish oat landraces based on morphological and phenological traits
Dvoncova D, Havrlentova M, Hlinkova A, Hozlar P (2010) Effect of fertilization and variety on the beta-glucan content in the grain of oats. Żywność Nauka Technologia Jakość 17(3):108
Dziurka K, Dziurka M, Warchoł M, Czyczyło-Mysza I, Marcińska I, Noga A et al (2019) Endogenous phytohormone profile during oat (Avena sativa L.) haploid embryo development. In Vitro Cell Dev Biol Plant 55:221–229
Endo RM, Brown CM (1964) Barley yellow dwarf virus resistance in oats. Crop Sci 4:279–283
Esposito F, Arlotti G, Bonifati AM (2005) Antioxidant activity and dietary fibre in durum wheat bran by-products. Food Res Int 38:1167–1173
Fan M, Zhang Z, Wang F, Li Z, Hu Y (2009) Effect of nitrogen forms and levels on β-glucan accumulation in grains of oat (Avena sativa L.) plants. J Plant Nutr Soil Sci 172(6):861–866
Fernández-González J, Isidro y Sánchez J (2025) Optimizing fully-efficient two-stage models for genomic selection using open-source software. Plant Methods 21(1):9
Foresman BJ, Oliver RE, Jackson EW, Chao S, Arruda MP, Kolb FL (2016) Genome-wide association mapping of barley yellow dwarf virus tolerance in spring oat (Avena sativa L.). PLoS One 11(5):e0155376
Frey KJ (1986) Genetic resources and their use in oat breeding. In: Proceedings of the second International oats conference. Springer, Dordrecht, pp 7–15
Frey KJ, Holland JB (1999) Nine cycles of recurrent selection for increased groat-oil content in oat. Crop Sci 39(6):1636–1641
Frey KJ, Cox TS, Rodgers DM, Bramel-Cox P (1984) Increasing cereal yields with genes from wild and weedy species. In: Chopra VL et al (eds) Genetics: new frontiers: proceedings of the XV International congress of genetics/editors. Oxford & IBH Publishing Co, New Delhi
Fu YB (2018) Oat evolution revealed in the maternal lineages of 25 Avena species. Sci Rep 8(1):1–12
Fu YB, Peterson GW, Scoles G, Rossnagel B, Schoen DJ, Richards KW (2003) Allelic diversity changes in 96 Canadian oat cultivars released from 1886 to 2001. Crop Sci 43(6):1989–1995
Gadisovich MB (2017) Comparative characteristics of productivity elements among film and huskless forms of oat. International Journal of Green Pharmacy (IJGP) 11(03)
Gazal A, Dar ZA, Zaffar G, Lone AA, Abidi I, Shabir A et al (2014) Trends in breeding oat for nutritional grain quality-an overview. J Appl Nat Sci 6(2):904–912
Gill SS, Tuteja N (2010) Reactive oxygen species and antioxidant machinery in abiotic stress tolerance in crop plants. Plant Physiol Biochem 48(12):909–930

Givens DI, Davies TW, Laverick RM (2004) Effect of variety, nitrogen fertiliser and various agronomic factors on the nutritive value of husked and naked oats grain. Anim Feed Sci Technol 113(1–4):169–181

Gnanesh BN, Mitchell Fetch J, Menzies JG, Beattie AD, Eckstein PE, McCartney CA (2013) Chromosome location and allele-specific PCR markers for marker-assisted selection of the oat crown rust resistance gene Pc91. Mol Breed 32:679–686

Gorash A, Armonienė R, Mitchell Fetch J, Liatukas Ž, Danytė V (2017) Aspects in oat breeding: nutrition quality, nakedness and disease resistance, challenges and perspectives. Ann Appl Biol 171(3):281–302

Grass L, Burris JS (1995) Effect of heat stress during seed development and maturation on wheat (Triticum durum) seed quality. I. Seed germination and seedling vigor. Can J Plant Sci 75(4):821–829

Hampshire J (2003) Investigations of vitamin E contents in oat varieties and industrial milling oats. Dtsch Lebensmitt Rundsch 99(6):222–231

He X, Skinnes H, Oliver RE, Jackson EW, Bjørnstad Å (2013) Linkage mapping and identification of QTL affecting deoxynivalenol (DON) content (Fusarium resistance) in oats (Avena sativa L.). Theor Appl Genet 126:2655–2670

Helsel DB, Frey KJ (1978) Grain yield variations in oats associated with differences in leaf area duration among oat lines 1. Crop Sci 18(5):765–769

Herrmann M, Roderick HW (1996) Characterisation of new oat germplasm for resistance to powdery mildew. Euphytica 89(3):405–410

Herrmann MH, Yu J, Beuch S, Weber WE (2014) Quantitative trait loci for quality and agronomic traits in two advanced backcross populations in oat (Avena sativa L.). Plant Breed 133(5):588–601

Heuschele DJ, Case A, Smith KP (2019) Evaluation of fast generation cycling in oat (Avena sativa). Cereal Res Commun 47(4):626–635

Hite RE, Sherwood RT, Marshall HG (1977) Adult plant resistance to powdery mildew in dal oats. Plant Dis Rep 61:273–277

Hizbai BT, Gardner KM, Wight CP, Dhanda RK, Molnar SJ, Johnson D et al (2012) Quantitative trait loci affecting oil content, oil composition, and other agronomically important traits in oat. Plant Genome 5(3)

Holthaus JF, Holland JB, White PJ, Frey KJ (1996) Inheritance of β-glucan content of oat grain. Crop Sci 36(3):567–572

Humphreys DG, Mather DE (1996) Heritability of β-glucan, groat percentage, and crown rust resistance in two oat crosses. Euphytica 91:359–364

Humphreys DG, Smith DL, Mather DE (1994) Nitrogen fertilizer and seeding date induced changes in protein, oil and β-glucan contents of four oat cultivars. J Cereal Sci 20(3):283–290

Hunter H (1924) Oats: their varieties and characteristics; a practical handbook for farmers, Seedmen, and students, vol 2. Ernest Benn

Hussain A, Khan S, Muhammad D, Bhatti MB (1993) Yield and quality of various varieties of oats. Pakistan J Sci Ind Res 36:258–260

Iannucci A, Codianni P, Cattivelli L (2011) Evaluation of genotype diversity in oat germplasm and definition of ideotypes adapted to the mediterranean environment. Int J Agron 2011:1

Ingalls JR, Thomas JW, Benne EJ, Tesar M (1965) Comparative response of wether lambs to several cuttings of alfalfa, birdsfoot trefoil, bromegrass and reed canarygrass. J Anim Sci 24(4):1159–1164

Jones IT, Jones ERL (1978) Mildew of oats. In: UK cereal pathogen virulence survey, pp 59–63

Joyce SA, Kamil A, Fleige L, Gahan CG (2019) The cholesterol-lowering effect of oats and oat beta glucan: modes of action and potential role of bile acids and the microbiome. Front Nutr 6:482770

Jung HG, Allen MS (1995) Characteristics of plant cell walls affecting intake and digestibility of forages by ruminants. J Anim Sci 73(9):2774–2790

Katsiotis A, Drossou A, Bladenopoulos K (2004) Genetic diversity of Avena landraces accessed by AFLPs. In: Proceedings 7th International oat conference. MTT

Kaur R, Kapoor R, Vikal Y, Kaur K (2018) Assessing genetic diversity in dual purpose oat (Avena sativa L.) cultivars based on morphological and quality traits. Int J Curr Microbiol App Sci 7(5):1574–1586

Kebede AZ, Admassu-Yimer B, Bekele WA, Gordon T, Bonman JM, Babiker E et al (2020a) Mapping of the stem rust resistance gene Pg13 in cultivated oat. Theor Appl Genet 133:259–270

Kebede G, Minta M, Feyissa F, Mohammed K, Dejene M, Geleti D, Yakob Y (2020b) Verification of forage oat (Avena Sativa) varieties in the Central Highland of Ethiopia. Livest Res Results 294

Kibite S, Edney MJ (1998) The inheritance of β-glucan concentration in three oat (Avena sativa L.) crosses. Can J Plant Sci 78(2):245–250

Kirkkari A-M, Peltonen-Sainio P, Rita H (2001) Reducing grain damage in naked oat through gentle harvesting. Agric Food Sci Finland 10:223–229

Kiviharju E, Pehu E (1998) The effect of cold and heat pretreatments on anther culture response of Avena sativa and A. sterilis. Plant Cell Tissue Organ Cult 54:97–104

Kiviharju E, Moisander S, Laurila J (2005) Improved green plant regeneration rates from oat anther culture and the agronomic performance of some DH lines. Plant Cell Tissue Organ Cult 81:1–9

Kiviharju E, Moisander S, Tanhuanpää P (2017) Oat anther culture and use of DH-lines for genetic mapping. In: Oat: methods and protocols, pp 71–93

Kouřimská L, Sabolová M, Horčička P, Rys S, Božik M (2018) Lipid content, fatty acid profile, and nutritional value of new oat cultivars. J Cereal Sci 84:44–48

Kriger OV, Kashirskikh EV, Babich OO, Noskova SY (2018) Oat protein concentrate production. Foods Raw Mater 6(1):47

Krishna A, Ahmed S, Pandey HC, Kumar V (2014) Correlation, path and diversity analysis of oat (Avena sativa L.) genotypes for grain and fodder yield. J Plant Sci Res 1(2):1–9

Krull CF, Borlaug NE (1970) The utilization of collections in plant breeding and production

Kujur MJ, Mehta AK, Bilaiya SK, Patil P (2017) Estimation of genetic diversity among genotypes of fodder oat based on principal component analysis. Int J Bio-Resour Stress Manag 8(6):807–810

Kumar N, Rana M, Kumar B, Chand S, Shiv A, Wani SH, Kumar S (2021) Genomic selection for wheat improvement. In: Physiological, molecular, and genetic perspectives of wheat improvement, pp 175–207

Kumari T, Jindal Y (2019) Genetic diversity and variability analysis in oats (Avena sp) genotypes. Electron J Plant Breed 10(1):1–8

Kynast RG, Riera-Lizarazu O (2011) Development and use of oat–maize chromosome additions and radiation hybrids. In: Plant chromosome engineering: methods and protocols, pp 259–284

Lásztity R (1998) Oat grain – a wonderful reservoir of natural nutrients and biologically active substances. Food Rev Intl 14(1):99–119

Li Y, Wang Z, Beier RC, Shen J, Smet DD, De Saeger S, Zhang S (2011) T-2 toxin, a trichothecene mycotoxin: review of toxicity, metabolism, and analytical methods. J Agric Food Chem 59(8):3441–3453

Liu H, Zwer P, Wang H, Liu C, Lu Z, Wang Y, Yan G (2016) A fast generation cycling system for oat and triticale breeding. Plant Breed 135(5):574–579

Lodhi MY, Marghazani IB, Hamayun K, Marri MJ (2009) Comparative performance study of different oat varieties under agro-climatic conditions of Sibi. J Anim Plant Sci 19(1):34–36

Loskutov IG (2001) Influence of vernalization and photoperiod to the vegetation period of wild species of oats (Avena spp.). Euphytica 117(2):125–131

Loskutov IG (2004) Taxonomy and evaluation of genus Avena L. In: Peltonen-Sainio P, Topi-Hulmi M (eds) Proceedings 7th international oat conference. MTT

Loskutov IG (2008) On evolutionary pathways of Avena species. Genet Resour Crop Evol 55(2):211–220

Luikham E, Anal PS, Anal PS (2015) Effect of phosphorus and potash on yield, quality and economics of oat fodder (Avena sativa L.). Agric Sci Digest Res J 35(2):161–162

Mäkelä N, Rosa-Sibakov N, Wang YJ, Mattila O, Nordlund E, Sontag-Strohm T (2021) Role of β-glucan content, molecular weight and phytate in the bile acid binding of oat β-glucan. Food Chem 358:129917

Marcińska I, Nowakowska A, Skrzypek E, Czyczyło-Mysza I (2013) Production of double haploids in oat (Avena sativa L.) by pollination with maize (Zea mays L.). Central Eur J Biol 8:306–313

Marshall HG (1992) Breeding oat for resistance to environmental stress. Oat Sci Technol 33:699–749

Marshall HG, Shaner GE (1992) Genetics and inheritance in oat. Oat Sci Technol 33:509–571

Martínez-Villaluenga C, Peñas E (2017) Health benefits of oat: current evidence and molecular mechanisms. Curr Opin Food Sci 14:26–31

Matzk F (1996) Hybrids of crosses between oat and Andropogoneae or Paniceae species. Crop Sci 36(1):17–21

Maunsell C, Macleod M, Nute G, Wade T (2004) AFENO: Avian feed efficiency from naked oats. http://randd.defra.gov.uk/Document.aspx?Document=LS3623_1301_FRP.doc.Accessed 29 Sep 2016

May WE, Mohr RM, Lafond GP, Johnston AM, Craig Stevenson F (2004) Effect of nitrogen, seeding date and cultivar on oat quality and yield in the eastern Canadian prairies. Can J Plant Sci 84(4):1025–1036

McCartney CA, Stonehouse RG, Rossnagel BG, Eckstein PE, Scoles GJ, Zatorski T et al (2011) Mapping of the oat crown rust resistance gene Pc91. Theor Appl Genet 122:317–325

McKenzie RIH, Martens JW, Green GJ (1971) The oat stem rust problem. In: Mutation breeding for disease resistance

Mcmullen MS, Patterson FL (1992) Oat cultivar development in the USA and Canada. Oat Sci Technol 33:573–612

Menon R, Gonzalez T, Ferruzzi M, Jackson E, Winderl D, Watson J (2016) Oats – from farm to fork. Adv Food Nutr Res 77:1–55

Mohler V, Zeller FJ, Hsam SL (2012) Molecular mapping of powdery mildew resistance gene Eg-3 in cultivated oat (Avena sativa L. cv. Rollo). J Appl Genet 53:145–148

Montilla-Bascón G, Rispail N, Sánchez-Martín J, Rubiales D, Mur LA, Langdon T et al (2015) Genome-wide association study for crown rust and powdery mildew resistance in an oat collection of commercial varieties and landraces. Front Plant Sci 6(103):10–3389

Moudrý J, Sterba Z, Moudrý J Jr, Bárta J (2004) The comparison of production ability of naked and husked oats. In: Peltonen-Sainio P, Topi-Hulmi M (eds) Proceedings 7th international oat conference. MTT

Naeem M, Khan MA, Chohan MSM, Khan AH, Salahuddin S (2002) Evaluation of different varieties of oats for green fodder yield potential. Asian J Plant Sci 1:142

Nan J, Ling Y, An J, Wang T, Chai M, Fu J et al (2023) Genome resequencing reveals independent domestication and breeding improvement of naked oat. GigaScience 12:giad061

Newell MA, Asoro FG, Scott MP, White PJ, Beavis WD, Jannink JL (2012) Genome-wide association study for oat (Avena sativa L.) beta-glucan concentration using germplasm of worldwide origin. Theor Appl Genet 125:1687–1696

Nishiyama I (1929) The genetics and cytology of certain cereals. Jpn J Genetics 5(1–2):1–48

Noga A, Skrzypek E, Warchoł M, Czyczyło-Mysza I, Dziurka K, Marcińska I et al (2016) Conversion of oat (Avena sativa L.) haploid embryos into plants in relation to embryo developmental stage and regeneration media. In Vitro Cell Dev Biol Plant 52:590–597

Nowakowska A, Skrzypek E, Marcińska I, Czyczyło-Mysza I, Dziurka K, Juzoń K et al (2015) Application of chosen factors in the wide crossing method for the production of oat doubled haploids. Open. Life Sci 10(1)

Ohm HW, Shaner G (1992) Breeding oat for resistance to diseases. Oat Sci Technol 33:657–698

Okon S, Kowalczyk K (2012) Deriving isolates of powdery mildew (Blumeria graminis DC. f. sp. avenae Em. Marchal.) in common oat (Avena sativa L.) and using them to identify selected resistance genes. Acta Agrobot 65(2)

Orr W, Molnar SJ (2008) Development of PCR-based SCAR and CAPS markers linked to β-glucan and protein content QTL regions in oat. Genome 51(6):421–425
Pant D, Mishra SN, Verma JS (1992) Effects of biparental mating in forage oats. Indian J Genetics Plant Breed 52(3):270–274
Pathan S, Damame S, Sinare B (2020) Effect of different cutting management on growth, yield, quality and economics of dual purpose oat, barley and wheat. Forage Res 46:182–186
Patterson FL, Schafer JF (1978) Registration of Clintland 60 and Clintland 64 oats 1 (Reg. No. 280 and 281). Crop Sci 18(2):354–355
Paudel D (2018) Rapid and simultaneous determination of nutritional constituents of united states grown oats using near infrared reflectance spectroscopy (NIRS). South Dakota State University
Pearson CJ, Ison RL (1997) Agronomy of grassland systems. Cambridge University Press
Peltonen-Sainio P (1994) Yield component differences between naked and conventional oat. Agron J 86(3):510–513
Peltonen-Sainio P (1997) Groat yield and plant stand structure of naked and hulled oat under different nitrogen fertilizer and seeding rates. Agron J 89(1):140–147
Peter Lawson & Son (1852) Synopsis of the vegetable products of Scotland in the Museum of the Royal Botanic Gardens of Kew. Private Press of P. Lawson and Son
Peterson DM (2004) Oat – a multifunctional grain. In: Pelton-Sainio P, Topi-Hulmi M (eds) Proceedings 7th international oat conference. MTT Agrifood Research, Helsinki, pp 21–26
Peterson DM, Wood DF (1997) Composition and structure of high-oil oat. J Cereal Sci 26(1):121–128
Peturson B (1944) Adult plant resistance of some oat varieties to physiologic races of crown rust. Can J Res 22(6):287–289
Poehlman JM, Kingsolver CH (1950) Disease reaction and agronomic qualities of oats selections from a Columbia x Victoria-Richland cross 1. Agron J 42(10):498–502
Ponitka A, Slusarkiewicz-Jarzina A (2009) Regeneration of oat androgenic plants in relation to induction media and culture conditions of embryo-like structures. Acta Soc Bot Pol 78(3):209–213
Quiles MJ (2006) Stimulation of chlororespiration by heat and high light intensity in oat plants. Plant Cell Environ 29(8):1463–1470
Ralla T, Salminen H, Edelmann M, Dawid C, Hofmann T, Weiss J (2018) Oat bran extract (Avena sativa L.) from food by-product streams as new natural emulsifier. Food Hydrocoll 81:253–262
Rauf M, Yoon H, Lee S, Shin MJ, Ko HC, Lee MC, Choi YM (2019) Evaluation of major dietary ingredients in diverse oats (Avena sativa L.) germplasm. J Crop Sci Biotechnol 22(5):495–507
Rines HW, Dahleen LS (1990) Haploid oat plants produced by application of maize pollen to emasculated oat florets. Crop Sci 30(5):1073–1078
Rines HW, Riera-Lizerazu O, Nunez VM, Davis DW, Phillips RL (1997) Oat haploids from anther culture and from wide hybridizations. In: Jain SM, Sopory SK, Veilleux RE (eds) In vitro haploid production in higher plants, vol 4. Kluwer Acad Publishers, Dordrecht, pp 205–221
Rocha Sr DJA, Flaresso Sr JA, Neto Sr JS, Cordova Sr UA (2023) Genomic selection for herbage yield in forage oats (Avena sp.). bioRxiv, 2023-11
Roy AK, Agrawal RK, Chand S, Ahmad S, Kumar RV, Mall AK, Bhardwaj NR, Mawar R, Singh DN, Kantwa SR, Faruqui SA (2020) Database of forage crop varieties: 2020. AICRP on forage crops & utilization, ICAR-IGFRI, Jhansi, pp 362
Sadras VO, Mahadevan M, Williams MM, Hoppo SD, Zwer PK (2019) Phenotypic plasticity of grain and hay quality in varieties and advanced lines from the Australian oat breeding program. Eur J Agron 102:23–32
Sahu M, Tiwari A (2020) Genetic variability and association analysis of oat (Avena sativa L.) genotypes for Green forage yield and other components. Curr J Appl Sci Technol:133–141
Saini P, Gani M, Saini P, Bhat JA, Francies RM, Negi N, Chauhan SS (2019) Molecular breeding for resistance to economically important diseases of fodder oat. In: Disease resistance in crop plants. Springer, Cham, pp 199–239

Sánchez-Martín J, Rispail N, Flores F, Emeran AA, Sillero JC, Rubiales D, Prats E (2017) Higher rust resistance and similar yield of oat landraces versus cultivars under high temperature and drought. Agron Sustain Dev 37(1):1–14

Sang S, Chu Y (2017) Whole grain oats, more than just a fiber: role of unique phytochemicals. Mol Nutr Food Res 61(7):1600715

Sangoi L (2001) Understanding plant density effects on maize growth and development: an important issue to maximize grain yield. Ciência rural 31:159–168

Semman U, Eba B, Dinkale T (2018) Performance evaluation of improved oat varieties/accessions at the highland of Guji zone, bore, Ethiopia. J Biol Agric Healthcare 8(17)

Sidhu PK, Davies PA (2009) Regeneration of fertile green plants from oat isolated microspore culture. Plant Cell Rep 28:571–577

Sidhu PK, Howes NK, Aung T, Zwer PK, Davies PA (2006) Factors affecting oat haploid production following oat× maize hybridization. Plant Breed 125(3):243–247

Sikora P, Tosh SM, Brummer Y, Olsson O (2013) Identification of high β-glucan oat lines and localization and chemical characterization of their seed kernel β-glucans. Food Chem 137(1–4):83–91

Simons MD (1972) Polygenic resistance to plant disease and its use in breeding resistant cultivars. J Environ Qual 1(3):232–240

Simons MD, Caldecott RS, Frey KJ (1962) On the attempted induction of resistance in oats to crown rust race 264 by artificial irradiation. Plant Dis Rep 46:88–91

Singh D, Chauhan A, Chaudhary A (2018) Relative performance of oat forage varieties for seed production, economics and fodder yield under Central Gujarat conditions. Forage Res 44(3):185–191

Slafer GA, Molina-Cano JL, Savin R, Araus JL, Romagosa I (2002) Barley science: recent advances from molecular biology to agronomy of yield and quality. CRC Press

Stadler LJ (1929) Chromosome number and the mutation rate in Avena and Triticum. Proc Natl Acad Sci USA 15(12):876

Stevens EJ, Armstrong KW, Bezar HJ, Griffin WB, Hampton JG (2004) Fodder oats an overview, vol 33. Food & Agriculture Org, pp 11–18

Stewart D, McDougall G (2014) Oat agriculture, cultivation and breeding targets: implications for human nutrition and health. Br J Nutr 112(S2):S50–S57

Sun CS, Lu T-G, Söndahl MR (1991) Anther culture of naked oat and the establishment of its haploid suspension cell. Acta Bot Sin 33:417–420

Suriyasak C, Harano K, Tanamachi K, Matsuo K, Tamada A, Iwaya-Inoue M, Ishibashi Y (2017) Reactive oxygen species induced by heat stress during grain filling of rice (Oryza sativa L.) are involved in occurrence of grain chalkiness. J Plant Physiol 216:52–57

Surje DT, De DK (2014) Correlation coefficient study in oat (Avena sativa L.) genotypes for fodder and grain yield characters. J Agric Sci Technol 1(1):89–93

Suttie JM, Reynolds SG (2004) Fodder oats: a world overview (no. 33). Food & Agriculture Org

Swamy BM, Sarla N (2008) Yield-enhancing quantitative trait loci (QTLs) from wild species. Biotechnol Adv 26(1):106–120

Syed SJ, Gadhe KS, Shaikh RP (2020) Studies on quality evaluation of OAT milk. J Pharm Phytochem 9(1):2275–2277

Szpicer A, Onopiuk A, Półtorak A, Wierzbicka A (2020) The influence of oat β-glucan content on the physicochemical and sensory properties of low-fat beef burgers. CyTA J Food 18(1):315–327

Tamm I (2003) Genetic and environmental variation of grain yield of oat varieties. Agron Res 1(1):93–97

Tanhuanpää P, Manninen O, Beattie A, Eckstein P, Scoles G, Rossnagel B, Kiviharju E (2012) An updated doubled haploid oat linkage map and QTL mapping of agronomic and grain quality traits from Canadian field trials. Genome 55(4):289–301

Thies F, Masson LF, Boffetta P, Kris-Etherton P (2014) Oats and CVD risk markers: a systematic literature review. Br J Nutr 112(S2):S19–S30

Thomas H, Powell W, Aung T (1980) Interfering with regular meiotic behaviour in Avena sativa as a method of incorporating the gene for mildew resistance from a. barbata. Euphytica 29(3):635–640
Tumino G, Voorrips RE, Morcia C, Ghizzoni R, Germeier CU, Paulo MJ et al (2017) Genome-wide association analysis for lodging tolerance and plant height in a diverse European hexaploid oat collection. Euphytica 213:1–12
Valentine J (1995) Naked oats. In: Welch RW (ed) The oat crop production and utilization, London, pp 504–532
Varma P, Bhankharia H, Bhatia S (2016) Oats: a multi-functional grain. J Clin Prevent Cardiol 5(1):9–17
Wang T, Du YL, He J, Turner NC, Wang BR, Zhang C et al (2017) Recently-released genotypes of naked oat (Avena nuda L.) out-yield early releases under water-limited conditions by greater reproductive allocation and desiccation tolerance. Field Crop Res 204:169–179
Wang T, Ma YQ, Huang XX, Mu TJ, Li YJ, Li XK et al (2021) Overexpression of OsUGT3 enhances drought and salt tolerance through modulating ABA synthesis and scavenging ROS in rice. Environ Exp Bot 192:104653
Wani SH, Wang D, Singh GP (eds) (2022) QTL mapping in crop improvement: present Progress and future perspectives. Academic Press, Elsevier
Warchoł M, Czyczyło-Mysza I, Marcińska I, Dziurka K, Noga A, Kapłoniak K et al (2019) Factors inducing regeneration response in oat (Avena sativa L.) anther culture. In Vitro Cell Dev Biol Plant 55:595–604
Webster FH, Wood P (2011) Oat utilization: past, present and future. In: Oats: chemistry and technology, pp 347–361
Weng X, Wang L, Wang J, Hu Y, Du H, Xu C, Xing Y, Li X, Xiao J, Zhang Q (2014) Grain number, plant height, and heading date is a central regulator of growth, development, and stress response. Plant Physiol 164(2):735–747
Whitehead A, Beck EJ, Tosh S, Wolever TM (2014) Cholesterol-lowering effects of oat β-glucan: a meta-analysis of randomized controlled trials. Am J Clin Nutr 100(6):1413–1421
Winkler L, Bockelman H, Bonman JM, Klos KE (2015) The genetic diversity of modern oat cultivars relative to historic cultivars and landraces held in the National Small Grains Collection
Würschum T, Maurer A, Longin CFH, Fußaeus C (2015) Genomic selection in oat: a novel method for breeding. Plant Breed 134(3):262–266
Yadav AK, Yadav VK, Viswakarma DN, Ram CN, Yadav V, Soni SK (2011) Genetic divergence in oats genotypes. Plant Arch 11(2):1013–1016
Yan H, Zhang H, Zhou P, Ren C, Peng Y (2023) Genome-wide association mapping of QTL underlying Groat protein content of a diverse panel of oat accessions. Int J Mol Sci 24(6):5581
Youngs VL, Forsberg RA (1979) Protein-oil relationships in oats 1. Crop Sci 19(6):798–802
Yu J, Herrmann M (2006) Inheritance and mapping of a powdery mildew resistance gene introgressed from Avena macrostachya in cultivated oat. Theor Appl Genet 113(3):429–437
Zhao J, Kebede AZ, Menzies JG, Paczos-Grzęda E, Chong J, Mitchell Fetch JW et al (2020) Chromosomal location of the crown rust resistance gene Pc98 in cultivated oat (Avena sativa L.). Thcor Appl Genet 133:1109 1122
Zhou S, Tong L, Liu L (2019) Oats. In: Bioactive factors and processing technology for cereal foods, pp 185–206
Zhu S, Rossnagel BG, Kaeppler HF (2004) Genetic analysis of quantitative trait loci for groat protein and oil content in oat. Crop Sci 44(1):254–260

Breeding Approaches for Maize Improvement to Enhance Its Forage Potential

Narkhede Gopal Wasudeo, Manchikatla Arun Kumar, G. Harish Kumar, Shabir Hussain Wani, and Mehdi Rahimi

1 Introduction

Maize (*Zea mays*) is a fundamentally important crop for the world, both as a major food crop and as a forage crop in livestock production. This fact increasingly holds importance as a basis for developing high-quality animal feeds from maize varieties, optimally selected for forage production. Forage maize is of primary relevance for livestock nutrition where pasture is limited or varies in quantity and quality. However, to fulfill the demands of modern animal husbandry in forage, it necessitates breeding for traits that impact quality, yield, and digestibility directly (Chand et al. 2022; Kumar et al. 2023; Talla et al. 2023).

Maize (*Zea mays*) is a very important crop to the world, not only as a staple food source but also as a forage crop in livestock production. The increasing demand for quality and sustainable feed for animals has supported the need for developing optimized maize varieties for the forage. Most importantly, forage maize plays a crucial part in livestock nourishment, especially where there is limited or erratic availability of pasture. For maize to be in a position to fully satisfy the current forage demands of modern animal husbandry, breeding efforts should focus on traits that contribute directly to forage quality, yield, and digestibility (Chand et al. 2022; Kumar et al. 2023; Talla et al. 2023).

N. G. Wasudeo · M. A. Kumar · G. H. Kumar
Department of Genetics and Plant Breeding, School of Agriculture, SR University, Warangal, Telangana, India

S. H. Wani
National Seed Project (NSP), Sher-e-Kashmir University of Agricultural Sciences and Technology – Kashmir (SKUAST-K), Srinagar, Jammu and Kashmir, India

M. Rahimi (✉)
Department of Biotechnology, Institute of Science and High Technology and Environmental Sciences, Graduate University of Advanced Technology Kerman, Kerman, Iran

S. H. Wani, S. S. Gosal (eds.), *Accelerated Plant Breeding, Volume 5*,
https://doi.org/10.1007/978-3-032-04527-0_11

Among the key breeding approaches discussed in this chapter are conventional breeding, hybridization methods, molecular breeding, and genomic selection applications. It also discusses the effect of some traits like plant height, leaf size, and water-use efficiency in forage maize varietal improvement. Emphasis has been placed on utilizing diverse genetic resources such as landraces and wild relatives of maize to introgress desirable traits and increase genetic variability in breeding programs. The chapter also discusses the challenges of improving maize for dual-purpose grain and forage and how to balance these objectives against diverse environmental conditions and changing climate patterns. It explores how interdisciplinary collaboration can optimize maize improvement among plant breeders, agronomists, and livestock nutritionists (Jiang et al. 2024; Caradus and Chapman 2024; Santiago López et al. 2018; Zhang et al. 2022a, b).

In the recent past, many advances have been achieved in breeding approaches to enhance forage potential in maize, issues faced, and innovations in modern breeding programs on maize. Maize breeders have been very successful in many aspects over the years in improving the grain yield for their crops, and with more interest seen in the use of maize as a forage crop, they are slowly converting to traits such as biomass production, nutrition quality, disease tolerance, and environmental stress. This chapter looks into the amalgamation of traditional breeding methods with modern biotechnological tools, thereby providing a more complete understanding of how genetic improvement can bring about the development of superior forage maize varieties (Yadav et al. 2015; Andorf et al. 2019; Kumar et al. 2023).

This chapter summarizes a comprehensive overview of the cutting-edge breeding approaches in vogue for augmenting maize's forage potential and indicates future pathways for developing more efficient, productive, and sustainable maize varieties as per the demands of modern agricultural systems. The research has the potential to profoundly affect livestock production systems worldwide through the introduction of maize varieties with promising traits of combining high yields in forage with excellent nutritional quality and contribute towards food security and sustainable agriculture (Franco et al. 2021; Kumar et al. 2023; Jiang et al. 2024).

2 Current Status of Maize Breeding

The maize evolutionary breeding study, which is particularly centered on forage traits, has finally driven it toward amplifying the feeding quality of the plants. It is known that forage maize is implicitly required in agriculture, especially livestock farming, for its high-productivity and nutritional values (Chand et al. 2022).

2.1 *Evolution of Maize Breeding with a Focus on Forage Traits*

2.1.1 Early Breeding and Landraces

- **Pre-1900s:** Maize breeding started as selecting local landraces, mostly open-pollinated varieties. These landraces adapted to the local climates and soils and were not specifically bred for forage characteristics. Farmers selected maize varieties according to their general performance, which included yield and the ability to withstand drought, and gave no consideration to nutritional quality or digestibility for livestock.
- **Forage Use:** Initially, maize was bred for grain, but as animal husbandry became more intensive, its valorization as a silage crop started. In winter, silage made from maize is utilized for feeding cattle when no fresh pasture is available (Capstaff and Miller 2018; Kumar et al. 2023).

2.1.2 Introduction of Hybrid Maize (Mid-1900s)

- **1930s–1950s:** The development of hybrid maize revolutionized maize breeding. The introduction of hybrid varieties, particularly after the discovery of heterosis (hybrid vigor), led to increased yields, disease resistance, and overall plant vigor. These early hybrids were primarily bred for grain production.
- **Forage focus emergence:** While hybrid maize was largely focused on grain yield, breeders began to recognize the importance of the entire plant as a source of forage. Forage hybrids started to emerge as breeders realized that improved maize varieties could produce both high grain and high-quality silage (Kumar et al. 2023; Singh et al. 2023; Farooqi et al. 2022).

2.1.3 Advances in Forage-Specific Traits (1960s–1980s)

- **Selection for digestibility:** Breeding for improved digestibility of forage became more prominent. Maize was being bred not only for yield but also for traits like lignin content, fiber digestibility, and protein concentration in the silage. Higher digestibility means better feed conversion for livestock, which in turn improves productivity.
- **Forage yield improvements:** Another important focus was on increasing the total biomass produced by maize plants, especially with an emphasis on the total amount of silage. This was important for dairy and beef production, as more silage meant more feed during the winter months.
- **Introduction of quality traits:** Traits like starch content, water-soluble carbohydrates, and fiber composition became focal points. High sugar content and low lignin were desirable traits because they improved both palatability and

digestibility. The focus was to create maize varieties that would result in high-energy silage that was easily fermentable (Dash et al. 2024; Prasanna et al. 2021; Benavente and Giménez 2021).

2.1.4 Molecular and Biotechnological Advances (1990s–2000s)

- **Genomics and marker-assisted selection:** Advances in genomics and the use of molecular markers revolutionized maize breeding. Marker-assisted selection (MAS) allowed for the identification and selection of genetic markers associated with desirable forage traits (e.g., digestibility, plant height, ear-to-stem ratio). This helped breeders select plants with superior forage qualities more efficiently.
- **Genetically engineered crops:** Genetic engineering has innovated in transgenic maize development, which increased efficiency among minor traits in fighting pests and diseases, drought-tolerance, and nutrient utilization. Genetically modified (GM) maize varieties are the commonest herbicide-resistant or Bt protein-producing insect-resistant ones.
- **Fiber reduction:** The ability of maize to create silage that can be consumed further by livestock is one of the developments in the breeding of maize to reduce plant fiber in silage. They developed forage types with reduced lignin content to improve feed efficiency, making the cell walls less tough. (Brummer and Wang 2020; Dwivedi et al. 2020; Okoye et al. 2023).

2.1.5 Recent Trends and Ongoing Improvements (2010s–Present)

- **High-quality forage hybrids**: Research efforts have become increasingly directed at developing maize hybrids for forage production. These hybrids would typically exhibit the following characteristics: high silage yield, good digestibility, and a well-balanced nutrient composition. Some latest hybrids are bred for a more ideal ear-to-stem ratio improving the nutritional balance between starch and fiber in the silage.
- **Climate resilient:** The move to breeding for the future and emerging fruits restructured as maize varieties maturing for the environment. And at the end of the day, while those factors only lengthened the distance, forage maize can also be grown under unfavorable conditions and still yield the best silage.
- **Nutrient optimization:** Furthermore, this optimization of maize plants for forage is becoming more specific. For example, some will be breeding with higher concentrations of those important nutrients such as protein and even some vitamins so that the maize silage can promote better animal health and productivity.
- **Precision breeding:** Modern breeding technologies such as CRISPR and other gene-editing technologies can offer opportunities to quicken the development of maize varieties with improved forage qualities. Specific genetic improvements can be made to traits such as digestibility, fiber content, and overall forage yield

(Maitra et al. 2020; Blessing et al. 2022; Sowiński 2024; Toker et al. 2024; Namata et al. 2025) by using these techniques.

2.1.6 Future Directions in Forage Maize Breeding

- **Multi-trait breeding:** To combine high biomass yield, digestibility, disease resistance, and environmental stress tolerance, future breeding efforts would focus on multi-trait improvements. Fermentation characteristics can also be enhanced in forage maize varieties for efficient silage preservation.
- **Efficiency and sustainability:** In the face of increasing challenges from a rising world population in feeding them, maize breeders may be directed into making maize production more sustainable by developing nitrogen use efficiencies and reducing water requirements while optimizing the energy balance associated with maize silage.
- **Integration with livestock systems:** More integrated approaches to breeding maize specifically for livestock systems will likely evolve, where breeders and livestock specialists collaborate to optimize silage quality based on the specific nutritional needs of different types of livestock (Bryan et al. 2024; Katoch 2022; Iqbal et al. 2018).

2.2 Overview of Maize Production, Consumption, and Breeding Advancements Globally

Maize (*Zea mays* L.), commonly known as corn, is one of the most important cereal crops in the world, playing a vital role in food security, livestock feed, industrial applications, and bio-energy production. Its production, consumption, and breeding have evolved significantly due to global demand, technological advancements, and challenges such as climate change and resource management (Santiago López et al. 2018; Talla et al. 2023; Chand et al. 2022).

2.2.1 Maize Production: Global Overview

Maize is grown across the globe, with major production areas in North America, South America, Asia, and Africa. The United States is the leading producer, followed by China, Brazil, Argentina, and Ukraine. Together, these countries account for the largest share of global maize output.

- **United States**: Accounts for about 35% of global maize production. The majority of US maize is used for animal feed, ethanol production, and exports.
- **China**: Primarily produces maize for domestic consumption, including food, feed, and industrial uses.

- **Brazil and Argentina**: Both have large-scale maize production for export, with Brazil increasingly becoming a major player in the global maize trade.
- **Ukraine**: Known for producing maize for export, especially to European and Asian markets.

2.2.2 Production Trends by Region

- **North America** (United States and Canada): Highly mechanized and technologically advanced maize production systems. The United States dominates both production and export.
- **Latin America** (Brazil, Argentina, Mexico): Increasingly competitive in maize production, with Brazil and Argentina becoming key exporters.
- **Africa**: Maize is a staple crop, but yields are lower due to challenges such as limited access to modern farming practices, pests, and droughts. Efforts are being made to increase productivity through better breeding and agronomic practices.
- **Asia** (China, India): Maize production is growing, but still lags behind other regions in terms of yield and efficiency; though it's increasing as the demand for animal feed grows (Andorf et al. 2019; Jiang et al. 2024; Yadav et al. 2015; Singh et al. 2023; Kumar et al. 2023).

Maize Consumption: Trends and Uses

2.2.3 Human Consumption

- **Direct human consumption**: Maize is a key staple in many countries, especially in Africa, Latin America, and parts of Asia. It is consumed as flour (for tortillas, polenta, and other traditional foods), maize meal, and popcorn.
- **Processed products**: Maize is used in a wide range of food products, including snacks, breakfast cereals, and corn syrup, as well as in the production of maize oil.

2.2.4 Animal Feed

- A significant portion of global maize production is used as animal feed, particularly in the United States, where it is a major component of livestock diets. This is a key driver of demand, as global meat and dairy consumption rises.
- In countries like Brazil and Argentina, maize is essential in the livestock sector, especially in beef and poultry production.

2.2.5 Industrial Uses

- **Ethanol production:** A significant portion of corn grown in the United States is converted into biofuels. Corn in biofuels has evolved in amounts made as a result of the Renewable Fuel Standard Act, but debates have erupted over how the scheme affects the environment and issues of food security. Forage maize is a source of cellulosic biofuel production (Mehta et al. 2025).
- **Corn starch and its products:** Corn is also used to make starch raw material which has various applications in food and pharmaceuticals as well as industrial processes.

2.2.6 Export Market

Major exporters of maize include the United States, Brazil, and Argentina. These countries make international markets, with Asia, Europe, and Africa being the prime importing regions. However, China is also emerging as a significant importer to meet its rising demand for animal feed (Singh et al. 2020; Vanhevel et al. 2024; Yadav et al. 2015; Chakraborty et al. 2023).

Advancements in Maize Breeding

Maize breeding has undergone significant advances due to the need to increase the yield, pest resistance, drought tolerance, and nutritional quality of maize. Major breeding research topics include the following:

- **Genetically modified (GM) maize:** Bt maize, a variety of maize resistant to the European corn borer, herbicide-tolerant maize (for example, Round-up Ready), has changed maize farming, especially in the USA and Argentina. GM crops brought a revolution in pesticides and yield improvement.
- **Drought-mitigating varieties:** Climate change is majorly threatening maize production, especially in arid and semi-arid conditions. Drought-proof maize varieties serve as a remedy to this problem, and efforts have been spearheaded by the International Maize and Wheat Improvement Centre (CIMMYT) for Africa and Latin America.
- **Nutritionally-enhanced maize:** Biofortification research is directed at improving the nutritional value of maize, particularly by increasing the levels of essential vitamins and minerals (e.g., pro-vitamin A, zinc). It is especially helpful in tackling micronutrient deficiency in areas where maize is a major staple diet (Chand et al. 2022; Ciampitti and Lemaire 2022; Chen 2024; Taube et al. 2020).

2.2.7 CRISPR and Gene Editing

Gene editing technologies like CRISPR/Cas9 provide new avenues for optimizing maize faster and more precisely; thus, varieties could be developed with better resistance to diseases and pests or enhanced nutritional profiles. (Ndudzo et al. 2024; Wang et al. 2022; Haq et al. 2022; Saini et al. 2023; Young et al. 2019; Jiang et al. 2022; Varotto 2023).

2.2.8 Hybrid Varieties

- Most likely, hybrid maize varieties popularly adopted for having much higher yields and being highly adaptable to multiple environments are hybrids combining very favorable traits from different parent lines. The productivity gains are associated with using hybrids especially in developing countries.

2.2.9 Climate-Resilient Crops

As climate change is causing fluctuating weather patterns and seasons, resulting in reduced maize production, breeding for heat tolerance, water efficiency, and pest resistance becomes critical in the future. Thus, scientists develop maize varieties to thrive well in the changing climates and adverse conditions surrounding such changes (Dash et al. 2024; Washburn et al. 2020; Fan et al. 2018).

2.2.10 Challenges and Future Prospects

Climate Change

Environmental factors such as rising temperatures, shifting precipitation patterns, and extreme weather events, such as droughts and floods, can be great threats to maize production. Research has been increasingly focused on developing varieties that can withstand these factors.

2.2.11 Pest and Disease Pressure

Various pests and diseases, which include fall armyworm and maize lethal necrosis disease, are important threats to global maize production, especially devastating to Africa.

2.2.12 Soil Degradation and Water Management

The continuous planting of maize is capable of depleting soils as well as overusing water. Thus, optimal sustainable agricultural practices and excellent irrigation techniques are important steps in securing future productivity.

2.2.13 Global Trade and Market Dynamics

- The geopolitical landscape, including trade policies and the impact of biofuel mandates (e.g., US ethanol policies), influences maize markets. Trade disruptions, such as those caused by the Russia-Ukraine war, can also have a significant impact on global maize prices and availability (Capstaff and Miller 2018; Chand et al. 2022; Wang et al. 2016).

2.2.14 Key Traits for Forage Improvement

Forage improvement refers to enhancing the quality, quantity, and resilience of forage crops (plants used to feed livestock) to ensure optimal productivity, sustainability, and animal health.

(a) **Morphological Traits**

A. **Leaves Characteristics**

2.2.15 Leaf Architecture in Maize

- **Erect vs. drooping leaves**: Maize typically exhibits slightly inclined leaves, but the orientation of leaves can vary due to genetic factors, environmental influences, and agricultural practices. Erect leaves tend to be more vertically oriented compared to broader or drooping leaves. The leaf angle can significantly affect the plant's ability to capture light efficiently and how the light is distributed across the plant.
- **Leaf area index (LAI)**: LAI refers to the total leaf area per unit ground area. A high LAI indicates a large amount of leaf area relative to the ground, which enhances light capture for photosynthesis. However, a very high LAI can also lead to self-shading, where lower leaves receive insufficient light, potentially reducing overall efficiency (Ciampitti and Lemaire 2022; Chen 2024; Taube et al. 2020; Fan et al. 2018).

2.2.16 Impact on Light Interception

- **Erect leaves**: Maize plants with more erect leaf angles tend to intercept light more effectively, especially under dense planting conditions. The vertical leaf arrangement ensures that light is more evenly distributed across the plant, reducing shading of lower leaves and promoting better photosynthetic activity throughout the canopy. This can be particularly advantageous in high-density planting systems, where plants are close together and competition for light is intense.
- **Leaf area index (LAI)**: As LAI increases, light interception improves up to a certain point. An optimal LAI allows the plant to capture a high amount of light without excessively shading the lower leaves, thus maintaining high photosynthetic efficiency throughout the canopy. However, beyond a critical LAI (typically around 4–6 for maize), further increases in leaf area can lead to diminished returns, as lower leaves get too shaded and the canopy becomes inefficient in light capture.

2.2.17 Forage Yield Implications

- **Optimal leaf architecture for forage yield:** Erect leaves and moderate LAI (not too high) contribute to higher forage yield by maize plants. Erect leaves can maximize the benefit of light interception without over-shading by lower leaves, thus improving photosynthesis as well. The moderate LAI enables the yield to remain high without causing much overcrowding in the canopy, which otherwise may affect light use efficiency.
- **Effect of erect leaves:** Among maize cultivars, those having more erect leaves (for instance, vertical leaves type) tend to do better in terms of forage yield of growth. The leaves make the plant save resources from self-shading in dense stands and channel them toward biomass production leading to high forage yield potential. This type of architecture is also associated with better air movement within the canopy which minimizes moisture, thereby reducing disease infection and improving overall plant health.
- **High LAI and forage yield:** An appropriate LAI is essential for maximizing forage yield. If the LAI is too low, the plant will not intercept enough light for vigorous growth, but if it is very high, there will be excess self-shading, especially on lower leaves, which can reduce the overall photosynthetic efficiency and potential yield (Sargent et al. 2022; He et al. 2024; Wan et al. 2024).

2.2.18 Environmental and Agronomic Considerations

- **Spacing and density:** Erect leaves and optimized LAI are particularly important in high-density planting systems. When plants are closely spaced, competition for light increases, and a more erect leaf angle helps reduce shading among adjacent ones, resulting in better overall light interception and utilization.

- **Climate and water availability:** For a certain high light intensity but reduced amount of water or nutrients, maintenance of LAI is important. Erect leaves may also be more efficient in terms of water use, as they can minimize the risk of over-transpiration associated with a whole-up-leaf position.
- **Genetic selection and breeding:** The breeding endeavor has concentrated more on the architecture of the leaf for improved light interception and high yield for maize to become genetically modified maize varieties. These efficient and optimized LAIs will yield better forage yield potential under intensive production systems (Momesso et al. 2019).

B. **Root Characteristics**

The roots of maize (Zea mays L.) have a significant role in their ability to uptake nutrients, sustain soil health, and support overall plant growth. Root characteristics, such as root mass, root length densities, and the arising distribution of roots, affect the efficiency of the uptake of nutrients, absorption of water, and soil structure.

2.2.19 Root Mass

Root mass indicates the overall weight of roots and practically goes hand in hand with the capacity of a plant to take in nutrients and water. The many roots of maize take up soil extensively, eventually establishing a well-formed system through which they will help plant growth.

- **Nutrient uptake:** A larger root mass can increase the capacity of the plant to absorb water and nutrients, particularly nitrogen, phosphorus, and potassium, which are important nutrients required for maize growth. The larger the root mass, the more likely the plant is able to access nutrients from a wider range of the soil.
- **Soil health:** The root mass has an upward effect concerning aeration of soil and infiltration of water into the soil. Growing in the soil, the roots of maize build channels favorable for the movement of air and water, thus modifying the soil structure. Furthermore, the dead root biomass, after decomposition, serves towards an organic amendment of the soil, thereby improving the fertility as well as microbial activities within the soil (Cooper and Messina 2023).

2.2.20 Root Length Density (RLD)

It is expressed as total root length per unit volume of soil (for example, cm of root per cubic centimeter of soil). It is an important variable in the way a plant can exploit the soil:

- **Potential nutrient uptake:** An increase in RLD in the available volume will lead to an increase in percentage surface area available for nutrient absorption. For instance, maize—which has a very dense and extensive root network—can obtain nutrients from deeper or compacted layers of soil, especially in soils with low nutrient availability. This trait is of special importance in nutrient-deficient or low moisture soils.

- **Soil Health:** Root systems are also denser than ordinary root systems and can improve soil structure through aggregate formation and increased soil porosity, enabling roots to penetrate better and water to infiltrate, thereby minimizing soil compaction. Extensive rooting, in addition, contributes organic matter to soil, improving soil texture and fertility in the long run (Ciampitti and Lemaire 2022; Taube et al. 2020; Ma et al. 2023).

2.2.21 Root Depth

It is equally a very important factor incepting maize to reach water and nutrients, especially in soils that have poor moisture and nutrient content in extreme drought conditions. Deep-rooting maize plants can reach subsurface water and nutrients where shallow-rooting maize plants cannot reach.

- Nutrient uptake comprising deep roots allows maize penetration below soil levels where some nutrients like calcium and magnesium are more abundant. With deep roots in drought-ravaged regions, maize would reach subterranean water or "moist zones" and thus reduce the increased stress exerted on it through waves of water scarcity.
- Deep-rooted plants prevent erosion by providing soil stabilization and allow development of deeper soil structure. Further, through the process of penetrating and breaking up compacted layers, deep roots create possible channels for other roots to spread further and improve the physical properties of soils.

2.2.22 Root Distribution and Architecture

Root distribution is deployed in the spread of roots in how they become embedded in the soil, while root architecture comprises the kind of the roots, for example, primary roots and secondary roots, as well as fibrous roots. Fibrous root system of maize typically spreads very widely within the upper layers of soil, though some varieties develop deeper roots when conditions are good enough.

- A root system more evenly distributed will result in efficiency and increase nutrient acquisition from an area of soil. Near-surface roots are effective at taking nutrients such as nitrogen, which are more available in the surface soil, but deep roots can reach nutrients that are less mobile or in lower concentrations within the soil.
- The root architecture of maize affects soil aggregation as well as porosity. Roots which grow horizontally or branch profusely in the topsoil are then responsible for producing stable soil aggregates because of improved soil structure and

erosion prevention (Ciampitti and Lemaire 2022; Taube et al. 2020; Lamptey et al. 2018).

2.2.23 Root Exudates and Microbial Interactions

Maize has roots exudation of these different types of metabolic wastes or organic compounds. Some of these end up in soil, while others interact with soil microbes. Examples of these exudates are sugars, amino acids, and organic acids that can support the beneficial soil microorganisms like mycorrhizal fungi and nitrogenous bacteria.

- Root exudates solubilizing mineral elements, such as phosphorus, favorably enhancing mycorrhizal fungi, which improve the uptake of nutrients for the plant. Also, few root exudates may modify the pH of the rhizosphere, making nutrients more bio-abundant.
- Healthy soil root exudates play an important role in the lives of soil biodiversity, supporting healthy microbial communities. Healthy soil microbes take part in the cycle of nutrients, organic matter decomposition, disease suppression, and most importantly, improve the soil's fertility and structure (Wu et al. 2023; Afridi et al. 2024).

2.2.24 Impact on Soil Erosion and Water Management

Soil roots reduce soil erosion by anchoring the soil and stabilizing and strengthening the soil, particularly in slope areas. The root system also serves to improve water retention and infiltration through the establishment of pathways through which water moves in the interior of the soil profile.

- A well-developed root system will reduce runoff and soil erosion to retain most of the topsoil. As roots break through compacted soil layers, they make channels which allow water to move inward at greater depths for a more improved water-holding capacity of soil.
- Benefits of improved water retention accrue to nutrient uptake during the statutory periods of such dry periods when water tends to be limited. This means that a root system which supports better water management can do comparatively well in maintaining nutrient uptake under such constraining conditions (Carr et al. 2021; Zhu et al. 2024; Handiso et al. 2023).

2.3 *Physiological Traits*

2.3.1 Photosynthetic Efficiency and Drought Tolerance

Photosynthetic efficiency and drought tolerance are the two most important aspects for forage quality, especially in limiting environmental factors like drought with respect to maize (*Zea mays*).

2.3.2 Photosynthetic Efficiency and Forage Quality

- Photosynthesis is the primary process in plants for energy production; the efficiency of this process determines overall growth, biomass accumulation, and the nutrient content of maize.
- For maize, photosynthetic efficiency has direct repercussions on the quantity and quality of its forage collected. Thus, a higher rate of photosynthesis would give rise to a greater biomass which could then accumulate nutrients, including carbohydrates, proteins, and lipids, required for livestock nutrition.
- Maize cultivars with superior photosynthetic efficacy utilize available light and carbon dioxide more effectively, consequently offering higher yields as well as improved forage quality. This is significant as maize is one of the major forage sources for animals, while efficient photosynthesis improves the overall digestibility and nutrient availability of that feed.
- Factors that indirectly affect photosynthetic efficiency include leaf architecture, chlorophyll content, and the ability to maintain stomatal conductance under varying environmental situations. Some of the maize varieties are bred to possess better chlorophyll production and increased leaf area index while being optimized in their photosystem activity, thus resulting in enhanced photosynthetic efficiency.

2.3.3 Drought Tolerance and Forage Quality

- Effects of drought stress on maize growth and productivity severely hinder its ability to produce good quality forage. Drought tolerance in maize encompasses a wider range of mechanisms such as better water use efficiency (WUE), stomatal regulation, osmotic adjustment, and deep-root systems that have access to moisture from deeper layers of the soil.
- During drought periods, maize plants may close their stomata to prevent evaporation and conserve water. However, most improved strains of drought-resistant maize can resist such conditions and maintain higher photosynthetic activity in limited water periods.

- Maize varieties with tolerance to drought are also likely to maintain leaf turgor and photosynthesis in dry conditions, hence, reduced yield losses and possibly improved quality of forage for the nutrition of livestock.
- Moreover, maize varieties with high drought tolerance will maintain fiber and mineral content in their forage. However, these two components might be compromised under water stress.
- Drought stress adversely impacts the maize crop at all stages of growth, especially through yield and jeopardizing quality forage production.
- Drought tolerance mechanisms in maize include aspects such as improved water use efficiency (WUE), stomatal regulation, osmotic adjustments, and deep root systems enabling it to access water from deeper layers of soil.
- During drought, maize plants may close their stomata to conserve water, thus limiting CO2 uptake, leading to reduced photosynthesis and yield quality.
- Improvement in drought tolerance gives maize the ability to tolerate these conditions and maintains hay better than that of plants competing at a lower level of photosynthetic activity during the same period of limited water.
- Drought-tolerant varieties of maize usually have better maintenance of leaf turgor and photosynthesis under drying conditions, thus resulting in less yield loss and better quality of forage in terms of nutrient content (for example, protein or energy).
- They can also maintain the fiber quality and minerals in the forage, which are important in livestock digestion, but can be reduced when under water stress.

2.3.4 Integrated Role in Forage Quality

- Photosynthetic efficiency is saying that there is drought tolerance in impacts considering maize forage quality, for drought conditions can directly diminish photosynthetic activity. The maize plant that shows the combined effects will have increased yield potential under stress conditions with good nutritional quality.
- Improved drought tolerance characters in maize hybrids usually feature high photosynthetic stability under water-limiting conditions for maintaining forage production and quality.
- Forage digestibility and nutrient composition also depend on these criteria: photosynthesis enters into the synthesis of complex sugars and proteins; drought tolerance is what lets plants be more resistant during periods of water stress. So, these drought-tolerant maize varieties can maintain higher digestible nutrients-in-like sugars, starch, and protein-in-form forage even during not very much water.

2.3.5 Implications for Breeding and Management

- The breeding of maize for higher efficiency in photosynthetically fixed carbon and drought tolerance is of very high significance for forage production in water-scarce areas.
- Agronomic practices such as optimized irrigation, conservation tillage, and soil fertility will complement the genetic traits of maize in enhancing photosynthetic efficiency and drought tolerance in forage quality enhancement.
- Precision farming technologies (drought monitoring, stress detection) help maize farmers manage crop for the optimal quality of forage by knowing the right hybrid varieties used for the crop and tailoring all practices grown against environmental stressors (Rad et al. 2021; Habermann et al. 2019; Gelley et al. 2020; Bashir et al. 2021; Bhattarai et al. 2020; Qi et al. 2021).

2.3.6 Nutritional Content

Maize (corn) in particular forms one of the important feeds for livestock being rich in energy, especially from starch, and usually having good digestibility. Furthermore, its desired effect in nutrition has made it a vital component for the balanced formulation of rations for animals.

2.3.7 Starch Concentration in Maize

- The starch occupies the main bulk in maize composition, as maize is predominantly a carbohydrate crop. Its typical starch content ranges from 60% to 70% of the dry matter in maize grain. Thus, it can be a very concentrated energy source for livestock.
- The starch in maize is quickly fermentable for readily available energy for both monogastric animals (pigs, poultry) and ruminant animals (cattle, sheep), although digestibility varies according to processing methods (grinding, cooking, steam flaking).

2.3.8 Organic Matter Digestibility (OMD)

Instinctively or organically, however, the OMD of maize is closely tied to different varietal, agronomic, and processing aspects, falling within 80–90%. This high digestibility accounts for the fact that maize is an excellent source of nutrients. In ruminants, maize silage or grain is very digestible; starch ferments rapidly in the rumen. However, the ruminal microbial population and the physical form of the maize affect the starch fermentation degree. Non-ruminants, like pigs, poultry, etc., have the highest starch digestibility, i.e., up to 95%. Starch is broken down in the small intestine; consequently, the energy intake is very high.

2.3.9 Crude Protein and Fiber Content (Related to Digestibility)

- Crude protein content: It is most evident that maize has lower protein content among other feeds, usually 7–9% dry matter, thus balancing protein enough when used along with more protein-rich constituents like soybean meal.
- Crude fiber: Moderately fibrous as maize grain has 2–3% in the dry matter rubrics, so it further enhances its digestibility as observed in higher fiber in maize silage or cob meal.

2.3.10 Amino Acid Profile

The amino acid composition of maize is significant in livestock nutrition. Lysine is the limiting amino acid in maize, and it is usually supplemented with other feedstuffs such as soybean meal to improve the amino acid profile of monogastric diets.

2.3.11 Fat Content

Maize has about 3–5% fat (in grain), which is part of the energy content, but it digests generally high. This fat is mostly in the form of unsaturated fatty acids and is helpful for livestock.

2.3.12 Mineral Content

Maize is a source of essential minerals such as phosphorus, potassium, and magnesium. On the other hand, it is poor in calcium, which may be automatically supplemented in the rations supplied for some livestock.

2.3.13 Maize Silage Versus Grain

- **Maize silage:** The change in nutritional profile is observed when maize is harvested for silage. The starch content is found to be slightly low due to fermentation but has high digestibility, about 70–80% OMD, and is, therefore, a highly acceptable forage material for the ruminants.
- **Maize grain:** Maize grain, on the contrary, is a much-preferred form for high-energy diets, especially for monogastric animals, because of the starch content (Bryan et al. 2024; Katoch 2022; Iqbal et al. 2018; Chand et al. 2022; Capstaff and Miller 2018).

3 Breeding Strategies

3.1 *Conventional Breeding Techniques*

Methods of breeding maize (*Zea mays* L.) for the improvement of its forage quality typically involve the selection and cross-breeding of maize varieties characterized by desired traits to determine good livestock feed qualities such as high yield, enhancement of nutritive quality, better digestibility, and resistance to pests and diseases.

3.1.1 Selection

The most frequently employed method of breeding, selection, is concerned with recognizing and choosing maize plants characterized by one or more traits, such as yield for forage, as well as nutritional quality (higher protein content, digestibility), resistance to pests and diseases, and improved drought tolerance. The main types of selection include the following:

- **Mass selection:** Selecting the best individuals from a population for their desirable forage traits and using their seeds for the next generation. This approach is commonly used to improve traits like drought tolerance and yield.
- **Individual plant selection:** Selecting the best-performing plants based on their forage yield, palatability, and other traits. This method is more labor-intensive but can lead to greater genetic gains.
- **Progeny testing:** Using the offspring of selected plants to evaluate their performance, ensuring that the selected traits are passed on to the next generation.

3.1.2 Hybridization

This crossing brings together two genetically distinct maize lines with a view to harnessing the best from both, thereby improving overall forage quality and yield. Hybridization is mainly needed for producing hybrids of high yield. In maize, hybrids are extensively used for both grain and forage production.

- **Simple crosses:** The most commonly used maize hybrids, formed by crossing two inbred lines. Simple cross hybrids are chiefly used for an increase in yield and vigor.
- **Double crosses:** This is a cross between two single-cross hybrids. This technique can help in increasing genetic diversity and improving disease resistance in forage maize.

3.1.3 Backcrossing

Backcrossing helps to bring in an attribute into a value without altering its overall performance. For example, if a high-yield maize variety is disease susceptible, the breeder can backcross it with a resistant variety, thereby maintaining all desirable forage traits of the original variety while introducing disease resistance from the other variety.

3.1.4 Recurrent Selection

This is a method of repeatedly selecting individuals with superior forage traits from a population to gradually improve the population's overall performance. Recurrent selection helps increase traits such as drought tolerance, forage yield, and quality over multiple generations. It is particularly effective in improving polygenic traits (those controlled by many genes), such as digestibility and overall forage quality.

3.1.5 Gene Pool Enhancement (Wide Crosses)

Wide crossing usually includes a process of crossing a maize plant with other species such as species or some varieties of maize that do not display genetic similarities. These may introduce new traits such as pest or disease resistance or better nutritional content into the maize gene pool. This procedure is especially effective when aiming at increasing genetic diversity or breeding new maize varieties with novel traits.

3.1.6 Quantitative Trait Locus (QTL) Mapping

Although QTL mapping is advanced in comparison to traditional breeding methods, it may well form a part of a conventional approach to decipher the genetic loci critical for important traits with particular reference to forage yield and quality. This understanding of the genetics of the traits will arm breeders with better knowledge in the selection and hybridization of plants for desirable characters.

3.1.7 Agro-Morphological Characterization

Agro-morphological traits of maize such as plant height, leaf size, stem diameter, and flowering time are evaluated, and they are critical traits for forage production. Breeders look for plants with robust growth, high leaf area (for better photosynthesis), and good biomass production, which would enhance their forage potential.

3.1.8 Pedigree Selection

Pedigree selection is concerned with the tracking of the ancestry of individual plants for selecting those that exhibit consistent performance across generations. This is particularly important in breeding such varieties where agronomic performance, disease resistance, and forage quality are desired.

3.1.9 Inbreeding Depression Management

Inbreeding depression is a reduced fitness of progeny as a result of inbreeding. Apart from transposing certain stable traits closer to end-users, inbreeding should be regulated judiciously to avoid detrimental aspects such as a decrease in yield or vigor. Breeding systems such as hybrids and recurrent selection would alleviate or reduce the effects of inbreeding depression (Ciampitti and Lemaire 2022; Chen 2024; Taube et al. 2020; Dash et al. 2024).

3.1.10 Benefits and Challenges

- **Advantage:** These traditional methods are also very useful for improving such important traits for forage maize as yield, digestibility, and resistance to disease and pest attacks. They are cost-effective techniques that do not involve advanced technologies and therefore are within reach of many breeders.
- **Challenges:** Most of the time, conventional breeding procedures take much longer to achieve positive changes and may sometimes require up to several generations before the desired improvement can be attained. Furthermore, the augmentation or improvement of such complex characters as digestibility and disease resistance would require a long and often intricate selection process and multiple annual tests (Prasanna et al. 2021; Singh et al. 2020; Hossain et al. 2022; Vanhevel et al. 2024; Washburn et al. 2020; Cooper and Messina 2023; Fan et al. 2018; Dwivedi et al. 2020; He et al. 2024).

3.2 Modern Biotechnological Approaches

Maize (*Zea mays* L.) is one of the indispensable crops not only for human consumption but also for livestock feed. Its forage productivity (which really refers to the ability of maize to generate a high-value feed for animals) can be further enhanced by modern biotechnological applications, intending to improve many aspects, such as yield, nutritional quality, disease resistance, and stress tolerance, subsequently increasing the forage potential of maize.

3.2.1 Genomic Selection and Marker-Assisted Breeding (MAB)

Genomic selection uses molecular markers to identify the genes responsible for favorable traits to hasten the breeding process. This technology helps breeders choose maize varieties that are superior in forage production, high biomass, digestibility, and pest or disease resistance.

3.2.2 Transgenic (GM) Crops

Genetic modification (GM) is a genetic engineering technique involving the insertion of specific genes for the introduction of desirable traits in maize. Transgenic crops can be created for forage maize with crops that have improved nutritional quality, pest resistance, or resistance to stress.

3.2.3 Improvement of Fiber and Digestibility

The modification of the lignin biosynthesis pathway through genetic engineering to reduce the lignin content in maize could be a step towards improvement for forage. Lignin is a major constituent that affects forage digestibility, and thus, a reduction in it would increase the nutritional value of maize to livestock.

3.2.4 Improved Nitrogen Use Efficiency

GM maize can be designed for better and improved nitrogen utilization, which requires less use of fertilizer and has improved quality in the fodder.

3.2.5 CRISPR-Cas9 Gene Editing

CRISPR-Cas9 places a precise editing tool for genes in maize, for example, and can directly change traits for enhanced forage potential.

- **Improvement of digestibility and fiber content**: CRISPR manipulates genes related to synthesis of cell wall components, which leads to reduced fiber content; hence, it would be easier for livestock to digest forage.
- **Stress tolerance**: Gene editing can also be applied to improve maize's tolerance to environmental stresses such as drought, salinity, and extreme temperatures, ensuring the forage can remain productive under altered climatic conditions.

3.2.6 Synthetic Biology

Synthetic biology involves designing and constructing new biological parts or systems, or redesigning existing biological systems. For maize, this approach could involve the following:

- **Improving metabolic pathways:** The genetic amending of maize for enhanced photosynthesis or carbon fixation promises to boost biomass production of maize, thus supplying forage material.
- **Increased nutrient content:** Enhancing nutrients like increasing concentration of proteins, minerals, or vitamins will also be feasible if synthetic biology is applied to maize forage, thus enhancing its nutritional profile for livestock (Brummer and Wang 2020; Dwivedi et al. 2020; Okoye et al. 2023; Benavente and Giménez 2021; Ndudzo et al. 2024; Wang et al. 2022; Haq et al. 2022; Saini et al. 2023; Young et al. 2019; Jiang et al. 2022; Varotto 2023; Sargent et al. 2022; He et al. 2024; Wan et al. 2024).

3.2.7 Microbial Inoculants and Endophyte Symbiosis

Using beneficial microorganisms, bacteria and fungi, as an adjunct to maize production has great potential for improving forage potential. They may help maize plants to absorb more nutrients from the soil, increase resistance against diseases, or improve forage quality (Ali et al. 2021).

- **Endophytic fungi:** These fungi can grow within maize plants without causing harm and contribute to improving the growth and stress resistance of plants. Some of the endophytes are also capable of producing substances that may increase palatability or nutrition in maize forage.

3.2.8 Metabolomics and Phenotyping

Metabolomics looks into the metabolites inherent in the metabolism process, and it can be utilized to evaluate the nutritional as well as functional properties of maize forage. Identification of specific metabolites related to improved forage quality would enable breeders to develop better maize cultivars through phenotyping and metabolic profiling (Mattoo et al. 2023).

3.2.9 Polyploidy and Hybridization

Polyploidy (the condition of having more than two sets of chromosomes) in maize can boost the size and yield of the plant, which can then reflect more biomass for forage. Hybrid maize varieties, often bred using advanced genetic tools, can also outperform traditional varieties in terms of forage yield, quality, and disease resistance (Batiru and Lübberstedt 2024).

3.2.10 Molecular Breeding for Abiotic Stress Resistance

It reduces the potential of forage by growing maize in those areas where water is restricted or the soil is not very good. Through molecular breeding methods, it is possible to develop maize varieties that could withstand or scavenge from abiotic stresses, such as drought, high temperature, or salinity of soil (Brummer and Wang 2020; Okoye et al. 2023).

3.2.11 High-Throughput Screening and Digital Phenotyping

The breakthroughs in high-throughput screening solve the problem of rapidly evaluating numerous advanced maize genotypes for their important forage traits. When combined with digital phenotyping based on sensors and imaging technologies for phenotyping health, biomass, and others, breeders will act with greater effectiveness in selecting superior functioning varieties for forage purposes (Teodoro et al. 2024).

3.2.12 Tissue Culture and Somatic Embryogenesis

Such techniques are used for propagating maize plants with desired characteristics. Somatic embryogenesis, for instance, is a technique for regenerating plants through somatic cells and can thus be used for rapid multiplication of plants having improved forage characteristics (McFarland and Kaeppler 2024).

4 Intercropping Systems

4.1 Benefits of Intercropping Maize with Legumes

Inclusion of intercropping maize with legumes offers manifold advantages, primarily enhancing the forage potential of maize (Table 1).

Table 1 Examples of successful intercropping practices that enhance forage potential

	Location	Legumes used	Key findings	References
Maize-legume intercropping review	Southern Mexico	Groundnut (*Arachis hypogaea*), velvet bean (*Mucuna pruriens*)	Increased maize yield by up to 53% compared to monocropping; improved soil health and reduced erosion	Pierre et al. (2022))
Resource use efficiency in Ghana	Northern Ghana	Various legumes (specifics not detailed)	Enhanced crop productivity and economic returns; maize+cowpea achieved highest system equivalent yield (SEY) of 5230 kg/ha	Kermah et al. (2017))
Integrated Pest management strategy	India	Cowpea (*Vigna unguiculata*), green gram (*Vigna radiata*), black gram (*Vigna mungo*), red gram (*Cajanus cajan*)	Maize+cowpea reduced pest infestation significantly compared to sole maize; highest benefit-cost ratio of 4.11	Saritha and Visalkshi (2024)
Forage legumes with maize	Various locations (not specified)	Diverse forage legumes (specifics not detailed)	Improved soil fertility and increased maize yields; intercropping systems showed better water use efficiency (WUE)	Saritha and Visalkshi (2024)
Legume/maize intercropping in China	Sichuan Province, China	Soybean (*Glycine max*), pea (*Pisum sativum*)	Increased crop yield and quality with optimized nitrogen application; improved nutrient use efficiency (NUE)	Zhang et al. (2022a, b)
Edible-legume intercropping systems	Various locations (not specified)	Pigeon peas, cowpeas, sesame (specifics not detailed)	Enhanced soil health and crop yields through natural agroecological services; improved sustainability of farming systems	Jalloh et al. (2024))

4.1.1 Improved Nutrient Cycling

- **Nitrogen fixation:** Leguminous plants such as beans, peas, or clover fix atmospheric nitrogen in the soil. This enriches the soil with nitrogen, which is essential for maize growth, making synthetic fertilizers unnecessary. Not only do legumes supply nitrogen for themselves, but they also benefit the maize and increase the overall fertility of the soil.
- **Reduced dependence on fertilizers:** This nitrogen can be used either completely or in part to supplement chemical nitrogen fertilizers, thus reducing the input cost and environmental pollution associated with chemical nitrogen fertilizers.

4.1.2 Enhanced Forage Quality

- **Increased protein content:** Legumes are rich in proteins, and their intercropping with maize can improve the protein content of the resulting forage. The forage then becomes more nutritious for livestock, improving both growth and milk production.
- **High Quality Silage:** Harvesting maize and legumes together comes with very good silage that can be used. The legumes will provide important nutrients and accessibility to improve the fermentation of silage, enabling it to be more digestible and better livestock feed.

4.1.3 Improved Soil Health

- **Erosion control:** The roots of legumes in the system can help reduce soil erosion and stabilize soil, thereby preventing runoff and preserving soil structure in areas prone to soil degradation.
- **Higher organic matter:** The decomposition of legume residues into organic matter contributes to soil organic matter, indirectly improving soil structure, water retention, and microbial activity, all of which enhance maize growth.

4.1.4 Weed Suppression

- **Competition with weeds**: Growing legumes with maize can serve as a partial means of suppressing weeds. Legumes, especially if planted fast or with a canopy, reduce the amount of soil exposed to sunlight and the ground area available for weeds, reducing herbicide needs.

4.1.5 Enhanced Pest and Disease Resistance

- **Diversified planting**: Intercropping with legumes can disrupt the life cycles of pests that are specific to maize. Legumes may act as a repellent to certain pests or attract beneficial insects, such as pollinators and predators of maize pests, thereby reducing the need for chemical pesticides.
- **Reduced disease pressure**: A more diverse crop system reduces the likelihood of disease buildup that is common in monoculture farming systems. The leguminous plants can also help break pest and disease cycles, reducing overall pressure on maize crops.

4.1.6 Increased Biodiversity

- **Enhanced biodiversity**: By diversifying the crop system, intercropping maize with legumes contributes to higher biodiversity on the farm. This supports beneficial organisms in the soil and surrounding environment, improving the overall resilience of the farming system.

4.1.7 Improved Land Use Efficiency

- **Maximized use of space**: Intercropping increases the overall productivity of the land. While maize grows tall and takes up space vertically, legumes typically grow shorter and can fill in the lower spaces. This leads to more efficient land use, as multiple crops are harvested from the same area.
- **Efficient use of resources**: With legume and maize plants occupying different niches in terms of nutrient and water requirements, intercropping can optimize the use of resources like water and light, enhancing productivity and forage yield.

4.1.8 Climate Resilience

- **Drought tolerance**: Some legumes have deep root systems that can access water in deeper soil layers, making the intercrop more resilient to drought conditions. Maize, which can be vulnerable to dry spells, benefits from the legumes' water-saving qualities.
- **Carbon sequestration**: The increased organic matter from legumes helps in carbon sequestration, contributing to soil carbon storage, which can mitigate climate change effects over the long term.

4.1.9 Economic Benefits

- **Increased yield**: The combined forage yields from maize and legumes can be higher than monoculture systems. Additionally, intercropping maize with legumes can increase the income potential for farmers by diversifying their sources of forage and potentially selling both the maize and legume crops.
- **Risk diversification**: Intercropping provides insurance against market fluctuations in the price of maize. In case of a poor maize season, the farmer still benefits from the sale or use of legumes (Toker et al. 2024; Blessing et al. 2022; Javanmard et al. 2020; Maitra et al. 2020; Iqbal et al. 2018; Sowiński 2024).

4.2 Environmental Considerations

4.2.1 Balancing Increased Productivity with Environmental Impacts

Balancing increased productivity with environmental impacts in maize forage breeding, particularly focusing on nitrogen use efficiency (NUE), is a critical issue for both improving agricultural sustainability and ensuring food and feed security (Momesso et al. 2019).

4.2.2 Understanding Nitrogen Use Efficiency (NUE) in Maize Forage Breeding

- **NUE definition**: NUE is the ratio of the amount of nitrogen (N) in the harvested crop to the total amount of nitrogen applied to the soil. Improving NUE in maize is key to reducing the need for synthetic fertilizers, which can have detrimental environmental effects such as nutrient runoff, water pollution, and greenhouse gas emissions.
- **Maize forage production**: Maize is often grown as a forage crop, especially for livestock feed. High-yielding maize varieties that can efficiently use nitrogen fertilizers are essential for optimizing both productivity and sustainability (Wani et al. 2021).

4.2.3 Genetic Improvements for High NUE in Maize Forage Breeding

Identifying Key Traits

- **Root architecture:** Stronger and deeper root systems can help maize plants access nitrogen more efficiently from the soil.
- **Nitrogen assimilation pathways:** Improving the ability of maize plants to take up, assimilate, and metabolize nitrogen is crucial. Genetic variations in enzymes like nitrate reductase and glutamine synthetase can affect this process.
- **Efficient use of fertilizer:** Some maize varieties are more efficient at converting applied nitrogen into biomass and forage yield, reducing the need for high inputs.
- **Marker-assisted selection (MAS)**: Using molecular markers linked to NUE traits can speed up the breeding process for selecting maize varieties that are both productive and nitrogen-efficient (Ciampitti and Lemaire 2022; Taube et al. 2020; Vanhevel et al. 2024; Prasanna et al. 2021).

4.2.4 Agronomic Practices for Sustainable Maize Forage Production

- **Optimizing fertilizer application**: Even with high NUE varieties, optimal fertilizer application is critical. Precision agriculture practices, such as variable rate nitrogen application and soil testing, can help minimize nitrogen losses and maximize plant uptake.
- **Crop rotation and intercropping**: Growing maize as part of a diversified crop rotation system or using intercropping (e.g., maize with legumes) can improve soil nitrogen content, reduce the need for synthetic fertilizers, and improve overall soil health.
- **Reduced till or no-till practices**: These practices can help preserve soil nitrogen and reduce nitrogen losses through leaching or volatilization (Koenig et al. 2023).

4.2.5 Environmental Impacts of Nitrogen Use in Maize Forage

- **Nitrogen runoff and leaching**: Excess nitrogen that is not taken up by maize can be lost to the environment, leading to water pollution (e.g., algal blooms in waterways) and contamination of groundwater with nitrates.
- **Greenhouse gas emissions**: The use of nitrogen fertilizers in agriculture can contribute to the emission of nitrous oxide (N_2O), a potent greenhouse gas. Breeding maize varieties that require less nitrogen or are more efficient at utilizing applied nitrogen can help mitigate these emissions (Wani et al. 2021).

4.2.6 Integrating Sustainability Goals into Breeding Programs

- **Breeding for multi-functional traits**: Beyond improving NUE, maize forage varieties can be bred to enhance traits such as drought resistance, pest resistance, and higher biomass yield. This comprehensive approach can improve the overall sustainability of maize forage systems.
- **Life cycle assessment (LCA)**: Conducting an LCA of different maize forage varieties and breeding strategies can help quantify the environmental impact of breeding decisions. This allows breeders to compare trade-offs between productivity and environmental sustainability (Karnatam et al. 2023).

4.2.7 Collaboration with Farmers and Stakeholders

- **Farmer education**: It is essential to engage farmers in understanding the importance of NUE and adopting practices that minimize fertilizer use, such as split applications of nitrogen or using organic fertilizers alongside synthetic ones.
- **Policy support**: Government policies that support sustainable agricultural practices (e.g., through subsidies for low-emission fertilizers or research grants for

NUE-focused breeding) can incentivize farmers to adopt these techniques (McCluskey and Tracy 2021).

4.2.8 The Future of Maize Forage Breeding for Sustainability

- Advances in genomics, such as genome-wide association studies (GWAS) and CRISPR technology, could accelerate the development of maize varieties with improved NUE and other beneficial traits.
- The focus on precision breeding and integrated agricultural systems that combine genetic, agronomic, and environmental factors will be crucial for balancing productivity with minimizing environmental impacts (Washburn et al. 2020; Fan et al. 2018; Dwivedi et al. 2020; He et al. 2024; Hossain et al. 2022; Prasanna et al. 2021).

4.2.9 Strategies to Mitigate Negative Environmental Effects

Improving maize forage yield while mitigating its negative environmental impacts requires a combination of sustainable farming practices and technological innovations.

Soil Health Management

- **Cover cropping**: Use of cover crops like legumes (e.g., clover or vetch) during the offseason helps improve soil structure, increase organic matter, and reduce soil erosion. This can enhance nutrient cycling, reduce the need for synthetic fertilizers, and prevent soil degradation.
- **Crop rotation**: Rotating maize with other crops, such as legumes or small grains, reduces soil nutrient depletion and pest build-up. It helps in breaking pest and disease cycles while maintaining or improving soil fertility.
- **Conservation tillage**: Reducing tillage preserves soil structure, prevents erosion, and maintains soil carbon. This practice improves water retention and can reduce the energy input associated with plowing.

Genetic Improvement and Biotechnology

- **Genetically improved varieties**: Use drought-resistant, pest-resistant, or high-yielding maize varieties to maximize productivity while reducing resource input. These varieties can increase resilience to environmental stresses, leading to better yields with less environmental impact.
- **Biotechnology for forage quality**: Using genetically modified maize that improves nutritional quality or increases biomass yield can help make maize

forage more productive and sustainable (Okoye et al. 2023; Varotto 2023; Benavente and Giménez 2021).

5 Future Directions in Maize Forage Breeding

Climate change poses significant challenges to global agricultural systems, including maize forage production, which is vital for livestock feed and food security in many regions. These challenges arise from the intensifying effects of changing weather patterns, including temperature shifts, rainfall variability, and the increased frequency of extreme weather events (Chen 2024; Dwivedi et al. 2020; Brummer and Wang 2020; Okoye et al. 2023; Benavente and Giménez 2021).

5.1 Temperature Increases

Maize, like other crops, is sensitive to temperature changes, particularly heat stress during critical stages of growth such as flowering and grain formation. Climate change has led to rising temperatures globally, and this can have the following impacts on maize forage production:

- **Reduced growth and yield**: High temperatures can reduce maize plant growth and decrease the forage quality, as it leads to earlier senescence (aging) and a reduction in the plant's ability to photosynthesize.
- **Increased water demand**: Warmer conditions increase evapotranspiration rates, leading to higher water demand for maize cultivation. In areas where water availability is already limited, this can exacerbate drought stress, resulting in stunted growth and lower forage yields.
- **Shifting growing seasons**: Temperature fluctuations may cause shifts in planting and harvesting seasons. This can create a mismatch in forage supply and demand, impacting livestock feeding schedules.

5.2 Changes in Rainfall Patterns

The frequency and intensity of rainfall events are changing due to climate change. Maize relies on consistent water availability during its growing season, and any disruptions in rainfall patterns can severely affect its production:

- **Drought**: Prolonged periods of drought, especially during critical growth stages like flowering and grain filling, can lead to reduced biomass and lower forage quality. Drought stress limits the amount of green biomass available for livestock feed.

- **Excessive rainfall and flooding**: On the other hand, excessive rainfall and flooding can lead to waterlogging of soils, which restricts root growth, reduces nutrient uptake, and harms plant health, further reducing forage yields.
- **Unpredictable rainfall**: The variability in rainfall patterns also complicates crop planning, as farmers may not know when to plant or irrigate. This unpredictability can lead to inefficient use of resources, such as water and fertilizers.

5.3 Pests and Diseases

Warmer temperatures and shifts in moisture patterns can alter the dynamics of pests and diseases, which are already significant threats to maize production. Climate change may exacerbate the prevalence of certain pests and pathogens, which can cause additional stress to maize crops:

- **Pests**: Increased temperatures can favor the spread of insect pests like the corn earworm and rootworm, which damage both the maize plants and the quality of the forage produced.
- **Diseases**: Fungal diseases like gray leaf spot or maize rust can thrive under warm and wet conditions. Such diseases can compromise plant health and reduce both the quantity and quality of maize forage.

5.4 Reduced Forage Quality

The nutritional value of maize forage is affected by climate change, which is a vital input in livestock health and productivity. Thus, below is the relevant factor for lower forage quality:

- **Heat**: A longer growing period's heat stress leads majorly to lower protein content in maize forage and higher fiber content, thus lower digestibility for animals.
- **Water stress**: Drought leads to a declining band of protein and carbohydrate in maize as it increases lignin, making forage less palatable and more unfit for animal nutrition.
- **Growing seasons**: Due to the diminution of the growing season caused by increasing temperatures, maize may not ultimately achieve its potential optimal growth, whereby the forage is of lower quality.

5.5 *The Need for Resilient Maize Varieties*

Given the present circumstances of climate change, it is imperative to develop varieties of maize that are robust against different environmental stresses. These production types will ensure that forage production remains consistent and livestock gets enough feed even under changing climate conditions. Important traits for resilient maize include the following:

- **Drought resistance:** Tolerant of dry spells, they seem to be important varieties. They develop deeper roots and improve water use efficiency, and they reduce leaf senescence under dry conditions.
- **Heat resistance:** Heat-resistant maize varieties are those which do not lose their yield during or become poor feed quality due to increasing temperatures. These are for very important regions where the temperature is gradually rising.
- **Pest and disease resistance:** While genetic resistance commonly helps maize plants against diseases and pests, changes, for example, in integrated pest management and genetic resistance can help conserve pesticide use and reduce crop losses from pests and diseases.
- **Encourage improved nutrition:** Maize improved on these traits may yield maize varieties that have increased protein levels along with better digestibility to improve forage quality, especially under drought or heat stresses, where standard maize fails to maintain the nutritional profile.

6 Conclusion

Among these breeding strategies that have been suggested for maize improvement are those that aim at improved forage productivity, to meet the increasing demand for better-quality livestock feed. The chapter elaborately describes genetic diversity, breeding innovation, and advanced technologies like genomic selection and marker-assisted breeding, which act as strengthening strategies for the creation of maize varieties with superior forage attributes. By focusing on specific traits such as digestibility, yield, nutritional content, and resilience to environmental stresses, major changes can be made for enhancing maize as a valued forage crop. Furthermore, multi-trait selection strategies and evaluation of the genetic potential of local and exotic germplasm for optimizing maize varieties to agroclimatic regions can complement improvements in maize varieties. Sustained research and breeding program cooperation, along with modern tools, have great promise for the enhancement of forage productivity in maize and, consequently, their contributions toward more sustainable livestock production systems.

References

Afridi MS, Kumar A, Javed MA, Dubey A, de Medeiros FHV, Santoyo G (2024) Harnessing root exudates for plant microbiome engineering and stress resistance in plants. Microbiol Res 279:127564

Ali M, Ahmad Z, Ashraf MF, Dong W (2021) Maize endophytic microbial-communities revealed by removing PCR and 16S rRNA sequencing and their synthetic applications to suppress maize banded leaf and sheath blight. Microbiol Res 242:126639

Andorf C, Beavis WD, Hufford M, Smith S, Suza WP, Wang K, Woodhouse M, Yu J, Lübberstedt T (2019) Technological advances in maize breeding: past, present and future. Theor Appl Genet 132:817–849

Bashir N, Athar HUR, Kalaji HM, Wróbel J, Mahmood S, Zafar ZU, Ashraf M (2021) Is photoprotection of PSII one of the key mechanisms for drought tolerance in maize? Int J Mol Sci 22(24):13490

Batiru G, Lübberstedt T (2024) Polyploidy in maize: from evolution to breeding. Theor Appl Genet 137(8):182

Benavente E, Giménez E (2021) Modern approaches for the genetic improvement of Rice, wheat and maize for abiotic constraints-related traits: a comparative overview. Agronomy 11(2):376

Bhattarai B, Singh S, West CP, Ritchie GL, Trostle CL (2020) Effect of deficit irrigation on physiology and forage yield of forage sorghum, pearl millet, and corn. Crop Sci 60(4):2167–2179

Blessing DJ, Gu Y, Cao M, Cui Y, Wang X, Asante-Badu B (2022) Overview of the advantages and limitations of maize-soybean intercropping in sustainable agriculture and future prospects: a review. Chilean J Agric Res 82(1):177–188

Brummer EC, Wang ZY (2020) Biotechnology and molecular approaches to forage improvement. In: Forages: the science of grassland agriculture, vol 2, pp 567–579

Bryan G, Winichayakul S, Roberts N (2024) Nutritional enhancement of animal feed and forage crops *via* genetic modification. J R Soc N Z 55(2):327–342

Capstaff NM, Miller AJ (2018) Improving the yield and nutritional quality of forage crops. Front Plant Sci 9:535

Caradus JR, Chapman DF (2024) Evaluating pasture forage plant breeding achievements: a review. N Z J Agric Res:1–75

Carr TW, Balkovič J, Dodds PE, Folberth C, Skalský R (2021) The impact of water erosion on global maize and wheat productivity. Agric Ecosyst Environ 322:107655

Chakraborty M, Munshi SK, Haque A, Azad MAK, Islam T, Alam M, Shiddiky MJ (2023) Current biotechnological approaches in maize improvement. In: Maize improvement: current advances in yield, quality, and stress tolerance under changing climatic scenarios. Springer International Publishing, Cham, pp 137–180

Chand S, Indu Singhal RK, Govindasamy P (2022) Agronomical and breeding approaches to improve the nutritional status of forage crops for better livestock productivity. Grass Forage Sci 77(1):11–32

Chen S (2024) Advances in molecular breeding of forage crops: technologies, applications and prospects. Agriculture 14(2):279. https://doi.org/10.3390/agriculture14020279

Ciampitti IA, Lemaire G (2022) From use efficiency to effective use of nitrogen: a dilemma for maize breeding improvement. Sci Total Environ 826:154125

Cooper M, Messina CD (2023) Breeding crops for drought-affected environments and improved climate resilience. Plant Cell 35(1):162–186

Dash PK, Ganesan KN, Manivannan N, Kumar SV, Senthil N, Pushpam R (2024) Revolutionizing livestock sustainability: pioneering breeding strategies for superior forage biomass and quality. Horizon 11(4):1093–1107

Dwivedi SL, Goldman I, Ceccarelli S, Ortiz R (2020) Advanced analytics, phenomics and biotechnology approaches to enhance genetic gains in plant breeding. Adv Agron 162:89–142

Fan X, Bi Y, Zhang Y, Jeffers D, Yin X, Kang M (2018) Improving breeding efficiency of a hybrid maize breeding program using a three heterotic-group classification. Agron J 110(4):1209–1216

Farooqi MQU, Nawaz G, Wani SH, Choudhary JR, Rana M, Sah RP et al (2022) Recent developments in multi-omics and breeding strategies for abiotic stress tolerance in maize (*Zea mays* L.). Front Plant Sci 13:965878

Franco JG, Berti MT, Grabber JH, Hendrickson JR, Nieman CC, Pinto P, Van Tassel D, Picasso VD (2021) Ecological intensification of food production by integrating forages. Agronomy 11(12):2580. https://doi.org/10.3390/agronomy11122580

Gelley CH, Ashworth AJ, Keyser PD, Nave RL, Rhinehart JD (2020) Water-use efficiency of forage crops in the South-Eastern United States. Agronomy 10(9):1377

Habermann E, Dias de Oliveira EA, Contin DR, Delvecchio G, Viciedo DO, de Moraes MA, Prado RM, Costa KAP, Braga MR, Martinez CA (2019) Warming and water deficit impact leaf photosynthesis and decrease forage quality and digestibility of a C4 tropical grass. Physiol Plant 165(2):383–402

Handiso MA, Hemacho AH, Bongido BL, Anjulo MM (2023) Effect of conservation tillage on soil and water conservation, yield and yield components of maize (*Zea mays* L.) in south Ari District, Southern Ethiopia. Discover Agric 1(1):10

Haq SIU, Zheng D, Feng N, Jiang X, Qiao F, He JS, Qiu QS (2022) Progresses of CRISPR/Cas9 genome editing in forage crops. J Plant Physiol 279:153860

He B, Pan S, Zhao J, Zou X, Liu X, Wu S (2024) Maize improvement based on modern breeding strategies: Progress and perspective. ACS Agric Sci Technol 4(3):274–282

Hossain F, Muthusamy V, Bhat JS, Zunjare RU, Kumar S, Prakash NR, Mehta BK (2022) Maize breeding. In: Fundamentals of field crop breeding. Springer Nature Singapore, Singapore, pp 221–258

Iqbal MA, Hamid A, Ahmad T, Siddiqui MH, Hussain I, Ali S, Ali A, Ahmad Z (2018) Forage sorghum-legumes intercropping: effect on growth, yields, nutritional quality and economic returns. Bragantia 78(1):82–95

Jalloh AA, Mutyambai DM, Yusuf AA, Subramanian S, Khamis F (2024) Maize edible-legumes intercropping systems for enhancing agrobiodiversity and belowground ecosystem services. Sci Rep 14(1):14355

Javanmard A, Machiani MA, Lithourgidis A, Morshedloo MR, Ostadi A (2020) Intercropping of maize with legumes: a cleaner strategy for improving the quantity and quality of forage. Cleaner Eng Technol 1:100003

Jiang Y, Sun K, An X (2022) CRISPR/Cas system: applications and prospects for maize improvement. ACS Agric Sci Technol 2(2):174–183

Jiang C, You Y, Lai X, Zhang Z, Gao W, Ma R, Yang X (2024) Maximizing food equivalent unit yield for forage maize production without notably compromising dry matter yield and feed quality in a semi-arid region. Ind Crop Prod 218:118942

Karnatam KS, Mythri B, Un Nisa W, Sharma H, Meena TK, Rana P, Vikal Y, Gowda M, Dhillon BS, Sandhu S (2023) Silage maize as a potent candidate for sustainable animal husbandry development perspectives and strategies for genetic enhancement. Front Genet 14:1150132

Katoch R (2022) Biotechnological techniques for nutritional quality improvement in forages. In: Nutritional quality Management of Forages in the Himalayan region. Springer Singapore, Singapore, pp 465–492

Kermah M, Franke AC, Adjei-Nsiah S, Ahiabor BD, Abaidoo RC, Giller KE (2017) Maize-grain legume intercropping for enhanced resource use efficiency and crop productivity in the Guinea savanna of northern Ghana. Field Crop Res 213:38–50

Koenig KM, Li C, Hunt DE, Beauchemin KA, Bittman S (2023) Effects of sustainable agronomic intensification in a forage production system of perennial grass and silage corn on nutritive value and predicted milk production of dairy cattle. J Dairy Sci 106(1):274–293

Kumar S, Singh P, Devi U, Yathish KR, Saujanya PL, Kumar R, Mahanta SK (2023) An overview of the current fodder scenario and the potential for improving fodder productivity through genetic interventions in India. Anim Nutr Feed Technol 23(3):631–644

Lamptey S, Yeboah S, Li L (2018) Response of maize forage yield and quality to nitrogen fertilization and harvest time in semi-arid Northwest China. Asian J Res Agric For 1:1–10

Ma R, Jiang C, Shou N, Gao W, Yang X (2023) An optimized nitrogen application rate and basal topdressing ratio improves yield, quality, and water-and N-use efficiencies for forage maize (*Zea mays* L.). Agronomy 13(1):181

Maitra S, Shankar T, Banerjee P (2020) Potential and advantages of maize legume intercropping system. In: Maize-production and use, pp 1–14

Mattoo AK, Cavigelli MA, Mišić DM, Gašić U, Maksimović VM, Kramer M, Kaur B, Matekalo D, Zivkovic JN, Roberts DP (2023) Maize metabolomics in relation to cropping system and growing year. Front Sustain Food Syst 7:1130089

McCluskey C, Tracy WF (2021) Engaging farmer stakeholders: maize producers' perceptions of and strategies for managing on-farm genetic diversity in the upper Midwest. Sustainability 13(16):8843. https://doi.org/10.3390/su13168843

McFarland FL, Kaeppler HF (2024) History and current status of embryogenic culture-based tissue culture, transformation and gene editing of maize (*Zea mays* L.). Plant Genome e20451

Mehta BK, Kumar S, Chauhan S, Ashiwal C, Asati R, Purwar M, Bansal S, Shashikumara P, Meena SK, Kumar A, Ahmed S (2025) Forage maize as source of cellulosic biofuel production. In: Forage crops in the bioenergy revolution: from fields to fuel. Springer Nature Singapore, Singapore, pp 195–214

Momesso L, Crusciol CA, Soratto RP, Vyn TJ, Tanaka KS, Costa CH, Neto JF, Cantarella H (2019) Impacts of nitrogen management on no-till maize production following forage cover crops. Agron J 111(2):639–649

Namata MJ, Xu J, Habyarimana E, Palakolanu SR, Wang L, Li J (2025) Genome editing in maize and sorghum: a comprehensive review of CRISPR/Cas9 and emerging technologies. Plant Genome 18(2):e70038

Ndudzo A, Makuvise AS, Moyo S, Bobo ED (2024) CRISPR-Cas9 genome editing in crop breeding for climate change resilience: implications for smallholder farmers in Africa. J Agric Food Res:101132

Okoye CO, Wang Y, Gao L, Wu Y, Li X, Sun J, Jiang J (2023) The performance of lactic acid bacteria in silage production: a review of modern biotechnology for silage improvement. Microbiol Res 266:127212

Pierre JF, Latournerie Moreno L, Garruña R, Jacobsen KL, Laboski CAM, Us Santamaría R, Ruiz-Sánchez E (2022) Effect of maize–legume intercropping on maize physio-agronomic parameters and beneficial insect abundance. Sustainability 14(19):12385

Prasanna BM, Cairns JE, Zaidi PH, Beyene Y, Makumbi D, Gowda M et al (2021) Beat the stress: breeding for climate resilience in maize for the tropical rainfed environments. Theor Appl Genet 134(6):1729–1752

Qi M, Liu X, Li Y, Song H, Yin Z, Zhang F, He Q, Xu Z, Zhou G (2021) Photosynthetic resistance and resilience under drought, flooding and rewatering in maize plants. Photosynth Res 148:1–15

Rad RD, Sharifabad HH, Torabi M, Azizinejad R, Salemi H, Heidari Soltanabadi M (2021) Impact of drought stress on biochemical responses, energy, and water productivity on maize forage (*Zea mays* L.). SN Appl Sci 3:1–11

Saini H, Thakur R, Gill R, Tyagi K, Goswami M (2023) CRISPR/Cas9-gene editing approaches in plant breeding. GM Crops Food 14(1):1–17

Santiago López U, Rosales Nieto CA, Santiago López E, Santiago López N, Preciado Rangel P, Palmo Gil A, Real D (2018) Yield of forage, grain and biomass in eight hybrids of maize with different sowing dates and environmental conditions. Revista Mexicana de Ciencias Pecuarias 9(1):86–104

Sargent D, Conaty WC, Tissue DT, Sharwood RE (2022) Synthetic biology and opportunities within agricultural crops. J Sustain Agric Environ 1(2):89–107

Saritha R, Visalakshi M (2024) Legume based profitable intercropping system for Management of Fall Armyworm in maize. Legum Res 47(9):1583–1587

Singh RK, Prasad A, Muthamilarasan M, Parida SK, Prasad M (2020) Breeding and biotechnological interventions for trait improvement: status and prospects. Planta 252:1–18

Singh VK, Pundir S, Chaturvedi D, Kaur A, Pandey A, Mandal S et al (2023) Enhancing maize (*Zea mays* L.) crop through advanced techniques: a comprehensive approach. In: New prospects of maize. IntechOpen

Sowiński J (2024) Intercropping maize (*Zea mays* L.) and field beans (*Vicia faba* L.) for forage, increases protein production. Sci Rep 14(1):16419

Talla VVR, Sandal SS, Walia P, Vidya A (2023) Quality enhancement in forage crops of cereals and legumes. Int J Plant Soil Sci 35(16):166–179

Taube F, Vogeler I, Kluß C, Herrmann A, Hasler M, Rath J, Loges R, Malisch CS (2020) Yield progress in forage maize in NW Europe - breeding progress or climate change effects? Front Plant Sci 11:1214

Teodoro PE, Teodoro LP, Baio FH, Silva Junior CA, Santana DC, Bhering LL (2024) High-throughput phenotyping in maize and soybean genotypes using vegetation indices and computational intelligence. Plant Methods 20(1):164

Toker P, Canci H, Turhan I, Isci A, Scherzinger M, Kordrostami M, Yol E (2024) The advantages of intercropping to improve productivity in food and forage production – a review. Plant Prod Sci 27(3):155–169

Vanhevel Y, De Moor A, Muylle H, Vanholme R, Boerjan W (2024) Breeding for improved digestibility and processing of lignocellulosic biomass in *Zea mays*. Front Plant Sci 15:1419796

Varotto S (2023) Current status and future prospective of genome editing application in maize. In: A roadmap for plant genome editing. Springer Nature, Cham, pp 165–182

Wan X, Wu S, Wei X, McConnell LL (2024) Plant biotechnology and molecular breeding for global food security. ACS Agric Sci Technol 4(5):521–523

Wang H, Li K, Hu X, Liu Z, Wu Y, Huang C (2016) Genome-wide association analysis of forage quality in maize mature stalk. BMC Plant Biol 6:1–12

Wang Y, Tang Q, Pu L, Zhang H, Li X (2022) CRISPR-Cas technology opens a new era for the creation of novel maize germplasm. Front Plant Sci 13:1049803

Wani SH, Vijayan R, Choudhary M, Kumar A, Zaid A, Singh V, Kumar P, Yasin JK (2021) Nitrogen use efficiency (NUE): elucidated mechanisms, mapped genes and gene networks in maize (*Zea mays* L.). Physiol Mol Biol Plants 27(12):2875–2891

Washburn JD, Burch MB, Franco JAV (2020) Predictive breeding for maize: making use of molecular phenotypes, machine learning, and physiological crop models. Crop Sci 60(2):622–638

Wu L, Weston LA, Zhu S, Zhou X (2023) Rhizosphere interactions: root exudates and the rhizosphere microbiome. Front Plant Sci 14:1281010

Yadav OP, Hossain F, Karjagi CG, Kumar B, Zaidi PH, Jat SL, Chawla JS, Kaul J, Hooda KS, Kumar P, Yadava P, Dhillon BS (2015) Genetic improvement of maize in India: retrospect and prospects. Agric Res 4:325–338

Young J, Zastrow-Hayes G, Deschamps S, Svitashev S, Zaremba M, Acharya A et al (2019) CRISPR-Cas9 editing in maize: systematic evaluation of off-target activity and its relevance in crop improvement. Sci Rep 9(1):6729

Zhang H, Shi W, Ali S, Chang S, Jia Q, Hou F (2022a) Legume/maize intercropping and N application for improved yield, quality, water and n utilization for forage production. Agronomy 12(8):1777

Zhang Z, Wei J, Li J, Jia Y, Wang W, Li J, Lei Z, Gao M (2022b) The impact of climate change on maize production: empirical findings and implications for sustainable agricultural development. Front Environ Sci 10:954940

Zhu Z, Li J, Zhu D, Gao Z (2024) The impact of maize canopy on splash erosion risk on soils with different textures under sprinkler irrigation. Catena 234:107608

Genome Editing Tools for Improving the Yield of Forage Crops

Varsha Gupta, Baishnab Charan Tripathy, and Neelima Gupta

1 Introduction

Forage crops refer to herbaceous legumes (Fabaceae) or grasses (Poaceae) grown in tropical grasslands and deserts (Muir et al. 2011). In tropical regions, popular grasses are *Brachiaria*, *Panicum* species, and Napier grass (*Pennisetum purpureum*). Livestock production is critically essential for smallholders' livelihoods, particularly in the developing parts of the world, such as sub-Saharan Africa, where women frequently tend to the livestock production systems (Njuki and Sanginga 2013). Countries grow different crops for their livestock, depending on their climate and the animals' needs. In temperate environments, the most common grasses for grazing include fescue, bentgrass, orchard grass, and ryegrass. For example, the *tabular festulolium* is a crossbreed of *Festuca* and *Lolium* that has been grown since the 1970s. The unique blend combines the unmatched forage excellence of *Lolium multiflorum* with the ability to endure stress and persist over time of *Festuca arundinacea* (Ghesquière et al. 2010). Varity of maize have been specifically developed for livestock feed. Legumes such as clover, medics, trefoil, and vetches are also widely cultivated. Among brassica forage species, kale and oilseed rape are commonly used, while beet is another important temperate forage crop. The most extensively cultivated perennial legume is lucerne (alfalfa), which can be grown either as a sole crop or in combination with tropical and temperate grasses.

V. Gupta (✉)
Department of Life Sciences, Chhatrapati Shahu Ji Maharaj University, Kanpur, Uttar Pradesh, India
e-mail: varshagupta@csjmu.ac.in

B. C. Tripathy
Biotechnology, Sharda University, Noida, Uttar Pradesh, India

N. Gupta
Hari Singh Gaur (A Central) University, Sagar, Madhya Pradesh, India

S. H. Wani, S. S. Gosal (eds.), *Accelerated Plant Breeding, Volume 5*,
https://doi.org/10.1007/978-3-032-04527-0_12

According to FAOSTAT data from 2017, livestock production has witnessed an astounding increase of more than 200% over the past 54 years. The United Nations predicts that by 2050, the world's human population will surpass 9 billion, resulting in a rise in the consumption of animal-based food items. However, maintaining a balance between cultivating field crops and grain for human consumption and forage crops for animal feed in such a scenario can be challenging. Cereal crops have been utilized as animal feed for many years due to their high dry matter production. However, due to their low protein content, they are not considered high-quality forage sources. Forage legume crops are known to provide superior-quality feed and fodder for livestock. Crop rotation and multi cropping is well established concept, it was found that when cereal crops are sown along with legume crops, it can lead to an improvement in yield, protein content, and other quality parameters (Zhang et al. 2015). Legume crops can fix atmospheric nitrogen (N) for usage by the plant by utilizing a symbiotic relationship between the plant and *Rhizobium*. This makes them an invaluable asset for producing forage with adequate nutritional value. Legume crops have a long-standing history of providing forage and feed for animals, with a wide range of as many as 60 varieties cultivated for this purpose. In this book chapter, we will discuss in detail the role of various novel technologies for improving the yield of forage crops.

2 Forage Grasses and Legumes

Grasses, which are a part of the Poaceae family, are the most important plant family. Grasses are a vital plant family, with economic and ecological significance (Lee et al. 2020; Huang et al. 2022). Grasslands, occupying nearly a quarter of the surface of the earth, serve as a source of food, housing, and building materials for humans and animals. They are also valuable carbon dioxide sinks and are critical in oxygen production and carbon storage (Lopez et al. 2022; Terrer et al. 2021). Grasses are more robust to dry and warm settings than trees, making grasslands a better carbon sink than forests in future climatic conditions (Dass et al. 2018). Grasses are required to improve their yield and resistance to stresses caused by climate change (Giridhar and Samireddypalle 2015). Plants face disparities in water availability, temperature, and soil composition, leading to stress conditions (Cushman et al. 2022). Genome editing techniques can provide stress resistance to plants (Bailey-Serres et al. 2019; Dhakate et al. 2022). Despite their difficult reproductive and genetic characteristics, efforts are made to use genome editing and genetic tools in forage grasses (Liu et al. 2020; Wang et al. 2022), which are a diverse group due to their high genetic variability. They can be classified based on their life cycle and ecotype. C4 plants are hot-season grasses, while C3 plants are cold-season grasses (Moser and Hoveland 1996; Moser et al. 2004).

Legumes are important food crops with high agricultural value due to their nutrient content, secondary metabolites, and ability to fix nitrogen (Le et al. 2007). Certain plant organic compounds, such as phytoalexins and chitinases, help defend against pests and pathogens (He and Dixon 2000). The Fabaceae plant family is well-studied and important in agriculture due to its capability to fix nitrogen through rhizobia (Doyle and Luckow 2003). While alfalfa remains the most popular and widely grown legume crop for forage and feed in many countries, farmers across different regions have varying preferences for legume crops that provide quality forage. For instance, cowpea is a nutritious fodder commonly used in sub-Saharan Africa and India. In these regions, grains are required as food, while haulms are given to livestock as nutritious fodder. In addition to commonly cultivated legume crops like soybeans and peanuts, other lesser-known legume crops, such as white clover, Austrian winter pea, and sweet clover, have also gained popularity in different regions in the United States as a source of quality forage. These crops provide high protein content, improve soil fertility, and reduce the need for synthetic fertilizers.

As the effects of climate change continue to unfold, the agricultural sector faces increasing challenges. It is crucial to evaluate the qualities of various fodder crops, such as the feasibility of sustainable growth and consistent yields, along with their digestibility and animal performance. It is important to identify and prioritize fodder crops which are resilient to the impacts of climate change while also providing high-quality forage. By conducting such assessments, we can ensure a diverse selection of these crops that adapt to changing climate conditions and contribute to the sustainable production of animal feed and forage. This step is crucial in developing a more resilient agricultural system capable of enduring the impacts of climate change.

Alfalfa is ranked as the fourth most widely grown and economically valuable crop, following corn, soybean, and wheat (Rashmi et al. 1997; Samac and Temple 2004). *Alfalfa is an important perennial legume widely grown to feed livestock and produce biofuel feedstock. The beauty of this crop is that it requires almost negligible nitrogen fertilizer for ideal growth while it also provides nitrogen to subsequent crops* (Radovic et al. 2009; Triboi and Triboi-Blondel 2014). *Alfalfa is a good source of proteins, minerals, enzymes, antioxidants, and vitamins A, C, K, and E, along with important compounds with pharmaceutical properties* (Bora and Sharma 2011). *It also contains two thiol redox compounds, homoglutathione (hGSH) and glutathione (GSH)* (Klapheck 1988; Baldacci-Cresp et al. 2012). *Alfalfa is a versatile plant that exhibits different ratios of hGSH/GSH in various parts like leaves, stems, and roots* (Pasternak et al. 2014). *This makes it a valuable and essential forage crop, which can also serve as a cover crop in grasslands to improve weed control. Moreover, alfalfa is suitable for producing recombinant pharmaceutical proteins* (Fu et al. 2015) *and is important in phytoremediation* (Nirola et al. 2016). *MAlfalfa belongings to edicago* genus, is a highly coveted and extensively cultivated autotetraploid, the perennial leguminous crop with excellent quality forage and high nutritional value. *Alfalfa*'s genetic makeup is primarily derived from two subspecies: subsp. sativa and subsp. *varia*, and subsp. *falcata* is also utilized either directly or in combination with other subspecies. In addition, *alfalfa*'s genetic versatility allows for the creation of many other species of the genus, particularly species of the perennial genus, through successful genetic hybridization.

Soybean, a plant that belongs to the legume family, is an annual crop cultivated for centuries for its high-quality oil content and seed protein. Initially, it was primarily grown as a forage crop in North East Asia and was later introduced in the mid-1800s to the United States. However, with the increasing demand for soybean meal and oil, its usage shifted to grain production in 1941. Recently, there has been a strong interest in soybeans as a forage crop due to their several advantages. For instance, as a forage crop, soybean has high protein content at the stage of seed-filling, making them an excellent livestock nutrition source. Additionally, it has a wide choice of growth stages suited for easy harvesting, making it a versatile crop. Moreover, soybeans are an efficient cover crop that reduces soil erosion. Lastly, soybeans are widely adaptable to different climatic zones, making them an ideal crop for farmers worldwide.

Cowpea is an annual legume crop that holds immense significance in the diet of millions of people residing in subtropical and tropical regions. This crop, originally from sub-Saharan Africa, is now grown in more than 100 countries worldwide. Cowpeas are an excellent choice for those seeking a nutritious legume, as they are known to contain substantial amounts of protein and energy. Due to its nutrient-rich profile, it is considered an ideal feed option for livestock, particularly during the winter and dry seasons when forage is scarce. Livestock farmers and animal feed manufacturers have found cowpeas to be a valuable addition to the diet of their animals. This is because cowpeas contain essential nutrients that can help promote the health and productivity of livestock. As a result, cowpeas have become a popular choice among those in the industry who are looking to maintain the well-being of their animals and ensure optimal production levels. Its high nutritional value is due to its impressive mineral and vitamin content, including iron, magnesium, zinc, and vitamins A, B, and C. Additionally, cowpea is a versatile crop that can be used in various forms, including as fresh or dry fodder, silage, or hay, making it a valuable resource for farmers and ranchers. Moreover, cowpea has the potential to become a valuable forage and fodder crop in the future as it can thrive in regions with sandy and less fertile soils and low rainfall. The use of cowpea as fodder is prevalent in Asia, particularly in India, where the green cowpea plants are either grazed or cut and combined with dry cereals to prepare animal feed. In mixed cropping systems, cowpea haulms are highly desirable as fodder since grain and fodder can be obtained from a single crop. With its adaptability to various soil types and climatic conditions, cowpea is a versatile legume crop gaining popularity worldwide.

3 Factors Affecting Productivity

Drought: Forage crops are highly dependent on water availability. Drought conditions can significantly reduce yields and affect the quality of forage.

Excessive rainfall: On the other hand, excessive rainfall can lead to waterlogging, affecting root health and nutrient absorption.

3.1 *Pests and Diseases*

Insects: Various pests, such as grasshoppers, armyworms, and aphids, can damage forage crops by feeding on leaves or roots.

Diseases: Fungal, bacterial, and viral diseases can impact forage crops, reducing yield and quality.

Nutrient depletion: Continuous cultivation of forage crops without proper nutrient management can deplete the soil of essential nutrients, impacting future yields.

Soil erosion: Erosion can lead to the loss of topsoil, affecting the structure and fertility of the soil.

Seed quality: Poor quality or contaminated forage seeds can lead to suboptimal germination and crop performance.

Seed availability: Limited availability of quality forage seeds, especially for specific varieties, can constrain farmers.

Overgrazing: Improper grazing management can lead to overgrazing, reducing the overall productivity of forage crops.

Harvest timing: Timing is crucial when harvesting forage crops. Harvesting too early or too late can affect both yield and nutritional quality.

Market demand and economic factors: Market fluctuations—Changes in market demand and prices for forage crops can impact the economic viability of forage production.

Input costs: Rising inputs, such as fertilizers and fuel, can affect the profitability of forage farming.

Extreme weather events: Augmented frequency and intensity of severe weather events can pose additional challenges for forage crop production.

Technology adoption: Limited technology adoption—Some farmers may face challenges in adopting modern agricultural technologies, such as precision farming or improved forage varieties. Addressing these challenges requires research, extension services, and effective farm management practices to ensure sustainable forage production.

3.2 *Abiotic Stresses*

Climate change causes extreme temperatures, unusual precipitation patterns, and soil degradation, impacting agriculture. Plants respond to stress by adjusting through acclimation or adaptation, involving physiological, metabolic, physical, and genetic changes. Acclimation helps plants respond quickly to ensure their survival, while adaptation is a lengthy process to preserve the population through evolution. Plants undergo epigenetic modifications in response to stressful events (Guarino et al. 2022). Nature never ceases to amaze us! Organisms have the immense capability to adapt and evolve in response to new and extreme conditions. This results in a stunning display of morphological, metabolic, and physiological changes that allow them to thrive where others may fail. It's genuinely fascinating to see how creative and resilient life can be! (He et al. 2018).

Plants face multiple abiotic stresses, such as salinity, drought, heat, and cold, which can profoundly impact their growth and development. The extent of this impact is directly related to crop yield and productivity. Although we will discuss each of these stresses separately in this chapter, they are frequently interconnected in nature, resulting in more significant effects than when they occur individually. As a result, plants must undergo multiple stressors to acclimate and survive.

3.2.1 Temperature Conditions

Climate change alters temperature conditions, affecting plant growth and crop yield, which in turn impacts food security. Temperature variations can cause two types of abiotic stress: heat and cold stress.

3.2.2 Heat Stress

Climate change is causing steady temperature increases, leading to heat waves worldwide. This causes heat stress for plants, which tend to react and survive through heat-avoidance or heat-tolerance mechanisms (Jagadish et al. 2021; Aleem et al. 2020). Plants have avoidance processes and heat-tolerance mechanisms to survive heat stress. Heat sensors are activated when the plasma membrane becomes more fluid and permeable. These sensors interact with calcium channels, which activate stress response mechanisms involving different enzymes and proteins (Sustek-Sanchez et al. 2023; Xu et al. 2020).

Plant cells have various ways to cope with heat stress. Calcium transport-related transmembrane proteins such as Annexin, Synaptotagmin A, and Cyclic Nucleotide-Gated Channels (CNGCs) are believed to act as heat sensors in plants. These proteins regulate the entry of Ca^{2+} from the apoplast to the cytosol and help accumulate heat-shock proteins (HSPs). HSPs prevent protein misfolding and aggregation that may occur due to heat stress. Heat stress leads to oxidative stress, which enhances the formation of reactive oxygen species (ROS). Plants use antioxidant enzymes such as peroxidase and catalase to prevent damage from oxidative stress. Additionally, metabolic changes occur due to heat stress that affects enzyme activity, carotenoids, and chlorophyll, leading to reduced photosynthesis and carbon fixation (Wu et al. 2022; Hasanuzzaman et al. 2020).

3.2.3 Low-Temperature Tolerance and Winter Hardiness

Most forage species require vernalization, which involves exposing seeds or early seedlings to low temperatures to flower. However, prolonged winter can be problematic for forage grasses as surviving the winter is a complex process influenced by various factors such as low temperatures, frost, waterlogging, desiccation, ice encasement, and snow cover. The primary cause of winter damage is ice encasement. To survive freezing temperatures, plants need to respond to environmental cues. Dehydrins (DHNs) are a class of proteins that help plants withstand harsh environmental situations such as drought, high salinity, low temperatures, and physical damage. DHNs align with plasma membrane intrinsic proteins to safeguard cellular membranes and DNA from harm. The ICE-CBF-COR signaling pathway is responsible for regulating plant response to cold, which comprises the Inducer of CBF Expression (ICE), C-repeat Binding Factors (CBFs), and the Cold-Regulated genes (CORs). During the winter season, DHNs and Group 2 Late Embryogenesis Abundant (LEA) proteins play a crucial role in the survival of plants. The CBF cold-responsive pathways regulate DHNs. CBF/DREB (CBFs, also known as DREB1 genes) are transcription factors that attach to DRE/CRT elements in COR genes' promoters. In *Elymus nutans* Griseb, genes that encode LEA14-A, COR14a, and dehydrin showed augmented transcriptional levels in the genotype with superior cold tolerance. The QTLs were identified in meadow fescue for winter survival, frost, and drought tolerance, which are co-located with CBFs and DHNs. The CBF6 gene is located in an important frost tolerance/winter survival QTL.

During cold acclimation, perennial grasses create fructans and raffinose family oligosaccharides. These non-structural storage carbohydrates are synthesized from sucrose and stored within vacuoles. Fructans are accumulated as a replacement for starch or in addition to it and are crucial for the plant's survival during winter stress and tissue regeneration in spring. The enzymatic action of fructosyltransferases, invertases, and fructan exohydrolases facilitates fructan accumulation (Chalmers et al. 2005; Wu et al. 2021).

3.2.4 Drought

Drought harms plant productivity and worsens with climate change. Plants have developed various defense mechanisms to survive prolonged dry spells in semi-arid areas, such as antioxidant defense systems and intricate gene expression programs. While annual crops complete their reproductive cycle, perennial crops rely on drought tolerance and avoidance strategies. Recent studies suggest that seed development pathways are linked to vegetative desiccation tolerance. Understanding this mechanism could help engineer drought-tolerant forage crops. Water scarcity triggers responses in plants at various levels that cause a reduction in aboveground biomass growth and accumulation, resulting in a yield penalty. Although water deficit-adapted genotypes maintain growth, ABA-mediated stomatal closure can negatively impact photosynthetic activity and halt growth. Drought stress is a common problem faced by plants, which results in the enhanced generation of reactive oxygen species (ROS), causing damage to cellular membranes and oxidative stress. However, Abscisic acid shows an important role in inducing the protective function of Dehydrin proteins (DHNs). These DHNs interact with intrinsic proteins in the plasma membrane to safeguard the membranes and DNA from damage. Moreover, salinity, drought, and intensive light stress result in the accumulation of the 3′-phosphoadenosine 5′-phosphate (PAP), which regulates nuclear stress response genes. The concentration of PAP, which is regulated by phosphatase SAL1, breaks down PAP into Adenosine monophosphate (AMP) and, hence, reduces the PAP levels.

Alfalfa's reaction to drought stress was first studied in the 1990s. Identification of genes associated with drought stress adaptation can aid breeding programs. MicroRNAs and their target genes were implicated during low oxygen, wound formation, high temperature, or oxidative stress. miR156 overexpression enhances alfalfa's drought and heat stress tolerance by silencing Squamosa Promoter Binding Protein-Like 13 (SPL13).

3.2.5 Salinity

Soil salinization is an environmental stress that limits plant growth by preventing water uptake and the absorption of essential elements. Salt stress primarily causes ionic and osmotic stress in plant cells, leading to changes in phytohormone signaling, cytoskeleton dynamics, cell wall structure, photosynthesis, and sugar signaling (Zhao et al. 2021). The competition of Na^+ with K^+ for binding sites necessary for cellular functions causes a toxic effect (Tester and Davenport 2003).

Certain molecules, such as the hyper-osmolality-gated calcium-permeable (OSCA) channel family, can detect excess sodium ions (Na^+) and signal it (Shabala et al. 2015; She et al. 2022). Calcium-sensing proteins, known as calcineurin B-like proteins (CBLs), interact further with serine/threonine protein kinases (CIPKs) to maintain ion transport and homeostasis. They regulate sodium ion transport, reactive oxygen species (ROS), and abscisic acid (ABA) signaling. When plants are exposed to salt stress, it leads to an increase in cyclic nucleotides and ROS, which activate mitogen-activated protein kinase (MAPK) cascades. This, in turn, increases the synthesis of osmolytes, helping to alleviate salt-induced osmotic stress. Additionally, salt stress causes an accumulation of ABA, activating SnRK2s to transduce signals and induce stress-responsive genes. Non-halophytic plants are more sensitive to salinity stress, while halophytic plants have adapted to salinized environments. Therefore, identifying the genetic components that underlie salinity tolerance is crucial for plant breeders (She et al. 2022; Vaziriyeganeh et al. 2021).

When plants are exposed to high salinity levels, they experience hyperosmotic pressure that leads to ion toxicity, especially due to the presence of Na^+. This results in an increase in the levels of cytosolic Ca^{2+}, ROS, ABA, and MAPK signaling. To counter these adverse effects, plants accumulate water-soluble compounds, such as betaines, sugars, polyols, and amino acids. Glycine betaine (GB) is particularly effective in protecting plants under salt, drought, and low-temperature stress conditions. By introducing foreign genes and overexpressing stress-related genes, salt tolerance in alfalfa can be increased. For instance, SeNHX1 and GmDREB1 help improve stability and photosynthesis while also protecting against sodium toxicity. Similarly, AtNDPK2 expression and CODA cDNA targeting of alfalfa chloroplasts under SWPA2 promoter enhance the plant's tolerance to various stresses.

Salinity stress can have detrimental effects on plants, such as high pH damage, osmotic shock, and toxication stress. However, alfalfa can withstand saline-alkaline stress due to its inherent antioxidant capacity. In fact, transcriptome analysis of alfalfa roots treated with ABA revealed that ABA-regulated differentially expressed isoforms play a crucial role in hormone signal transduction, plant immunity, transcriptional regulation, and anti-oxidative defense.

4 Genome Editing

Genome editing tools represent a highly advanced technology that enables scientists to make precise and targeted changes to an organism's DNA. These tools have revolutionized molecular biology, genetics, and biotechnology by providing a means to edit genes with exceptional accuracy. Notably, genome editing in crops holds significant promise for accelerating the development of new varieties with improved traits, ultimately contributing to global food security and sustainable agriculture. The potential benefits of genome editing in crop development are vast. This technique can significantly enhance crop yields, bolster drought resistance, and fortify nutrient content. Most importantly, it can be harnessed to develop crops that are remarkably resilient to pests and diseases, thereby decreasing the excessive requirement for damaging pesticides and herbicides. As such, genome editing has the potential to be a game-changer for the agricultural industry. Improved crop traits can lead to sustainable and efficient farming, benefiting all (Table 1). However, deploying these technologies is an important matter for regulatory considerations and public acceptance. Ethical and safety considerations are crucial in the responsible use of genome editing tools in agriculture.

4.1 CRISPR-Cas9

CRISPR-Cas9, one of the most extensively employed genome editing tool. A ribonucleoprotein complex consisting of a guide RNA and Cas9 enzyme targets specific DNA sequences. Cas9 produces double strand breaks at the designated site, and upon the cell's repair of the break, modifications or edits may be introduced to the genetic code. This technology enables the precise and targeted modification of an organism's DNA, thereby revolutionizing the fields of molecular biology, genetics, and biotechnology. Its high accuracy in gene editing has made it a powerful tool for scientific research. CRISPR-Cas9 is widely used in molecular biology, genetics, and biotechnology. This tool is derived from the bacterial immune system and involves an enzyme called Cas9 which can be programmed to target specific DNA sequences using a guide RNA. Once the targeted site is identified, Cas9 creates breaks in the DNA, and when the cell repairs the break, changes or edits can be

Table 1 Overview of genome editing tools and their potential applications

Genome editing tool	Purpose	Mode	Applications
CRISPR-Cas9	Crop improvement/stress tolerance	Drought tolerance, improved nutritional content, and disease resistance associated with gene editing	CRISPR is used to develop crops with increased resistance to pests and diseases, such as wheat resistant to powdery mildew and rice resistant to bacterial blight
TALENs (transcription activator-like effector nucleases)	Targeted genome editing in various crops	Specific modifications in genes associated with traits like herbicide resistance and improved yield	Create crops with enhanced resistance to specific herbicides, promoting effective weed control
ZFNs (zinc finger nucleases)	Crop improvement projects	Targeted changes in genes related to disease resistance and nutritional quality	Resistance of crops like maize to specific pathogens
Base editing	Allows converting a single DNA base pair to another without inducing double-strand breaks	Useful for introducing single-point mutations in genes associated with specific traits	Modify genes related to crop yield, nutritional content, and stress tolerance
CRISPR-Cas12 and CRISPR-Cas13	Employed for both DNA and RNA editing	Cas12 has been used for precise DNA editing, while Cas13 has been explored for RNA interference and virus resistance	Cas12 has been utilized for targeted modifications in genes associated with disease resistance in crops like tomatoes
Multiplex genome editing	Simultaneously editing multiple genes to achieve complex trait modifications	Improve traits like drought tolerance by targeting multiple water stress genes implicated in the plant	Crop improvement and stress tolerance
Gene silencing and RNA interference	Gene function is not manifested	To downregulate specific genes, leading to desirable traits such as reduced susceptibility to pests or improved shelf life	RNA interference has been utilized to develop crops that exhibit increased resistance to insect pests affecting yield

introduced to the genetic code. This technology enables scientists to modify an organism's DNA with high accuracy, allowing them to make precise and targeted changes to the genetic code. Overall, CRISPR-Cas9 has revolutionized the field of genetic research by providing a powerful tool for editing genes (Zhu et al. 2020; Wani et al. 2022; Bansal and Wani 2022; Tek et al. 2022; Wani and Hensel 2022; Choudhary et al. 2023; Wolabu et al. 2024; Fan et al. 2025).

4.2 TALENs

TALENs, short for Transcription Activator-Like Effector Nucleases, are a group of recombinant enzymes that have gained significant attention as an effective genetic tool for targeted genome editing. These enzymes contain two functional domains: a DNA-binding domain derived from naturally occurring TAL effectors and a nuclease domain. TALENs are highly efficient in modifying and editing specific genetic targets by creating double-strand breaks at precise locations within a DNA sequence. Due to their binding specificity, TALENs can be tailored to target nearly any genomic locus, making them valuable in genetic engineering applications such as creating disease models, developing gene therapies, and producing transgenic organisms. Furthermore, TALEN technology's versatility has paved the way for precise gene editing, correction, and silencing in multiple organisms. This binding specificity allows TALENs to be customized to target almost any genomic locus, with huge prospects for genomic research and ultimate improvement. The capability to customize TALENs for specific sequences provides a promising approach for specific and efficient genome editing in various research contexts.

4.3 ZFNs (Zinc Finger Nucleases)

Zinc Finger Nucleases, or ZFNs, are proteins created for editing genomes. They have two parts: the Zinc Finger DNA-binding and the nuclease domains. The Zinc Finger domain comprises small parts capable of recognizing and binding to unique and specific DNA sequences. The nuclease domain can break the DNA molecule at the site where the Zinc Finger domain is attached. ZFNs work by taking advantage of the natural way cells repair double-strand breaks in the DNA. The cell's repair machinery is activated when a double-strand break happens; repair can be done by either homology-directed repair (HDR) or non-homologous end joining (NHEJ). NHEJ links the broken ends together, often resulting in minor site changes. HDR uses a donor DNA molecule to make specific changes at the break site. ZFNs help make targeted genetic changes in many organisms, such as bacteria, yeast, plants, and animals.

4.4 Prime Editing

Prime editing is a more recent genome editing tool that allows the direct and precise editing of DNA without creating double-strand breaks. This technique uses a modified Cas9 enzyme and an engineered RNA guide specifying the desired edit. The process involves the synthesis of new DNA strands to replace the targeted sequence without causing breaks in the DNA.

4.5 *CRISPR-Cas12 and CRISPR-Cas13*

Similar to CRISPR-Cas9, Cas12 and Cas13 are other CRISPR-associated proteins with genome editing potential. Cas12 has DNA cleavage activity, while Cas13 targets RNA. These systems provide additional options for genome editing and/or RNA interference (Fig. 1).

5 Homologous Recombination (HR)

Homologous recombination is a natural cellular process involved in DNA repair. In genome editing, researchers can use HR to introduce specific changes by providing a DNA template with the desired sequence. This method is often used for precise gene insertion or replacement.

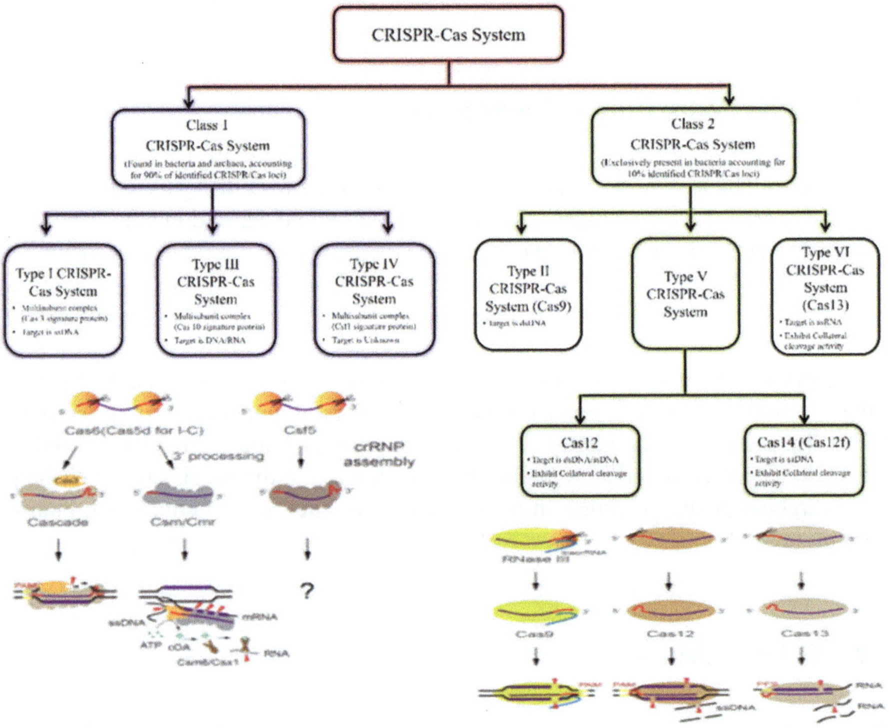

Fig. 1 The figure shows classification of different Cas variants with different Cas enzymes from different bacteria, adopted for multiple usage. Most reliable and validated are type II, type V & type VI. Cas12b (single crRNA-guided cis), Cas13a (trans-non-target nucleic acid), CasRx, and Cas14 are endonucleases while CasΦ is DNA cutter. Cas7-11 CRISPR/Cas systems are new CRISPR Class 1 effector resulting from fusion of Cas11 and Cas7 units whose target is the RNA:Spacer duplex without any collateral activity. (Figure adapted from https://encyclopedia.pub/entry/23773 and https://doi.org/10.3389/fmicb.2019.02471)

Genome editing tools have diverse applications, including functional genomics research, developing genetically modified organisms (GMOs), gene therapy, and studying genetic diseases (Table 1). Genome editing tools can precisely modify plant genetic material to improve crops. However, ethical and regulatory considerations must be considered for human germline editing and the environmental release of genetically modified organisms. These tools can enhance plant traits like disease resistance, stress tolerance, nutritional content, and yield. Here are some genome editing tools commonly employed for crop improvement:

Perennial grasses grown for feed in temperate climate zones often suffer from unfavorable environmental conditions. Genome editing using the CRISPR-Cas system can develop abiotic stress-resistant plants with precision and low off-target frequency mutations.

6 Genome Editing in Forage Crops: Prospects and Challenges

CRISPR-Cas9 is a revolutionary genome editing tool that has been effectively used to improve model and crop plants. However, forage plant improvements have not fully embraced this technology, despite its benefits. There is an opportunity to explore CRISPR-Cas9 to improve forage grasses and, ultimately, agriculture and livestock industry (Amombo et al. 2025). The EUGENIUS database in Europe has information on genetically modified organisms, including the green foxtail Line 193–31, which was modified using CRISPR-Cas9 to delay flowering. More details can be found in the GE *Setaria viridis* molecular characterization report in the EUGENIUS database.

Data shows that only a few of the members of the Poaceae family have been targeted using CRISPR-Cas tools for editing their genomes. These include grasses, all acclimatized for temperate regions. The model plant for C4 grasses, *S. viridis*, was subjected to genome editing three times. Most of the work has involved using CRISPR-Cas9 to remove a single gene (Table 2). The CRISPR-Cas9 system is highly efficient in generating knockouts by carrying out simple mutations (indels), resulting in an altered reading frame of the coding region. This system is mostly used for functional genomics and comprises two essential components – the Cas9 nuclease obtained from *Streptococcus pyogenes* and the short guide RNA (gRNA), which can target the DNA sequence of interest. By designing the precise gRNA, researchers could simultaneously mutate all gene alleles in a polyploid plant. This strategy has shown success in *Lolium arundinaceum* (allohexaploid) and *Panicum virgatum* L. (tetraploid) (Table 2) (Zhang et al. 2020).

Scientists have used the CRISPR-Cas9 system to specifically target and disable selected genes in forage grasses. These genes are linked to various traits such as flowering, tillering and branching, meiosis, and heat stress response, among others. The phytoene desaturase (PDS) gene was used as a marker to track the edited genes.

Table 2 Members of the Poaceae family who had been used for targeted mutagenesis with CRISPR-Cas9 technology for targeted mutagenesis. All these are adapted for temperate climates

Plant (common name)	Target for mutagenesis
Setaria viridis (green foxtail) (model plant for C4 grasses)	Floral organ number-2 (FON-2)
Setaria italic (foxtail millet)	Phytochrome C—PHYC Haploid induction (matrilineal-MTL)
Panicum virgatum (Switchgrass)	Teosinte branched 1 [tb1a & tb1b]) for tillering and branching
Lolium multiflorum (Italian ryegrass)	Meiosis
Lolium arundinaceum (tall fescue)	Heat stress response (17.9 kDa class II-HSP-17.9)

However, while researchers have reported on the success of the gene editing process, not much data is available for the phenotypic characteristics of the edited plants and how they perform in field trials. It is worth noting that different variants of the CRISPR-Cas system have also been used, such as CRISPR-Cas12a for *L. arundinaceum* and CRISPR-Cas9_Trex2 for *S. viridis* (Weiss et al. 2020; Zhang et al. 2021).

The CRISPR-Cas system has around 20 techniques that enable targeted modifications in the genome. Scientists have characterized various Cas enzymes, such as the Cas12a enzyme from the Lachnospiraceae bacterium. They have also modified the CRISPR-Cas9 system by co-expressing or fusing various proteins to the Cas9 nuclease. This technique allows for the inactivation of genes and the creation of loss-of-function mutants, as well as the obtaining of gain-of-function mutants. The Super Nova Tag (SunTag) system can potentiate transcriptional regulation. Scientists can modify a gene's promoter sequence, resulting in elevated expression levels. CRISPR-Cas9 is a highly effective gene-editing tool that can help scientists unlock the mysteries of the genome. An example of this tool in use can be seen with the overexpression of the ARGOS8 gene in maize, an auxin-regulated gene that plays an essential role in organ size 8. This modification has been shown to significantly enhance grain yield under drought-stress conditions (Shi et al. 2017). Knockout of the OsRR22 gene in rice via CRISPR-Cas9 resulted in salt tolerance without affecting agronomic traits in field trials. Genetic engineering can potentially improve yield and stress resistance (Zhang et al. 2019; Han et al. 2022).

Abiotic stress responses are complicated and involve multiple metabolic pathways. The genes responsible for these mechanisms are pleiotropic. Table 3 shows the identified genes related to abiotic stresses, which can be used to achieve stress tolerance.

Plants can detect changes in temperature in their surroundings through alterations in their cell membrane's structure and fluidity (Niu and Xiang 2018). When exposed to heat stress, the cell membranes become more fluid, activating pathways through heat sensors like CNGCs. Enhancing the expression of stress receptors results in a better response to stress. Overexpressing a heat sensor coding gene, such as the SYTA gene in *A. thaliana*, improved germination and seedling survival rates

Table 3 Genes implicated in various stress responses or sensing and the requirement of their upregulation or downregulation for stress tolerance

Plant	Gene	Role	Strategy for stress tolerance
Oryza sativa	OspsbA	Heat response	Upregulate
	OsCNGC14 OsCNGC16	Heat sensing	Upregulate
	OsPYL1/4/6	Heat response	Downregulate
	OsCOLD1	Cold sensing	Upregulate
	OsMYB30	Cold response	Downregulate
	OsSYT-5	Drought sensing	Upregulate
	OsDST	Drought response	Downregulate
	OsOSCA1.4	Salinity sensing	Upregulate
	OsbHLH024	Salinity response	Downregulate
Lolium arundinaceum	LaHsfA2c	Heat response	Upregulate
Solanum lycopersicum	SlMAPK3	Heat response	Downregulate
	SlPHYA SlPHYB1B2	Heat response	Downregulate
Elymus nutans	EnCOR410	Cold response	Upregulate
Agropyron cristatum	AcSnRK2.11	Cold response	Upregulate
	AcSnRK2.11	Drought response	Upregulate
	AcSnRK2.11	Salinity response	Upregulate
Arabidopsis thaliana	AtEGR2	Cold response	Downregulate
	AtCRPK1	Cold response	Downregulate
Cynodon dactylon	CdDHN4	Drought response	Upregulate
Triticum aestivum	TaSal1	Drought response	Downregulate
Hordeum vulgare	HvCBP20	Drought response	Downregulate
	HvITPK1	Salinity response	Downregulate
Zea mays	ZmDHN11	Salinity response	Upregulate

after heat stress exposure. The overexpression plants also exhibited enhanced expression of both HSPs and HSF while showing reduced membrane lipid peroxidation compared to non-overexpression lines. This suggests that increasing the expression of a sensor of heat stress can improve plant stress tolerance (Damaris and Yang 2021).

Another avenue for enhancing plant stress tolerance is through the upregulation of proteins implicated in the abiotic stress responses. Protein kinases, due to their ability to regulate posttranslational modifications of other proteins, are particularly suited for this task. They are crucial in the plant's response to abiotic and biotic stress.

Overexpressing SnRK2 genes (a kinase group specific to plants) in plants has improved their tolerance to abiotic stresses such as cold and drought. The overexpression of TaSnRK2.3, a wheat gene, in *Arabidopsis* plants resulted in better drought tolerance. Similarly, the overexpression of AcSnRK2.11, *Agropyron cristatum* gene, in *Nicotiana tabacum L.*, led to a significant increase in survival rates compared to wild-type plants following recovery periods from cold stress. These results suggest that the overexpression of these genes can be a potential strategy for enhancing plant stress tolerance. This can be achieved by upregulating stress-related genes like dehydrins.

To improve the capacity of plants to cope with non-living environmental factors, it is possible to suppress negative regulators of the stress response through genome editing. One promising target for downregulation is the class of enzymes that break down signaling molecules implicated in the stress response, such as inositol phosphatases. By lowering the activity of certain enzymes, the signaling molecules important in the stress response can build up, resulting in a more robust and effective stress response. This method can potentially increase the ability of crops and other plants to withstand various harsh environmental conditions, making them more resilient (Jia et al. 2019). The target phosphatase SAL1 could be selected for downregulation as it plays an important role in negatively regulating drought response (Chan et al. 2016). Using the CRISPR-Cas9 system, researchers have created TaSal1 knockout mutant wheat with less and smaller stomata (Table 4). "According to research, these plants tend to germinate and grow more efficiently in drought situations" (Abdallah et al. 2022). Another approach to creating plants that can withstand adverse environmental conditions is manipulating the expression of transcription factors associated with abiotic stress. Researchers showed that the transcription factor family of basic helix-loop-helix (bHLH) regulates abiotic stress responses in various plant species (Guo et al. 2021). It has been observed that the OsbHLH024 gene in rice has a negative effect on the plant's ability to tolerate salinity (Table 4). The plants that underwent mutation using the CRISPR-Cas9 system displayed better tolerance towards high salinity levels than the non-mutated plants.

Additionally, the plants that had their genes knocked out demonstrated a lower buildup of sodium ions and ROS (reactive oxygen species) but a higher level of potassium ions when compared to the control group of plants. Finally, the mutated plants exhibited an upregulated expression of genes responsible for encoding ion transporters in contrast to the non-mutated plants. (Alam et al. 2022). Downregulating homologs of the OsbHLH024 gene in grasses will effectively enhance their salinity stress tolerance, with the potential to significantly increase their survival and growth in saline soil conditions.

Recently, CRISPR-Cas has enabled targeted mutations in plant DNA through techniques like base and prime editing, although some technical issues remain unresolved (Anzalone et al. 2020; Hua et al. 2022). Scientists have developed a method to mutate the Cas9 nuclease to cut only one DNA strand, called a nickase. By activating the homology-directed repair (HDR) process instead of Non-Homologous End Joining (NHEJ), a wide range of mutations can occur. In the realm of genetics, mutations are known to take on various forms, ranging from single-nucleotide changes to more complex alterations such as chromosomal rearrangements. While mutations can have detrimental effects, it is important to recognize that they can

Table 4 Improvements recorded in various crops by overexpressing/downregulating or knocking out the gene for tolerance to specific stress, and work is ongoing for self-incompatibility in forage grasses

Plant	Gene	Editing was done by CRISPR-Cas9	Response
Arabidopsis thaliana	SYTA	Overexpressed	Higher germination; high seedling survival rate under heat stress
Arabidopsis thaliana	Wheat gene TaSnRK2	Overexpressed	Higher tolerance to drought stress
Nicotiana tabacum L.	Agropyron cristatum gene AcSnRK2.11	Overexpressed	Higher survival rates after recovery from cold stress
Wheat	Phosphatase SAL1 TaSal1	Downregulation	Better germination and growth under drought stress
Rice	Basic helix loop helix OsbHLH024	Knockout/ inactivation	Increased salinity tolerance Lower Na^+ and reactive oxygen species and higher K^+ Ion transporters upregulated
Forage plants	DUF247 for S and Z genes	Work in progress	Self-compatibility may be the outcome
Maize	ARG0S8 (auxin-related gene involved in organ size 8)	Overexpression	Improved yield of grains under drought stress
Rice	OsRR22 (type-B response regulator 22)	Knockout	Tolerance to high salt in growth chambers

also contribute to genetic diversity and evolution. By understanding the mechanisms and consequences of these genetic modifications, we can gain valuable insights into the underlying biology and develop innovative solutions for various health conditions. Thus, studying mutations is an essential component of advancing our knowledge and improving human health (Puchta et al. 2022; Villalobos-López et al. 2022).

The profound potential of CRISPR-Cas in enabling precise genetic modifications in crops, particularly in forage grasses, is substantial. This innovative technique facilitates the acquisition of sought-after traits significantly faster than conventional breeding techniques. Moreover, the ease of multiplexed genome editing allows for simultaneous targeting multiple sequences. Depending on the particular trait and species, there are certain obstacles that need to be addressed. These include evaluating the appropriateness of plant transformation techniques and ensuring the availability and quality of the reference genome. It is expected that these hurdles will be surmountable with advancements in technology.

Nevertheless, in the case of grasses, the confluence of genome editing with a breeding program warrants careful consideration of important biological attributes. It is worth noting that genome editing can potentially modify the reproductive characteristics of grasses. Forage grasses possess a sturdy self-incompatibility (SI) mechanism, making the inbreeding process nearly impossible. The SI system in grasses is governed by two multi-allelic genes, S and Z, for a long time (Lundqvist 1955; Cornish et al. 1979). It has been found that the S and Z loci are regulated by the specific genes called DUF247 (Manzanares et al. 2016; Herridge et al. 2022). Now that we know these genes' sequences and molecular functions, we can use genome editing techniques to generate self-fertile knockout lines by targeting them (Table 4). Potatoes have already been made self-compatible using a similar method (Ye et al. 2018). Developing male sterile lines is crucial for forage grasses. The research on maize has set an example by focusing on the male sterility one genes Ms1 and Ms45 through CRISPR-Cas9, which resulted in the production of male-sterile wheat lines for hybrid seed production. (Singh et al. 2018; Okada et al. 2019).

By knocking out the MTL gene, it's possible to induce haploids and produce fully homozygous doubled haploid lines. This technique has already been successfully implemented in *S. italica* (Cheng et al. 2021). Apomixis is a mode of asexual reproduction that can be observed in several grass species, such as *Poa pratensis*. This method is frequently employed in cultivating lawns, pastures, and leys. An important factor in enhancing the genetic stabilization of the hybrid vigor of parent lines is the induction of apomixis in other forage grasses. Scientists have made progress towards achieving this goal in rice by mutating genes related to meiosis through CRISPR-Cas. They produced clonal diploid gametes, which were then employed to trigger parthenogenesis by ectopically expressing BABY BOOM1 in the egg cell. Consequently, this resulted in the production of clonal progeny (Khanday et al. 2019; Zhu et al. 2020).

The CRISPR/Cas9 technology has been used in alfalfa to perform targeted mutagenesis. The Squamosa Promoter Binding Protein-Like 9 (SPL9) gene was chosen and successfully modified. Transgenic lines were pre-selected by droplet digital PCR (ddPCR) to screen large populations. However, because of the highly repeated clusters present in alfalfa's tetraploid genome, the editing efficiency of alfalfa was less than that of barrel medic. Despite this, CRISPR/Cas9-mediated modifications of the tetraploid alfalfa genome have been successfully performed, although the editing efficiency needs improvement.

The application of CRISPR/Cas9 technology has provided hope for genome editing in various plant species. In particular, studies have demonstrated the efficacy of this technology in legumes such as *Barrel medic*, *L. japonicus*, and *G. max*, as well as in several non-leguminous plant species. Notably, the application of CRISPR/Cas9 technology in alfalfa has been successful despite the challenges presented by its auto-tetraploid character. Specifically, using SPL9-silenced alfalfa plants did not reveal any visible phenotype, and ddPCR-based estimation of the reaction concentration per μl was a direct indicator of the genome editing rate. In addition, sequencing analysis confirmed the high specificity of the sgRNAs of SPL9 to the recognition site, with no off-target effects observed in the alfalfa genome. While conventional approaches such as transgene integration and gene stacking may be less suitable for alfalfa, the CRISPR/Cas9 technology has proven effective for targeted genome editing. RNA-Seq analysis was conducted to investigate the molecular mechanisms underlying various stresses in alfalfa, including frost, salinity, drought, resistance to aluminum, lead, waterlogging, and fall dormancy. Alfalfa has unique mechanisms to adapt to freezing temperatures. These mechanisms involve genes crucial in hormonal signal transduction, ubiquitin-mediated proteolysis pathways, and encoding membrane proteins. When it comes to salinity stress, it affects a variety of alfalfa genes, such as those encoding Sugar transporter ERD6-like 16, dihydroflavonol Reductase, transcription factor MYB59, and Inositol-1,4,5-Trisphosphate 5-Phosphatase. Additionally, 86 transcription factors responded to salinity stress, many belonging to GRAS, ARR, JUMONJI, and MYB families, preferentially upregulated in the tolerant alfalfa cultivar. The level of fall dormancy in alfalfa is determined by the genes responsible for auxin and ethylene signaling and carbohydrate transport. Beta-amylase, Ethylene Response Factor, Calcineurin B-like Interacting Protein Kinases, Glutathione Peroxidase, and Glutathione S-transferase are some of the genes that play an essential role in alfalfa's waterlogging stress resistance.

7 Advantages of Gene Editing

Advancements in gene engineering have opened up new possibilities for genome editing with the use of site-directed nucleases (SDNs), notably zinc finger nucleases (ZFNs) and transcription activator-like effector nucleases (TALENs). These cutting-edge techniques enable scientists to precisely modify DNA sequences, opening up new avenues for research and breakthroughs in the field of genetics (Shukla et al. 2009; Urnov et al. 2010; Mushtaq et al. 2021).

Genome editing is a technique that offers several advantages over transgenic methods. It provides precision, fewer off-target effects, streamlined production, multiplex possibility, and the ability to modify more traits. For example, genome editing can be implemented to simultaneously edit alpha-gliadin genes, lower gluten content in wheat (Sánchez-León et al. 2018), increase the production of gamma-aminobutyric acid in tomato (Li et al. 2018), or enhance the accumulation of provitamin D3 in tomato (Li et al. 2022). Maize is engineered to express the ARGOS8 gene, which contains no foreign DNA sequences. This could potentially exempt it from GMO regulation, but it ultimately depends on each country's policies. Genome editing has become a practical and transformative approach for improving forage crops, enabling precise modification of traits that directly enhance livestock nutrition and sustainability. For instance, CRISPR/Cas9-mediated editing of lignin biosynthesis genes in maize and alfalfa has successfully reduced lignin content, improving digestibility and nutrient availability for ruminants. In alfalfa, targeted gene edits have been used to increase protein concentration and reduce anti-nutritional factors, while in forage sorghum, genome editing has improved drought tolerance, ensuring stable yields under water-limited conditions. Similarly, soybean and cowpea have been modified to enhance nitrogen fixation and stress resilience, supporting both soil fertility and high-quality forage production. These applications not only improve feed efficiency and animal performance but also shorten breeding cycles compared to conventional methods, accelerating the development of climate-resilient cultivars. By integrating nutritional improvement with stress tolerance, genome editing offers a powerful strategy to revolutionize forage crop improvement and strengthen global food and protein security.

Genome editing is a sophisticated technique that enables the alteration of an organism's DNA with unparalleled precision. Because it does not involve the introduction of foreign DNA, genome editing is often thought to be less strictly regulated than traditional genetic modification techniques. In Japan, for example, a company called Sanatech Seed has successfully commercialized a high gamma-aminobutyric acid tomato created using genome editing. In the European Union, the Court of Justice has decided that organisms produced by mutagenesis techniques fall under the category of GMOs and are subject to the regulations outlined in Directive 2001/18/EC. This ruling applies to mutagenesis techniques introduced after 2001 when the GMO Directive was adopted. Site-directed nucleases are the tools used to modify plant genomes through genome editing. There are three scenarios in which this can occur: SDN-1, SDN-2, and SDN-3. The SDN-3 can currently produce transgenic plants. Alternatively, SDN-1 and SDN-2 do not require the integration of foreign DNA into the genome. Recently, the Court of Justice of the European Union (CJEU) ruled that the SDN-1 and SDN-2, including CRISPR-Cas, should be included under the Genetically Modified Organism (GMO) Directive. This decision has sparked much debate among academics and the biotech industry. Many experts argue that applying the GMO Directive to products created through a more precise method like random mutagenesis and transgenesis is excessive. It is interesting to note that some countries exempt certain products from GMO regulations based on their specific policies. For example, maize with an elevated expression of the ARGOS8 gene has no exogenous DNA sequences, which exempts it from GMO regulation in some countries.

The recent ruling on gene editing has led to a peculiar situation where two identical products with the same mutation could be regulated differently. As a result, detecting the origin of simple mutations, like single-nucleotide polymorphisms, has become challenging as no technology can determine it. The European Network of GMO Laboratories (ENGL) suggests that detecting methods for SDN-1 and SDN-2 products could be more reliable and challenging. This legal uncertainty has reduced the appeal of genome-editing research in the EU, and obtaining regulatory approval for gene-edited products has become as cumbersome as for GMOs. Although no regulatory approval applications involving gene editing have been made yet, some applications for authorization of products obtained with CRISPR-Cas9 in SDN-3 scenarios, such as maize DP-915635-4, have been submitted to member states and are under review by EFSA (Maize DP-915635-4, n.d.).

8 Conclusion and Future Perspectives

Genome editing holds immense promise for revolutionizing forage crop improvement by addressing challenges of low nutritional quality, poor digestibility, and vulnerability to climate stress. Its precision enables targeted modifications that enhance protein content, fiber digestibility, stress resilience, and nitrogen-use efficiency, thereby improving both livestock productivity and environmental sustainability. Looking ahead, integrating genome editing with advanced genomic tools, high-throughput phenotyping, and bioinformatics will accelerate the development of next-generation forage cultivars tailored to diverse agro-climatic conditions. However, widespread adoption will require overcoming regulatory, ethical, and public acceptance challenges, alongside ensuring equitable access to these technologies for farmers worldwide. With continued innovation, collaborative research, and responsible deployment, genome editing has the potential to become a cornerstone in developing climate-resilient, nutrient-rich forage crops, ultimately contributing to sustainable livestock production and global food security. Recent advancements in genome editing and transformation techniques can potentially improve forage crops by enhancing stress tolerance, increasing yield, and improving nutritional composition. However, regulatory uncertainty in the European Union has posed a significant challenge to commercializing these methods. Despite this challenge, modern techniques and proper breeding management strategies can help overcome these obstacles, allowing for the development of genome-edited forage grasses with significantly improved traits.

References

Abdallah NA, Elsharawy H, Abulela HA, Thilmony R, Abdelhadi AA, Elarabi NI (2022) Multiplex CRISPR/Cas9-mediated genome editing to address drought tolerance in wheat. GM Crops Food:1–17

Alam MS, Kong J, Tao R, Ahmed T, Alamin M, Alotaibi SS et al (2022) CRISPR/Cas9 mediated knockout of the OsbHLH024 transcription factor improves salt stress resistance in rice (Oryza sativa l.). Plants 11:1184. https://doi.org/10.3390/plants11091184

Aleem S, Sharif I, Amin E, Tahir M, Parveen N, Aslam R et al (2020) Heat tolerance in vegetables in the current genomic era: an overview. Plant Growth Regul 92:497–516. https://doi.org/10.1007/s10725-020-00658-5

Amombo E, Mumm RH, Nilahyane A, Hirich A, Hijri M, Ntui VO (2025) Enhancing crop nutrition in arid and semiarid regions of Africa through genome editing using CRISPR/Cas. Food Energy Secur 14(3):e70068

Anzalone AV, Koblan LW, Liu DR (2020) Genome editing with CRISPR–cas nucleases, base editors, transposases and prime editors. Nat Biotechnol 38:824–844. https://doi.org/10.1038/s41587-020-0561-9

Bailey-Serres J, Parker JE, Ainsworth EA, Oldroyd GED, Schroeder JI (2019) Genetic strategies for improving crop yields. Nature 575:109–118. https://doi.org/10.1038/s41586-019-1679-0

Baldacci-Cresp F, Chang C, Maucourt M, Deborde C, Hopkins J, Lecomte P et al (2012) Homoglutathione deficiency impairs root-knot nematode development in Medicago truncatula. PLoS Pathog 8:2471. https://doi.org/10.1371/journal.ppat.1002471

Bansal M, Wani SH (2022) Virus-mediated delivery of CRISPR/CAS9 system in plants. In: Genome editing: current technology advances and applications for crop improvement. Springer International Publishing, Cham, pp 197–203

Bora KS, Sharma A (2011) Phytochemical and pharmacological potential of Medicago sativa: a review. Pharm Biol 49:211–220. https://doi.org/10.3109/13880209.2010.504732

Chalmers J, Lidgett A, Cummings N, Cao Y, Forster J, Spangenberg G (2005) Molecular genetics of fructan metabolism in perennial ryegrass. Plant Biotechnol J 3:459–474. https://doi.org/10.1111/j.1467-7652.2005.00148.x

Chan KX, Mabbitt PD, Phua SY, Mueller JW, Nisar N, Gigolashvili T et al (2016) Sensing and signaling of oxidative stress in chloroplasts by inactivation of the SAL1 phosphoadenosine phosphatase. Proc Natl Acad Sci 113:E4567–E4576. https://doi.org/10.1073/pnas.1604936113

Cheng Z, Sun Y, Yang S, Zhi H, Yin T, Ma X et al (2021) Establishing in planta haploid inducer line by edited SiMTL in foxtail millet (Setaria italica). Plant Biotechnol J 19:1089–1091. https://doi.org/10.1111/pbi.13584

Choudhary JR, Bhavyasree RK, Sheoran S, Choudhary M, Chandra S, Kaswan V, Wani SH (2023) Forward and reverse genetics in crop breeding. In: Advanced crop improvement, volume 1: theory and practice. Springer International Publishing, Cham, pp 257–275

Cornish MA, Hayward MD, Lawrence MJ (1979) Self-incompatibility in ryegrass. Heredity 43:95–106. https://doi.org/10.1038/hdy.1979.63

Cushman JC, Denby KJ, Mittler R (2022) Plant responses and adaptations to a changing climate. Plant J 109:319–322. https://doi.org/10.1111/tpj.15641

Damaris RN, Yang P (2021) Protein phosphorylation response to abiotic stress in plants. In: Wu XN (ed) Plant phosphoproteomics: methods and protocols methods in molecular biology. Springer US, New York, pp 17–43

Dass P, Houlton BZ, Wang Y, Warlind D (2018) Grasslands may be more reliable carbon sinks than forests in California. Environ Res Lett 13:074027. https://doi.org/10.1088/1748-9326/aacb39

Dhakate P, Sehgal D, Vaishnavi S, Chandra A, Singh A, Raina S et al (2022) Comprehending the evolution of gene editing platforms for crop trait improvement. Front Genet 13. https://doi.org/10.3389/fgene.2022.876987

Doyle JJ, Luckow MA (2003) The rest of the iceberg. Legume diversity and evolution in a phylogenetic context. Plant Physiol 131:900–910. https://doi.org/10.1104/pp.102.018150

Fan S, Jia L, Wu J, Zhao Y (2025) Harnessing the potential of CRISPR/Cas in targeted alfalfa improvement for stress resilience. Int J Mol Sci 26(7):3311

Fu G, Grbic V, Ma S, Tian L (2015) Evaluation of somatic embryos of alfalfa for recombinant protein expression. Plant Cell Rep 34:211–221. https://doi.org/10.1007/s00299-014-1700-x

Ghesquière M, Humphreys MW, Zwierzykowski Z (2010) Festulolium. In: Boller B, Posselt UK, Veronesi F (eds) Fodder crops and amenity grasses. Springer, New York, pp 288–311

Giridhar K, Samireddypalle A (2015) Impact of climate change on forage availability for livestock. In: Sejian V, Gaughan J, Baumgard L, Prasad C (eds) Climate change impact on livestock: adaptation and mitigation. Springer India, New Delhi, pp 97–112. https://doi.org/10.1007/978-81-322-2265-1_7

Guarino F, Cicatelli A, Castiglione S, Agius DR, Orhun GE, Fragkostefanakis S et al (2022) An epigenetic alphabet of crop adaptation to climate change. Front Genet 13. https://doi.org/10.3389/fgene.2022.818727

Guo J, Sun B, He H, Zhang Y, Tian H, Wang B (2021) Current understanding of bHLH transcription factors in plant abiotic stress tolerance. Int J Mol Sci 22:4921. https://doi.org/10.3390/ijms22094921

Han X, Chen Z, Li P, Xu H, Liu K, Zha W et al (2022) Development of novel rice germplasm for salt-tolerance at seedling stage using CRISPR-Cas9. Sustainability 14:2621. https://doi.org/10.3390/su14052621

Hasanuzzaman M, Bhuyan MHMB, Zulfiqar F, Raza A, Mohsin SM, Mahmud JA et al (2020) Reactive oxygen species and antioxidant defense in plants under abiotic stress: revisiting the crucial role of a universal defense regulator. Antioxidants 9:681. https://doi.org/10.3390/antiox9080681

He XZ, Dixon RA (2000) Genetic manipulation of isoflavone 7-O-methyltransferase enhances biosynthesis of 4'-O-methylated isoflavonoid phytoalexins and disease resistance in alfalfa. Plant Cell 12:1689–1702. https://doi.org/10.1105/tpc.12.9.1689

He M, He C-Q, Ding N-Z (2018) Abiotic stresses: general defenses of land plants and chances for engineering multistress tolerance. Front Plant Sci 9. https://doi.org/10.3389/fpls.2018.01771

Herridge R, McCourt T, Jacobs JME, Mace P, Brownfield L, Macknight R (2022) Identification of the genes at s and z reveals grass self-incompatibility's molecular basis and evolution. Front Plant Sci 13. https://doi.org/10.3389/fpls.2022.1011299

Hua K, Han P, Zhu J-K (2022) Improvement of base editors and prime editors advances precision genome engineering in plants. Plant Physiol 188:1795–1810. https://doi.org/10.1093/plphys/kiab591

Huang W, Zhang L, Columbus JT, Hu Y, Zhao Y, Tang L et al (2022) A well-supported nuclear phylogeny of poaceae and implications for the evolution of C4 photosynthesis. Mol Plant 15:755–777. https://doi.org/10.1016/j.molp.2022.01.015

Jagadish SVK, Way DA, Sharkey TD (2021) Plant heat stress: concepts directing future research. Plant Cell Environ 44:1992–2005. https://doi.org/10.1111/pce.14050

Jia Q, Kong D, Li Q, Sun S, Song J, Zhu Y et al (2019) The function of inositol phosphatases in plant tolerance to abiotic stress. Int J Mol Sci 20:3999. https://doi.org/10.3390/ijms20163999

Khanday I, Skinner D, Yang B, Mercier R, Sundaresan V (2019) A male-expressed rice embryogenic trigger redirected for asexual propagation through seeds. Nature 565:91–95. https://doi.org/10.1038/s41586-018-0785-8

Klapheck S (1988) Homoglutathione: isolation, quantification and occurrence in legumes. Physiol Plant 74:727–732. https://doi.org/10.1111/j.1399-3054.1988.tb02044.x

Le BH, Wagmaister JA, Kawashima T, Bui AQ, Harada JJ, Goldberg RB (2007) Using genomics to study legume seed development. Plant Physiol 144:562–574. https://doi.org/10.1104/pp.107.100362

Lee S, Choi S, Jeon D, Kang Y, Kim C (2020) Evolutionary impact of whole genome duplication in poaceae family. J Crop Sci Biotechnol 23:413–425. https://doi.org/10.1007/s12892-020-00049-2

Li R, Li R, Li X, Fu D, Zhu B, Tian H et al (2018) Multiplexed CRISPR/Cas9-mediated metabolic engineering of g-aminobutyric acid levels in solanum lycopersicum. Plant Biotechnol J 16:415–427. https://doi.org/10.1111/pbi.12781

Li J, Scarano A, Gonzalez NM, D'Orso F, Yue Y, Nemeth K et al (2022) Biofortified tomatoes provide a new route to vitamin d sufficiency. Nat Plants 8:611–616. https://doi.org/10.1038/s41477-022-01154-6

Liu Y, Wang W, Yang B, Currey C, Fei S (2020) Functional analysis of the teosinte branched 1 gene in the tetraploid switchgrass (Panicum virgatum l.) by CRISPR/Cas9-directed mutagenesis. Front Plant Sci 11. https://doi.org/10.3389/fpls.2020.572193

Lopez B, Hines PJ, Ash C (2022) The unrecognized value of grass. Science 377:590–591. https://doi.org/10.1126/science.add6362

Lundqvist A (1955) Genetics of self-incompatibility in festuca pratensis huds. Hereditas 41:518–520. https://doi.org/10.1111/j.1601-5223.1955.tb03007.x

Manzanares C, Barth S, Thorogood D, Byrne SL, Yates S, Czaban A et al (2016) A gene encoding a DUF247 domain protein cosegregates with the s self-incompatibility locus in perennial ryegrass. Mol Biol Evol 33:870–884. https://doi.org/10.1093/molbev/msv335

Moser LE, Hoveland CS (1996) Cool-season grass overview. In: Moser LE, Buxton DR, Casler MD (eds) Cool-season forage grasses. https://doi.org/10.2134/agronmonogr34.c1

Moser LE, Burson BL, Sollenberger LE (2004) Warm-season (C4) grass overview. In: Burson LEM, Burson BL, Sollenberger LE (eds) Warm-season (C4) grasses. https://doi.org/10.2134/agronmonogr45.c1
Muir JP, Pitman WD, Foster JL (2011) Sustainable, low-input, warm- season, grass–legume grassland mixtures: mission (nearly) impossible? Grass Forage Sci 66:301–315. https://doi.org/10.1111/j.1365-2494.2011.00806
Mushtaq M, Pir HA, Zaid A, Wani SH (2021) CRISPR/Cas system: a powerful approach for enhanced resistance against rice blast. In: Biocontrol agents and secondary metabolites. Woodhead Publishing, pp 649–658
Nirola R, Megharaj M, Beecham S, Aryal R, Thavamani P, Vankateswarlu K et al (2016) Remediation of metalliferous mines, revegetation challenges and emerging prospects in semi-arid and arid conditions. Environ Sci Pollut Res 23:20131–20150. https://doi.org/10.1007/s11356-016-7372-z
Niu Y, Xiang Y (2018) An overview of biomembrane functions in plant responses to high-temperature stress. Front Plant Sci 9. https://doi.org/10.3389/fpls.2018.00915
Njuki J, Sanginga PC (2013) Gender and livestock: key issues, challenges and opportunities. In: Njuki J, Sanginga PC (eds) Women, livestock ownership and markets: bridging the gender gap in eastern and southern Africa. Routledge, New York
Okada A, Arndell T, Borisjuk N, Sharma N, Watson-Haigh NS, Tucker EJ et al (2019) CRISPR/Cas9-mediated knockout of Ms1 enables the rapid generation of male-sterile hexaploid wheat lines for use in hybrid seed production. Plant Biotechnol J 17:1905–1913. https://doi.org/10.1111/pbi.13106
Pasternak T, Asard H, Potters G, Jansen MA (2014) The hiol compounds glutathione and homo-glutathione di erentially a ect cell development in alfalfa (Medicago sativa L.). Plant Phys Biochem 74:16–23. https://doi.org/10.1016/j.plaphy.2013.10.028
Puchta H, Jiang J, Wang K, Zhao Y (2022) Updates on gene editing and its applications. Plant Physiol 188:1725–1730. https://doi.org/10.1093/plphys/kiac032
Radovic J, Sokolovic D, Markovic J (2009) Alfalfa-most important perennial forage legume in animal husbandry. Biotechnol Anim Husb 25:465–475. https://doi.org/10.2298/BAH0906465R
Rashmi R, Sarkar M, Vikramaditya T (1997) Cultivaton of alfalfa (Medicago sativa L.). Anc Sci Life 17:117–119
Samac DA, Temple SJ (2004) Development and utilization of transformation in Medicago species. In: Liang GH, Skinner DZ (eds) Genetically modified crops, their development, uses and risks. The Haworth Press, New York, pp 165–202
Sánchez-León S, Gil-Humanes J, Ozuna CV, Giménez MJ, Sousa C, Voytas DF et al (2018) Low-gluten, nontransgenic wheat engineered with CRISPR/Cas9. Plant Biotechnol J 16:902–910. https://doi.org/10.1111/pbi.12837
Shabala S, Wu H, Bose J (2015) Salt stress sensing and early signalling events in plant roots: current knowledge and hypothesis. Plant Sci 241:109–119. https://doi.org/10.1016/plantsci.2015.10.003
She K, Pan W, Yan Y, Shi T, Chu Y, Cheng Y et al (2022) Genome-wide identification, evolution and expressional analysis of OSCA gene family in barley (Hordeum vulgare l.). Int J Mol Sci 23:13027. https://doi.org/10.3390/ijms232113027
Shi J, Gao H, Wang H, Lafitte HR, Archibald RL, Yang M et al (2017) ARGOS8 variants generated by CRISPR-Cas9 improve maize grain yield under field drought stress conditions. Plant Biotechnol J 15:207–216. https://doi.org/10.1111/pbi.12603
Shukla VK, Doyon Y, Miller JC, DeKelver RC, Moehle EA, Worden SE et al (2009) Precise genome modification in the crop species zea mays using zinc-finger nucleases. Nature 459:437–441. https://doi.org/10.1038/nature07992
Singh M, Kumar M, Albertsen MC, Young JK, Cigan AM (2018) Concurrent modifications in the three homeologs of Ms45 gene with CRISPR-Cas9 lead to rapid generation of male sterile bread wheat (Triticum aestivum l.). Plant Mol Biol 97:371–383. https://doi.org/10.1007/s11103-018-0749-2

Sustek-Sanchez F, Rognli OA, Rostoks N, Sõmera M, Jaskūne K, Kovi MR et al (2023) Improving abiotic stress tolerance of forage grasses – prospects of using genome editing. Front Plant Sci 14:1127532. https://doi.org/10.3389/fpls.2023.1127532

Tek MI, Calis O, Fidan H, Shah MD, Celik S, Wani SH (2022) CRISPR/Cas9 based mlo-mediated resistance against Podosphaera xanthii in cucumber (Cucumis sativus L.). Frontiers Plant Science 13:1081506

Terrer C, Phillips RP, Hungate BA, Rosende J, Pett-Ridge J, Craig ME et al (2021) A trade-off between plant and soil carbon storage under elevated CO2. Nature 591:599–603. https://doi.org/10.1038/s41586-021-03306-8

Tester M, Davenport R (2003) Na+ tolerance and na+ transport in higher plants. Ann Bot 91:503–527. https://doi.org/10.1093/aob/mcg058

Triboi E, Triboi-Blondel AM (2014) Towards sustainable, self-supporting agriculture: biological nitrogen factories as a key for future cropping systems. In: Dent D (ed) Soil as world heritage. Springer, Dordrecht, pp 329–342. https://doi.org/10.1007/978-94-007-6187-2_32

Urnov FD, Rebar EJ, Holmes MC, Zhang HS, Gregory PD (2010) Genome editing with engineered zinc finger nucleases. Nat Rev Genet 11:636–646. https://doi.org/10.1038/nrg2842

Vaziriyeganeh M, Khan S, Zwiazek JJ (2021) Transcriptome and metabolome analyses reveal potential salt tolerance mechanisms contributing to maintenance of water balance by the halophytic grass puccinellia nuttalliana. Front Plant Sci 12. https://doi.org/10.3389/fpls.2021.760863

Villalobos-López MA, Arroyo-Becerra A, Quintero-Jiménez A, Iturriaga G (2022) Biotechnological advances to improve abiotic stress tolerance in crops. Int J Mol Sci 23:12053. https://doi.org/10.3390/ijms231912053

Wang H, Jia G, Zhang N, Zhi H, Xing L, Zhang H et al (2022) Domestication-associated PHYTOCHROME c is a flowering time repressor and a key factor determining setaria as a short-day plant. New Phytol 236:1809–1823. https://doi.org/10.1111/nph.18493

Wani SH, Hensel G (2022) Genome editing: current technology advances and applications for crop improvement. Springer, Cham, p 345

Wani SH, Choudhary M, Barmukh R, Bagaria PK, Samantara K, Razzaq A, Jaba J, Ba MN, Varshney RK (2022) Molecular mechanisms, genetic mapping, and genome editing for insect pest resistance in field crops. Theor Appl Genet 135(11):3875–3895

Weiss T, Wang C, Kang X, Zhao H, Elena Gamo M, Starker CG et al (2020) Optimization of multiplexed CRISPR/Cas9 system for highly efficient genome editing in setaria viridis. Plant J 104:828–838. https://doi.org/10.1111/tpj.14949

Wolabu TW, Mahmood K, Chen F, Torres-Jerez I, Udvardi M, Tadege M, Cong L, Wang Z, Wen J (2024) Mutating alfalfa COUMARATE 3-HYDROXYLASE using multiplex CRISPR/Cas9 leads to reduced lignin deposition and improved forage quality. Front Plant Sci 15:1363182

Wu S, Greiner S, Ma C, Zhong J, Huang X, Rausch T et al (2021) A fructan exohydrolase from maize degrades both inulin and Levan and co-exists with 1-kestotriose in maize. Int J Mol Sci 22:5149. https://doi.org/10.3390/ijms22105149

Wu X, Wang Y, Bian Y, Ren Y, Xu X, Zhou F et al (2022) A critical review on plant annexin: structure, function, and mechanism. Plant Physiol Biochem 190:81–89. https://doi.org/10.1016/j.plaphy.2022.08.019

Xu H-S, Guo S-M, Zhu L, Xing J-C (2020) Growth, physiological and transcriptomic analysis of the perennial ryegrass lolium perenne in response to saline stress. R Soc Open Sci 7:200637. https://doi.org/10.1098/rsos.200637

Ye M, Peng Z, Tang D, Yang Z, Li D, Xu Y et al (2018) Generation of self-compatible diploid potato by knockout of s-RNase. Nat Plants 4:651–654. https://doi.org/10.1038/s41477-018-0218-6

Zhang S, Shi Y, Cheng N, Du H, Fan W, Wang C (2015) De novo characterization of fall dormant and nondormant alfalfa (Medicago sativa L.) leaf transcriptome and identification of candidate genes related to fall dormancy. PLoS One 10:e0122170. https://doi.org/10.1371/journal.pone.0122170

Zhang A, Liu Y, Wang F, Li T, Chen Z, Kong D et al (2019) Enhanced rice salinity tolerance via CRISPR/Cas9-targeted mutagenesis of the OsRR22 gene. Mol Breed 39:47. https://doi.org/10.1007/s11032-019-0954-y

Zhang Y, Ran Y, Nagy I, Lenk I, Qiu J-L, Asp T et al (2020) Targeted mutagenesis in ryegrass (Lolium spp.) using the CRISPR/Cas9 system. Plant Biotechnol J 18:1854–1856. https://doi.org/10.1111/pbi.13359

Zhang L, Wang T, Wang G, Bi A, Wassie M, Xie Y et al (2021) Simultaneous gene editing of three homoeoalleles in self-incompatible allohexaploid grasses. J Integr Plant Biol 63:1410–1415. https://doi.org/10.1111/jipb.13101

Zhao S, Zhang Q, Liu M, Zhou H, Ma C, Wang P (2021) Regulation of plant responses to salt stress. Int J Mol Sci 22:4609. https://doi.org/10.3390/ijms22094609

Zhu H, Li C, Gao C (2020) Applications of CRISPR–cas in agriculture and plant biotechnology. Nat Rev Mol Cell Biol 21:661–677. https://doi.org/10.1038/s41580-020-00288-9

Correction to: Accelerated Breeding in Bajra Using Genomic Approaches

Ruchika Kashyap, Dinesh Kumar Saini, Sangram Singh Sandhu, Heena Puri, Shahbaz Ahmed, Jagmohan Singh, Karansher Sandhu, and Rahul Kapoor

Correction to: Chapter 4 in: S. H. Wani, S. S. Gosal (eds.), *Accelerated Plant Breeding, Volume 5*, https://doi.org/10.1007/978-3-032-04527-0_4

The book was inadvertently published with incorrect affiliation for two of the contributors associated with Chapter 4, Heena Puri, as "Department of Entomology, University of Nebraska–Lincoln, Lincoln, NE, USA". It has now been updated as "Insect Testing Scientist, Impetus Agriculture Inc., Saint Louis, MO, USA". Similarly, Karansher Sandhu listed as associated with "Department of Entomology, University of Nebraska–Lincoln, Lincoln, NE, USA" has now been replaced with "Department of Crop and Soil Sciences, Washington State University, Pullman, WA, USA".

The updated version of this chapter can be found at
https://doi.org/10.1007/978-3-032-04527-0_4

S. H. Wani, S. S. Gosal (eds.), *Accelerated Plant Breeding, Volume 5*,
https://doi.org/10.1007/978-3-032-04527-0_13

Index

S. H. Wani, S. S. Gosal (eds.), *Accelerated Plant Breeding, Volume 5*,
https://doi.org/10.1007/978-3-032-04527-0

MIX
Papier aus verantwortungsvollen Quellen
Paper from responsible sources
FSC® C105338

If you have any concerns about our products,
you can contact us on
ProductSafety@springernature.com

In case Publisher is established outside the EU,
the EU authorized representative is:
Springer Nature Customer Service Center GmbH
Europaplatz 3, 69115 Heidelberg, Germany

Printed by Libri Plureos GmbH
in Hamburg, Germany